Clay Minerals: From Paleoclimatic and Paleoenvironmental Indicators to Industrial Raw Materials

Clay Minerals: From Paleoclimatic and Paleoenvironmental Indicators to Industrial Raw Materials

Guest Editors

Elisa Laita Florian
Javier García Rivas
Isabel Abad

MDPI

Basel • Beijing • Wuhan • Barcelona • Belgrade • Novi Sad • Cluj • Manchester

Guest Editors
Elisa Laita Florian
Department of Geology
University of Jaén
Jaén
Spain

Javier García Rivas
Department of Mineralogy
and Petrology
Complutense University of
Madrid
Madrid
Spain

Isabel Abad
Department of Geology
University of Jaén
Jaén
Spain

Editorial Office
MDPI AG
Grosspeteranlage 5
4052 Basel, Switzerland

This is a reprint of the Special Issue, published open access by the journal *Minerals* (ISSN 2075-163X), freely accessible at: https://www.mdpi.com/journal/minerals/special_issues/541AM887JX.

For citation purposes, cite each article independently as indicated on the article page online and as indicated below:

Lastname, A.A.; Lastname, B.B. Article Title. *Journal Name* **Year**, *Volume Number*, Page Range.

ISBN 978-3-7258-7717-1 (Hbk)
ISBN 978-3-7258-7718-8 (PDF)
https://doi.org/10.3390/books978-3-7258-7718-8

Cover image courtesy of Elisa Laita Florian

Contents

About the Editors

Elisa Laita Florian

Elisa Laita Florian (Department of Geology, University of Jaén) specializes in clay mineralogy, geochemistry and sedimentary processes. Her work focuses on the characterization of clay minerals, with particular emphasis on their use as indicators of paleoenvironmental and paleoclimatic conditions and their relevance as industrial raw materials.

Her research interests include clay mineral assemblages, weathering and alteration processes, and the geochemical evolution of sedimentary basins. She has participated in several national and international research projects focused on the relationship between clay mineral assemblages and paleoenvironmental conditions, as well as on the evaluation of clay deposits for industrial purposes. Within industrial applications, her work has been particularly centered on the use of clays in ceramics. She has conducted studies on clay deposits from different geological settings, including several regions in Europe and other areas characterized by different climatic and tectonic conditions, allowing for the comparative analysis of clay formation processes. She has authored and coauthored numerous peer-reviewed publications in international journals, contributing to the advancement of knowledge on clay minerals through the integration of mineralogical, geochemical, and petrological approaches, bridging fundamental research with applied perspectives.

Javier García Rivas

Javier García Rivas (Department of Mineralogy and Petrology, Complutense University of Madrid) is a specialist on clay mineralogy, focusing on the mineralogical, crystal-chemical, and geochemical characterization of clay minerals, with particular emphasis on Mg-rich bentonites and fibrous clays such as palygorskite and sepiolite. His research integrates multiple analytical techniques, including X-ray diffraction, infrared spectroscopy, geochemical analysis and electron microscopy. His work focuses not only on pure geological issues but also on the possible applicability of natural bentonites in different fields, such as engineering barriers on radioactive waste repositories and even on archaeometrical studies of the Maya Blue pigment. His research activity has led to numerous authored and co-authored peer-reviewed publications in international journals, contributing to the advancement of knowledge in clay mineralogy.

Isabel Abad

Isabel Abad (Department of Geology, University of Jaén) is a specialist on the mineral transformations of phyllosilicates in low-temperature sedimentary and metamorphic environments in different contexts and on using electron microscopy tools to characterize geological processes at the nanoscale. Her research interests include the clay mineral-rich fault zones, seeking to explain their behavior on the basis of the texture and features of these minerals, and the processes of mineral neoformation in these deformation-dominated environments; she is also involved in the study of mineral transformations in sediments from saline environments rich in organic matter, both recent and fossil. Collaboration with other researchers has led to her participation in other lines of research and projects related to mineralogy and geochemistry. These include her recent participation in an Antarctic project related to the slope stability of volcanic pyroclastic buildings, where she is characterizing the mineralogy and geochemistry of the deposits, and her involvement in a Chilean project about Li-rich clays. Her research activity has led to numerous authored and coauthored peer-reviewed publications in international journals, contributing to the advancement of knowledge on clay minerals in a wide range of geological contexts.

Preface

This Reprint presents a collection of contributions focused on the origin, evolution, and applications of clay minerals in diverse geological settings. Its scope covers mineralogical, geochemical, and environmental perspectives, with the aim of advancing the understanding of clay formation processes and their significance as indicators of past environmental conditions and valuable industrial resources.

The motivation for this Reprint arises from the increasing recognition of clay minerals as key components linking Earth surface processes, paleoclimatic change, and resource potential.

Recent analytical developments and interdisciplinary approaches have opened new opportunities to better characterize clay minerals and to interpret their role in both geological settings and technological applications.

This Reprint seeks to provide a coherent and updated reference that bridges fundamental research with applied perspectives, supporting future studies and innovations in the field.

Elisa Laita Florian, Javier García Rivas, and Isabel Abad
Guest Editors

minerals

MDPI

Editorial

Editorial for Special Issue "Clay Minerals: From Paleoclimatic and Paleoenvironmental Indicators to Industrial Raw Materials"

Elisa Laita [1,2,*], Javier García-Rivas [3] and Isabel Abad [2]

1 Centre National de la Recherche Cientifique (CNRS), Institut des Sciences Analytiques et de Physico-Chimie Pour l'Environnement et les Matériaux (IPREM), Université de Pau et des Pays de l'Adour, UMR 5254, 64000 Pau, France
2 Departamento de Geología & Centro de Estudios Avanzados en Ciencias de la Tierra, Energía y Medio Ambiente (CEACTEMA), Universidad de Jaén, Campus Las Lagunillas sn, 23071 Jaén, Spain; miabad@ujaen.es
3 Departamento de Mineralogía y Petrología, Universidad Complutense de Madrid, C/José Antonio Nováis 2, 28040 Madrid, Spain; javier.garcia.rivas@ucm.es
* Correspondence: elisa.laita-florian@univ-pau.fr

Clays and clay minerals constitute important mineral resources from both scientific and industrial perspectives. The genesis of clay minerals takes place when low-temperature aqueous solutions interact with rocks on the Earth's surface. Certain factors, such as the environment, temperature, amount of water available, and the type of weathered rock, determine the clay minerals formed. Therefore, clay minerals can provide information about the paleoclimate or paleoenvironment under which they were formed. Additionally, clays are also materials of great industrial and economic interest. Currently, clays are used in many types of industries since they constitute important components used in the manufacturing of many products, such as plastics, paper, cement, absorbent materials, ceramic and refractory materials, among others. The physical and chemical characterization of different clays (e.g., kaolin, smectites, fibrous clays) is of great interest since their industrial applications are closely related to their structure and composition.

This Special Issue, entitled "Clay Minerals: From Paleoclimatic and Paleoenvironmental Indicators to Industrial Raw Materials", includes ten contributions that investigate the origin, mineralogical characteristics, environmental significance, and technological applications of clay minerals. The studies examine clay deposits formed in different geological settings and highlight their importance both as paleoenvironmental indicators and as valuable industrial resources. Several contributions focus on the genesis of clay minerals in sedimentary environments (contributions 1, 2, 6, 7, 8, 9, and 10). These studies analyze clay assemblages formed in settings such as playa-lake systems, marine evaporitic environments, and ancient sedimentary basins. Through mineralogical and geochemical analyses, the authors explore processes, such as detrital input, authigenic mineral formation, and mineral transformations, related to variations in hydrochemistry and depositional conditions. Some of these papers emphasize the use of clay minerals as paleoenvironmental and paleoclimatic indicators (contributions 1, 2, 7, 9, and 10). Variations in clay mineral assemblages and geochemical characteristics are used to reconstruct past environmental conditions, including weathering intensity, hydrochemical evolution, and climatic regimes affecting sediment deposition. In addition, some contributions also address the mineralogical and physicochemical characterization of clay deposits with potential industrial applications (contributions 3, 4, 5, 8, 9, and 10). Using techniques such as X-ray diffraction, electron microscopy, and spectroscopic analyses, these studies evaluate sepiolite,

 https://doi.org/10.3390/min16040342

bentonite, kaolin, and laterites for applications in ceramics, adsorption, and environmental technologies.

Overall, the papers included in the Special Issue highlight the multidisciplinary significance of clay minerals in both palaeoclimatic and palaeoenvironmental studies and industrial applications.

The paper by Laita et al. (contribution 1) investigates clay-rich and iron-bearing palaeosols developed at the Middle–Upper Jurassic unconformity of the External Prebetic and the Iberian Range (South Iberian Palaeomargin and East Iberian Palaeomargin) in order to reconstruct the palaeoclimatic conditions that affected the exposed carbonate platform during the Callovian–Oxfordian. Their results reveal a mineral assemblage dominated by kaolinite and iron oxides, such as goethite and hematite. The occurrence of kaolinite crystals formed in situ, together with concentric iron-coated grains, indicates the development of plinthitic palaeosols under intense chemical weathering during periods of subaerial exposure. Geochemical proxies also confirm a high degree of alteration consistent with warm and humid climatic conditions during the Callovian–Oxfordian transition. This contribution demonstrates how clay minerals can act as robust indicators of past climatic regimes and weathering intensity. By linking clay mineral genesis with the regional paleoenvironmental evolution in the western Tethys, the study highlights the importance of clay mineral assemblages as key archives for reconstructing ancient climate and surface processes.

The paper by del Buey et al. (contribution 2) investigates the origin and evolution of palygorskite in lacustrine environments, focusing on three recent hyperalkaline playa-lakes in Central Spain and their relationship with nearby Miocene lacustrine deposits. The authors address this issue through an integrated mineralogical and microanalytical approach to trace palygorskite from the source (Miocene marlstones and mudstones) to the sink (recent playa-lakes). The results demonstrate the coexistence of both inherited (detrital) and neoformed (authigenic) palygorskite in the studied lakes. Detrital palygorskite shows degradation, wider fibers, and near-ideal compositions, indicating inheritance from older lacustrine deposits. Authigenic palygorskite forms narrower fibers with compositions controlled by lake hydrochemistry, enriched in Al_2O_3, MgO, and Fe_2O_3. Iron-rich varieties occur with saponite, reflecting Mg competition during precipitation, also observed between smectite and palygorskite in Miocene mudstones. These findings highlight how physicochemical differences in closely spaced hyperalkaline lakes can strongly influence the crystallochemistry of authigenic palygorskite and demonstrate the value of palygorskite as a sensitive geochemical proxy for interpreting paleoenvironmental conditions in lacustrine systems.

The paper by Sousa et al. (contribution 3) examines the interaction between expandable clay minerals and glycerin-based drilling fluids, which have been proposed as an alternative to conventional inhibited fluids for drilling highly reactive formations. Despite their potential advantages, operational problems such as drill bit and drill pipe balling, caused by the agglomeration and accretion of cuttings, can still occur. The authors investigate how these interactions affect wellbore stability and the extent to which accretion develops during drilling in clay-rich formations. Bentonite pellets were used as an analog for reactive formations. These pellets are mainly composed of interstratified illite-smectite clay minerals, and they consistently display highly plastic behavior and a strong tendency to expand. Accretion tests using glycerin-based drilling fluids containing different types of expansion inhibitors show that accretion remained significant in all tested fluids, with values exceeding 58%, indicating that neither glycerin nor the tested inhibitors were effective in reducing clay expansion or preventing cuttings aggregation. These findings highlight the critical role of clay mineralogy in drilling performance and demonstrate

https://doi.org/10.3390/min16040342

how detailed mineralogical characterization of clays directly informs their technological performance and industrial applications.

The paper by Bakouan et al. (contribution 4), explores the valorization of two lateritic materials from Bririmian and Precambrian environments in Burkina Faso, focusing on how geological provenance influences their mineralogical and physicochemical properties. The findings reveal that while geological context gives information regarding mineralogical composition of the laterites (quartz, kaolinite, hematite, and goethite), it alone does not determine their suitability as adsorbents. Instead, key physicochemical traits such as specific surface area (41.15–58.65 m^2/g), cation exchange capacity (52.3–58.7 cmol(+)/kg^{-1}), and anion exchange capacity (64.6–86.5 cmol(−)/kg^{-1}) are critical. The results demonstrate high efficacy in removing arsenate and arsenite from aqueous solutions, achieving up to 99.7% removal at optimal conditions, alongside notable removal rates for Pb(II) and Cu(II). These results underscore the potential of the laterites as cost-effective adsorbents for heavy metals and anionic contaminants in groundwater and highlight the role of laterites as versatile industrial raw materials for environmental remediation.

The paper by Ionescu et al. (contribution 5) presents a comprehensive investigation of a Miocene carbonatic clay from Transylvania (Romania), a material historically utilized for over a century in traditional ceramics, bricks, and tiles. The authors explore the mineralogical and microstructural transformations of the clay upon firing between 700 °C and 1200 °C. Their results reveal the progressive transformation of primary minerals (illite, muscovite, feldspar, carbonate, Fe oxyhydroxides, and quartz) into high-temperature phases such as gehlenite, clinopyroxene, maghemite, hematite, mullite, and α-cristobalite, alongside a glassy matrix. These firing-induced phases serve as distinctive "ceramic markers," providing valuable indicators of thermal history. By correlating optical appearance, microstructure, and specific mineral associations, the authors propose a novel "ceramic thermometer" capable of estimating firing temperatures in ancient ceramic artifacts. This approach not only enhances the understanding of thermal behavior in clay-based materials but also bridges geological, archeological, and industrial perspectives. The study exemplifies how detailed mineralogical analyses can inform both heritage studies and modern ceramic production, highlighting the significance of clay as a versatile natural resource.

The paper by Yaremchuk et al. (contribution 6) investigates the mineralogical evolution of clays associated with Lower and Middle Miocene marine evaporite deposits in the Ukrainian Carpathian Foredeep, emphasizing the role of brine concentration in controlling clay mineral transformations. The authors characterize clay mineral assemblages across different evaporite facies (gypsum, halite, and potash) and document systematic variations linked to the geochemical conditions of deposition and diagenesis. In the gypsum facies, assemblages are dominated by smectite and illite, with occasional mixed-layer chlorite-smectite and illite-mectite, as well as chlorite. In halite facies, the composition varies with stratigraphic unit: Eggenburgian rock salt (Vorotyshcha Suite) contains illite, chlorite, and mixed-layer illite-smectite, whereas the Badenian rock salt (Tyras Suite) also includes smectite, corrensite, and mixed-layer chlorite-smectite. Potash facies are characterized primarily by illite and chlorite. These assemblages reflect progressive aggradational transformations of unstable clay minerals such as kaolinite and smectite into the more stable illite and chlorite under increasingly saline evaporitic conditions. The study also highlights the influence of organic matter sorption, which may slow mineral transformation processes. The authors document contrasting mineralogical patterns in the weathering zone of the evaporite deposits, where reduced ionic concentrations promote degradational processes (forming illite-smectite) and neoformation of kaolinite during desalination. This study demonstrates how clay minerals in evaporitic environments record both palaeoenvironmental conditions and post-depositional alteration.

https://doi.org/10.3390/min16040342

The paper by Bauluz et al. (contribution 7) provides a comprehensive mineralogical and geochemical characterization of the Monte Castelo kaolin deposits (Burela, NW Spain), derived from acid igneous rocks undergoing varying degrees of chemical weathering. The authors trace the transformation of primary aluminosilicates (albite, K-feldspar, and K-micas) into kaolinite as weathering intensifies. Their observations reveal that kaolinite preferentially forms booklets and fine-grained matrices, and develops intergrowths along muscovite cleavages, displaying bidimensional crystallographic continuity without detectable intermediate phases. This insight into mica-kaolinite relationships underscores the role of high water-to-rock ratios in controlling weathering processes. The study also addresses analytical challenges in evaluating kaolinite crystallinity, demonstrating that while quartz impurities disrupt Hinckley index calculations, the AGFI index provides a robust alternative, effectively mitigating interference from coexisting minerals. Geochemically, the study documents a general decline in most elemental contents with progressive weathering, except for Al, TiO_2, HREEs, Ta, Hf, Th, U, V, Cr, S, Zr, Mo, and Sn, highlighting selective enrichment processes. By linking nano-scale textures, mineral transformations, and chemical evolution, this study advances our understanding of kaolin genesis under natural weathering conditions.

The paper by García-Rivas et al. (contribution 8) provides a comprehensive mineralogical and crystal-chemical characterization of bentonite deposits from the Benavila outcrop in Portugal, representing the largest known occurrence of bentonite in continental Portugal. The study offers detailed insight into the mineralogical composition, structural characteristics, and geochemical features of these clay-rich materials. The results reveal that smectite is the dominant phase, accompanied by variable proportions of calcite and minor accessory minerals (tectosilicates and other phyllosilicates). Crystal-chemical analyses indicate that the smectites belong to intermediate compositions within the montmorillonite–beidellite series and display mixed cis-vacant and trans-vacant structural configurations, reflecting complex formation and diagenetic processes. In addition, the geochemical data suggests the influence of carbonate-rich fluids in the evolution of the deposit. The combination of all the data shows the possible pathway of alteration that gave rise to the deposit. This study exemplifies how advanced mineralogical and microanalytical techniques can be integrated to unravel the origin, evolution, and structural variability of clay minerals. Additionally, it contributes to a deeper understanding of these bentonites as a basis for further industrial applications.

The paper by Lescano et al. (contribution 9) explores how the geological formation environment influences the physicochemical properties of a lacustrine-derived and a hydrothermal-derived sepiolite from Spanish and Argentinian companies, a critical factor for both palaeoclimatic interpretations and industrial applications. The investigation reveals striking differences between the two sepiolites. The lacustrine-derived sepiolite exhibits a higher SiO_2/MgO ratio and contains amorphous silica impurities, whereas the hydrothermal-derived sepiolite aligns closely with sepiolite's theoretical stoichiometry and displays enhanced crystallinity. Morphologically, the lacustrine-derived sepiolite forms compact, aggregated fibrous structures, while the hydrothermal-derived sepiolite is characterized by disaggregated, needle-like fibers with nanometric diameters and high aspect ratios. Both sepiolites are predominantly hydrophilic, yet the hydrothermal-derived sepiolite shows suspended particles at the interface, hinting at a slightly higher hydrophobicity than the lacustrine-derived sepiolite. These observations underscore the pivotal role of formation environment in determining sepiolite's structural, chemical, and surface features, with direct implications for its dispersibility, adsorption capacity, and interfacial behavior. The study advances our understanding of sepiolite as a paleoenvironmental indicator and informs its optimized use as an industrial raw material.

https://doi.org/10.3390/min16040342

The paper by Zakusin et al. (contribution 10) provides a comprehensive mineralogical and adsorption properties characterization of smectite-palygorskite clays from the Dashkovskoye and Borshchevskoye deposits (Lower Carboniferous, Russia), offering critical insights into their formation and environmental history. The study reveals that Al-palygorskite from the Dashkovskoye deposit formed through sedimentation from suspended matter in a shallow-water basin, contrasting with the chemogenic genesis of palygorskites and the clastic/redeposited nature of smectites. In contrast, the Borshchevskoye palygorskites exhibit a more complex terrigenous origin, formed from redeposited chemogenic Al–palygorskites transported from the Dashkovskoye region, with Fe-palygorskites entering via upstream soil redeposition. Notably, the study demonstrates a depth-dependent trend, where deeper basin layers contain increasing smectite proportions and decreasing palygorskite content. Beyond their geological significance, all analyzed clays show substantial adsorption capacities (32–49 mg-eq/100 g), highlighting their potential for industrial applications.

In summary, the contributions published in the Special Issue "Clay Minerals: From Paleoclimatic and Paleoenvironmental Indicators to Industrial Raw Materials" confirm the importance of integrated mineralogical and geochemical approaches in improving our understanding of clay mineral formation and evolution. These studies highlight how the characterization of clay minerals can provide valuable insights into past environmental and climatic conditions, while also clarifying the processes controlling the transformation and stability of clay minerals in different geological settings. The papers included in this Special Issue also emphasize the relevance of clay minerals as industrial raw materials. Through the application of modern analytical techniques, these studies illustrate how the properties of clays are closely linked to their genesis and geological environment, which ultimately determine their technological behavior and potential industrial applications.

As Guest Editors of this Special Issue, we hope that readers will find these contributions both insightful and stimulating, providing new perspectives on the role of clay minerals as indicators of paleoenvironmental and palaeoclimatic conditions and as valuable materials for industrial use. Finally, we would like to express our sincere gratitude to the Minerals editorial staff for their continuous support and guidance, and to all reviewers for their valuable time and constructive comments that helped improve the quality of the published papers.

Conflicts of Interest: The authors declare no conflicts of interest.

List of Contributions:

1. Laita, E.; Abad, I.; Reolid, M. Palaeoclimatic Inferences from Clayey-Iron Palaeosols: A Weathering Event Recorded in the Middle–Upper Jurassic Unconformity (South Iberian Palaeomargin, Western Tethys). *Minerals* **2024**, *14*, 741. https://doi.org/10.3390/min14080741.
2. del Buey, P.; Sanz-Montero, M.E.; Rodríguez-Aranda, J.P.; Sánchez-Román, M.; Nieto, F. The Miocene Source-to-Sink Evolution of Fibrous Clay Minerals in Hyperalkaline Playa-Lakes, Duero Basin (Central Spain). *Minerals* **2025**, *15*, 50. https://doi.org/10.3390/min15010050.
3. Sousa, A.P.O.; Lima, M.C.S.; Costa, W.R.P.; Nascimento, R.C.A.M.; Delgado, J.M.P.Q.; Lima, A.G.B.; Amorim, L.V. Effect of Interaction Between Expandable Minerals and Glycerin-Based Fluids on the Occurrence of Accretion. *Minerals* **2025**, *15*, 245. https://doi.org/10.3390/min15030245.
4. Bakouan, C.; Chenoy, L.; Guel, B.; Hantson, A.-L. Physicochemical and Mineralogical Characterizations of Two Natural Laterites from Burkina Faso: Assessing Their Potential Usage as Adsorbent Materials. *Minerals* **2025**, *15*, 379. https://doi.org/10.3390/min15040379.
5. Ionescu, C.; Simon, V.; Hoeck, V.; Gál, Á. Thermal Behaviour of a Carbonatic Clay: A Multi-Analytical Approach. *Minerals* **2025**, *15*, 390. https://doi.org/10.3390/min15040390.

https://doi.org/10.3390/min16040342

6. Yaremchuk, Y.; Hryniv, S.; Peryt, T. Controls on the Transformation of Clay Minerals in the Miocene Evaporite Deposits of the Ukrainian Carpathian Foredeep. *Minerals* **2025**, *15*, 395. https://doi.org/10.3390/min15040395.
7. Bauluz, B.; Yuste, A.; Alvira, S.; García-Vicente, A. The Burela Kaolin Deposit (NW Spain): Genesis, Composition and Micro- and Nanotexture. *Minerals* **2025**, *15*, 416. https://doi.org/10.3390/min15040416.
8. García-Rivas, J.; Dias, M.I.; Paiva, I.; Fernandes, P.G.; Marques, R.; García-Romero, E.; Suárez, M. Mineralogical and Geochemical Characterization of the Benavila (Portugal) Bentonites. *Minerals* **2025**, *15*, 836. https://doi.org/10.3390/min15080836.
9. Lescano, L.; Marfil, S.A.; Castillo, L.A.; Barbosa, S.E. Influence of Geological Origin on the Physicochemical Characteristics of Sepiolites. *Minerals* **2025**, *15*, 950. https://doi.org/10.3390/min15090950.
10. Zakusin, S.; Zakusina, O.; Koroleva, T.; Morozov, I.; Chernov, M.; Krupskaya, V. Composition, Genesis, and Adsorption Properties of Smectite–Palygorskite Clays (Lower Carboniferous, Russia). *Minerals* **2026**, *16*, 70. https://doi.org/10.3390/min16010070.

https://doi.org/10.3390/min16040342

minerals MDPI

Article

Palaeoclimatic Inferences from Clayey-Iron Palaeosols: A Weathering Event Recorded in the Middle–Upper Jurassic Unconformity (South Iberian Palaeomargin, Western Tethys)

Elisa Laita *, Isabel Abad and Matías Reolid

Departamento de Geología & Centro de Estudios Avanzados en Ciencias de la Tierra, Energía y Medio Ambiente (CEACTEMA), Universidad de Jaén, Campus Las Lagunillas sn, 23071 Jaén, Spain; miabad@ujaen.es (I.A.); mreolid@ujaen.es (M.R.)

* Correspondence: elaita@ujaen.es

Abstract: The study of iron crusts containing iron-coated grains from different sections of the Prebetic (SE Iberia) and the overlying marine sedimentary rocks also containing iron-coated grains in the Prebetic and the Iberian Range (NE Iberia) allowed us to determine the palaeoclimatic and palaeoenvironmental conditions under which they originated. The iron crusts are mainly composed of clay minerals (kaolinite and illitic phases) and/or goethite and hematite. The kaolinite texture indicates that it is authigenic, whereas the illitic phases are probably detrital. The mineralogy and texture of the iron crusts allow us to classify them as plinthitic palaeosols. The iron-coated grains consisting of a nucleus and a cortex, both composed of a mixture of kaolinite, goethite, and hematite, originated in situ during the plinthite development. Reworking processes caused the fragmentation and incorporation of the iron-coated grains into the overlying ferruginous oolithic limestones and terrigenous-carbonated breccia. New marine iron-coated grains formed later in the ferruginous oolithic limestones. The high Chemical Index of Alteration and Chemical Index of Weathering values and the geochemical ratios (Ba/Sr, Rb/Sr, Sr/Cu, Ga/Rb) from iron deposits reflect intense weathering under warm and humid conditions in the South Iberian Palaeomargin during the Callovian–Oxfordian, which may also take place in the East Iberian Palaeomargin (Iberian Range).

Keywords: geochemical ratios; illitic phases; iron-coated grains; iron crusts; kaolinite; palaeoclimate; plinthite

1. Introduction

Clay minerals can be found in soils, sediments, and numerous sedimentary rocks at the Earth's surface. The genesis of clay minerals occurs predominantly in areas where chemical weathering and edaphic processes (i.e., related to soil) take place, which are strongly controlled by climatic factors such as temperature and precipitation [1,2]. Consequently, the study of clay minerals from palaeosols (i.e., soils developed in the landscapes of the past) gives key information for palaeoclimatic and palaeoenvironmental reconstructions [3–6].

Variations in clay mineral assemblages formed in palaeosols developed in Mesozoic successions have been used as palaeoclimatic and palaeoenvironmental proxies [6–9].

Intense chemical weathering processes occur under warm and humid conditions (subtropical to tropical), giving place to a very effective hydrolysis resulting in the genesis of oxisols (according to the USDA (United States Department of Agriculture) soil taxonomy) and bauxites. In these tropical soils, kaolinite is the main clay mineral together with Al hydroxides and Fe oxyhydroxides [1–3,6,9–11].

Palaeosols can contain not only authigenic clay minerals formed by chemical weathering but also clay minerals of different origins (detrital or diagenetic). This must be taken into account when using clay minerals as palaeoclimatic and palaeoenvironmental proxies. Detrital minerals are inherited and will, therefore, reflect the weathering processes developed in the source area, whereas diagenetic processes can modify the association of clay minerals (both detrital and authigenic), transforming them into new diagenetic phases and then invalidating the use of authigenic clay minerals as palaeoclimatic and palaeoenvironmental proxies [6,8,9,12–14].

The worldwide loss of carbonate productivity from the Late Callovian to Early Oxfordian has been attributed to climatic and oceanographic changes [15,16]. Oxygen isotope data point to a progressive cooling of the ocean waters from Middle Callovian to Early Oxfordian with the subsequent sea-level fall and the erosion of carbonate platforms resulting in common unconformities in wide areas, for example, in the Tethys [17–21]. Subsequently, a warming event occurred during the Middle Oxfordian [15,22–25]. The $\delta^{18}O$ variations also reflect precipitation regimes against a general warm climate for the Middle to Late Oxfordian [15,16].

During the Middle–Late Jurassic transition, the Tethyan Domain experienced important sea-level fluctuations due to the fragmentation of continental palaeomargins under an extensive rifting phase [21,26–28]. The fragmentation of continental palaeomargins and the sea level changes in the Western Tethys gave place to a stratigraphic unconformity recorded in different basins of western Europe and northern Africa [17–20,29–31].

Numerous regional studies describe the presence of this discontinuity in the External Prebetic of the Betic Cordillera (SE Iberia), which represents the Mesozoic sedimentation of the South Iberian Palaeomargin, with an iron crust containing iron-coated grains overlayed by ferruginous oolithic limestones [31]. This discontinuity has previously been characterized from a sedimentological, mineralogical, and geochemical point of view [21,32]. However, detailed characterization of the texture of the clay minerals assemblages in the iron crusts and the iron-coated grains is still lacking.

In this work, X-ray diffraction, optical and electron microscopy, and geochemical analysis have been carried out to undertake mineralogical characterization of the iron crust containing iron-coated grains to the Middle–Late Jurassic discontinuity of the External Prebetic, that represents proximal areas of the South Iberian Palaeomargin. New outcrops and those previously reported in the literature for the External Prebetic, are analyzed, including examples from the same discontinuity from the East Iberian Palaeomargin, represented by the Iberian Range. The analyses allow us: (1) to determine the origin of the clay minerals (authigenic, detrital, or diagenetic), (2) to evaluate the palaeoclimatic and palaeoenvironmental conditions under which these iron crust and iron-coated grains were generated during the Middle–Upper Jurassic transition in this area of the Betic Cordillera (SE Iberia), and (3) to compare the palaeoenvironmental conditions and processes occurring in the South Iberian Palaeomargin (External Prebetic) and the East Iberian Palaeomargin (Iberian Range).

2. Geological Setting

The Betic Cordillera, situated to the S and SE of the Iberian Peninsula, represents the western most sector of the European Alpine Orogen. Two main geological domains are differentiated in the Betic Cordillera: the Betic External Zones and the Betic Internal Zones [33]. The Betic External Zones comprise Mesozoic to Lower Neogene sedimentary rocks that were deposited in the South Iberian Palaeomargin (Western Tethys). The higher rank division of the Betic External Zones is into Prebetic and Subbetic [33,34]. Within the Jurassic paleogeography of the South Iberian Palaeomargin, the Prebetic represents a comparatively proximal epicontinental shelf, whereas the Subbetic represents distal epioceanic swell and through areas. According to stratigraphic and tectonic data, the Prebetic domain is subdivided into External and Internal Prebetic [33,35] (Figure 1). The Prebetic successions are mainly formed by shallow marine facies, with continental episodes and intervals of erosion, whereas pelagic facies are dominant in the Subbetic domain [33,36].

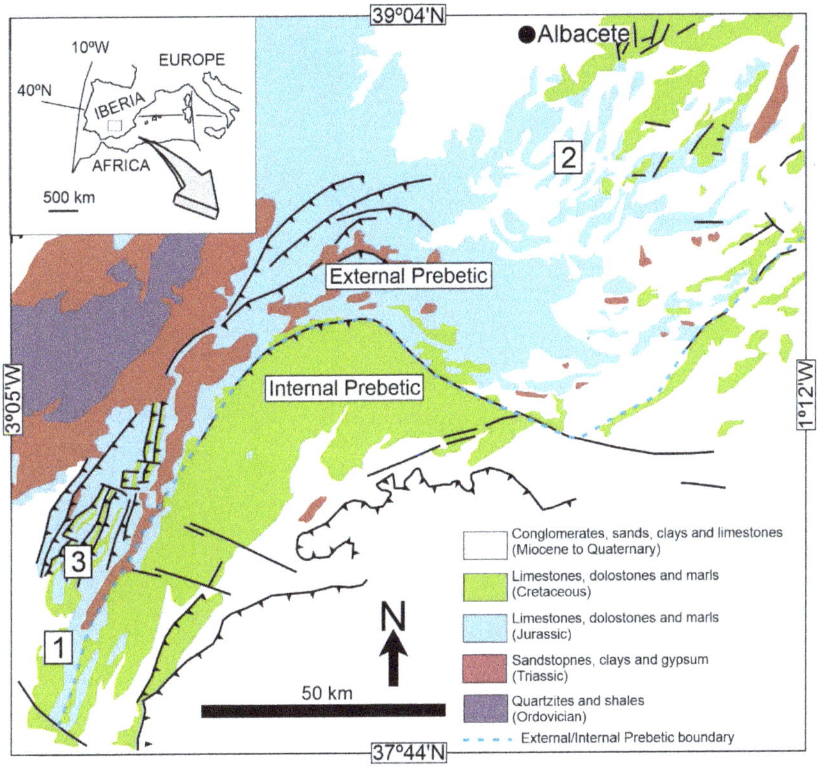

Figure 1. Geological map of the Betic Cordillera showing the Prebetic and its location on the Iberian Peninsula (SE Spain). The location of the previously studied outcrops by Reolid and Abad [21] and Reolid et al. [32] in the Central External Prebetic (1: RGCH, RGCHP, and CHO sections) and Eastern External Prebetic (2: PC1 to P4 sections) are represented together with the new outcrop analyzed in this study located in the north area of the Central External Prebetic (3: VI section). Modified from Reolid et al. [32].

The Middle–Upper Jurassic materials studied here are located in External Prebetic outcrops (Figures 1 and 2). During the Early–Middle Jurassic, the External Prebetic constituted a carbonate inner-self environment, which became a carbonate–siliciclastic mid-shelf environment during the Late Jurassic (Middle Oxfordian to Early–Middle Kimmeridgian) [37]. There is a discontinuity associated with the Middle/Upper Jurassic boundary [36,38].

In the Central External Prebetic, the Callovian/Oxfordian boundary is located between the carbonate deposits (Lower–Middle Jurassic) of the Chorro Formation [39,40] and the marl–limestone rhythmites (Middle–Upper Oxfordian) of the Lorente Formation [40,41]. The Chorro Formation encompasses approximately 400 m of Lower Jurassic dolomitized limestones, along with 50 m of Middle Jurassic oolithic limestones with megaripples and oncolithic–bioclastic facies [32]. The presence of *Protopeneroplis striata* [35,42], *Mesoendothyra croatica*, and *Trocholina palestiniensis* [38] confirms Middle Jurassic age and a shallow inner-shelf setting. The Upper Jurassic (Middle Oxfordian–Lower Kimmeridgian) Lorente Formation constitutes the onset of hemipelagic sedimentation within the southeastern epicontinental system. In the proximal sectors of the shelf, the Lorente Formation mainly consists of 70–100 m thick succession of marl–limestone rhythmites, as documented by Reolid [40] and Rodríguez-Tovar [43]. The Lorente Formation is commonly preceded by <1 m of spongiolithic limestones [44,45].

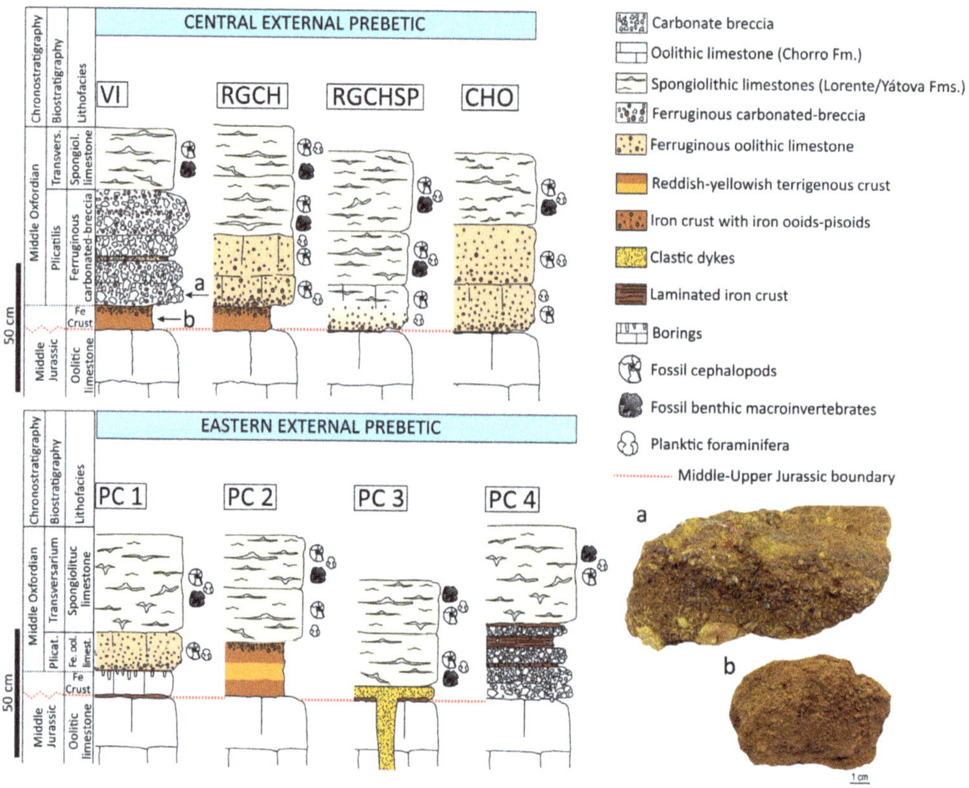

Figure 2. Sections from the central and eastern sectors of the External Prebetic showing the iron crusts and the ferruginous oolithic limestones in the Callovian–Oxfordian discontinuity (modified from Reolid and Abad [21]) and images of the ferruginous–carbonated breccia containing iron-coated grains (**a**) and the iron crust containing iron-coated grains (**b**) from the Villas (VI) section.

In the eastern sector, this discontinuity is located between the Chorro Formation and the spongiolithic limestones (Middle–Upper Oxfordian) of the Yátova Formation [46]. The spongiolithic limestones can reach around 16 m thick [44,45]. This stratigraphic succession of the Eastern External Prebetic is similar to that described in the Iberian Range [19,46].

Iron crusts containing iron-coated grains have been documented in relation to this Middle/Upper Jurassic discontinuity in the Sierra de Cazorla (Central External Prebetic) and Altos de Chinchilla (Eastern External Prebetic) [21,32] (Figure 2). These crusts are also observed in the Sierra de las Villas (north area of the Central External Prebetic), showing a palaeosols appearance (Figure 2). According to Reolid et al. [32], the discontinuity in the External Prebetic appears as a slightly irregular erosive surface with abundant iron oxides, which coincides with the palaeokarst previously described in the Eastern External Prebetic [21,47].

Ferruginous oolithic limestones (10–40 cm thick) occur over the iron crust in the central (Sierra de Cazorla) and eastern (Altos de Chinchilla) sectors of the External Prebetic [21,32] (Figure 2). Locally, these ferruginous oolithic limestones overlie the discontinuity surface directly and pinch out laterally [21,31,32,40]. This lithofacies correlates with the Arroyofrío Formation of the Iberian Range that preceded the spongiolithic limestone lithofacies of the Yátova Formation (SE Iberia) [19,46].

3. Samples and Methods

This work analyzes iron crusts and iron-coated grains from the External Prebetic sections previously studied by Reolid and Abad [21] and Reolid et al. [32]. Additionally, it includes a new outcrop located in the north area of the Central Prebetic (Sierra de las Villas, province of Jaén, SE Spain) (Figures 1 and 2). This new outcrop provides new sedimentary features, complementing the knowledge about the environmental conditions that occurred during the Middle–Upper Jurassic transition. In addition, this study includes samples from the equivalent ferruginous oolithic limestones from the neighbor domain, the Iberian Range, represented by the Arroyofrío Formation. With respect to previous works, this study is focused on the presence and palaeoenvironmental significance of clay minerals in these deposits.

Therefore, the studied outcrops are distributed in the Central External Prebetic, Eastern External Prebetic, and Iberian Range. The sections from the Central External Prebetic are located in the western nappes of the Sierra de Cazorla (province of Jaén, SE Spain): Riogazas-Chorro (RGCH, 37°52′22″ N, 3°0′26″ W), Riogazas-Chorro-SP (RGCHSP, 37°52′22″ N, 3°0′22″ W), and El Chorro (CHO, from 37°52′27″ N, 3°0′22″ W to 37°52′22″ N, 3°0′16″ W). The new outcrop in Sierra de las Villas (province of Jaén, SE Spain) is located 12 km north of the outcrops of Sierra de Cazorla: Villas (VI, 37°59′08″ N, 2°55′52″ W).

The sections from the Eastern External Prebetic are located in the Sierra del Chortal (province of Albacete, SE Spain): Pozo Cañada 1 (PC1, 38°47′36″ N, 1°43′25″ W), Pozo Cañada 2 (PC2, 38°47′37″ N, 1°43′25″ W), Pozo Cañada 3 (PC3, 38°47′45″ N, 1°43′18″ W) and Pozo Cañada 4 (PC4, 38°47′48″ N, 1°43′13″ W). This area is around 140 km northeast of the outcrops of the Central External Prebetic.

Finally, the iron-coated grains studied in the Iberian Range for comparing with the External Prebetic come from the Arroyofrío Formation of the Sierra de Albarracín (province of Teruel, E Spain): Gea de Albarracín (AF section, 40°24′45″ N, 1°19′56″ W). This outcrop is around 310 km northeast of the outcrops of the Central External Prebetic.

3.1. X-ray Diffraction Study

The matrix of the iron crust from Sierra de las Villas and the iron-coated grains separated from this matrix were analyzed by X-ray diffraction (XRD) from unoriented powders (whole rock analysis) to determine their mineralogical composition. The <2 μm fraction of the crust was extracted by centrifugation and was air-dried, and ethylene glycol-treated oriented aggregates were prepared by sedimentation on glass slides to determine the clay minerals present in the samples. To obtain the diffraction patterns, a PANalytical Empyrean diffractometer was used at the Scientific-Technical Instrumentation Center (CICT) of the University of Jaén (Spain), equipped with an X′Celerator solid-state linear detector and θ/θ goniometer, using CuKα radiation, 45 kV voltage and 40 mA current. The XRD patterns were acquired from 4° to 64° 2θ for the whole rock samples and from 4° to 32° 2θ for the <2 μm fraction. In both cases, a step increment of 0.01° 2θ and a counting time of 10s/step was used. The recording was made with XPowderX software (Version 2023.06.01) [48].

Relative proportions of the mineral phases were calculated using Reference Intensity Ratios (RIR) from the literature [49–51]. The RIR values were calculated following the procedure described by Hillier [52]. The Al^{3+} content in the goethite structure was calculated in the unoriented powder XRD patterns of the crust and the iron pisoids as an indicator of the origin of goethite [32,53], following the procedure proposed by Schulze [54] based on the position of de d_{110} and d_{111} peaks of goethite in the XRD patterns.

3.2. Optical and Electron Microscopy Study

Polished thin sections of the iron crust and the ferruginous–carbonated breccia from the Villas section were studied by optical microscopy to identify the minerals and characterize their texture. Polished thin sections of the ferruginous oolithic limestone from RGCH, CHO, and PC1 sections and the equivalent lithofacies of the AF section in the Iberian Range were also studied by optical microscopy to characterize the texture of iron-coated grains and compare them with those from the Villas section. The iron-coated grains in the crust, the ferruginous oolithic limestone, and the ferruginous–carbonated breccia, are described as follows: macropisoids (>5 mm), pisoids (5–1 mm), ooids (1000–100 μm) and micro-ooids (<100 μm) [55].

The polished thin sections of the Villas section (the crust and the ferruginous–carbonated breccia) and rock fragments of the iron crust and the iron-coated grains were examined by Scanning Electron Microscopy (SEM) at the CICT (University of Jaén, Spain) using a Merlin Carl Zeiss Gemni II SEM. The polished thin sections and the rock fragments were previously carbon-coated. Compositional images were obtained using two detectors: an angular selective backscattered electron detector (AsB) and an energy selective backscattered electron detector (EsB). Chemical information and mapping were also obtained using an energy-dispersive X-ray spectroscopy (EDS). The accelerating voltage for the AsB and the EDS was 15 kV with a beam current of 750 pA, whereas for the EsB, the accelerating voltage was 4 kV with a beam current of 750 pA. For the rock fragments, morphological images were obtained using a secondary (In-lens) detector. In this case, the accelerating voltage was 5 kV with a beam current of 750 pA. A polished thin section of the ferruginous oolithic limestone from the AF section and rock fragments of the iron crust from the PC2 section were also studied by SEM using the same analytical conditions as for the Villas section samples. The purpose of studying these samples was to compare the mineralogy and texture of the clay minerals in the iron-coated grains of the AF section and the iron crust of the PC2 section with those observed in the Villas section.

For those figures containing XRD patterns, optical and SEM images, mineral abbreviations are according to Whitney and Evans [56].

3.3. Geochemical Study

Elemental mappings of the ferruginous–carbonated breccia from the Villas section have been obtained by an X-ray-microfluorescence M4 Tornado Brucker at the CICT (University of Jaén, Spain). The major elements of the matrix of the iron crust and the iron-coated grains from the Villas section were analyzed separately using X-ray fluorescence (XRF) in a PANalytical Zetium spectrometer with a maximum power of 4 kW and provided with an X-ray tube of Rh anode. Trace element analyses were carried out using a PerkinElmer NexION 300D (PerkinElmer, Inc., Waltham, MA, USA) inductively coupled plasma-mass spectrometer (ICP-MS). A total of 100 mg of sample powders were digested using HNO_3 + HF in a Teflon-lined vessel at 180 °C and 200 p.s.i. for 30 min, evaporation to dryness, and then subsequently dissolved in 100 mL of 4 vol% HNO_3. Both the XRF and the ICP-MS analyses were carried out at the Scientific Instruments Centre of the University of Granada (CIC-UGR, Spain).

Palaeoclimatic and palaeoweathering proxies (Ba/Sr, Sr/Cu, Ga/Rb, and Rb/Sr) were calculated from the chemical analyses of the iron crust from the Villas section and the chemical analyses reported by Reolid et al. [32] for the iron crust in the RGCH section to infer the palaeoclimatic conditions.

To evaluate the degree of chemical weathering in the iron crust from the Villas section and those reported in the RGCH and PC2 sections [21,32], the Chemical Index of Alteration (CIA) and the Chemical Index of Weathering (CIW) were determined. These indexes were calculated by adapting the formula proposed by Nesbitt and Young [57]:

$$CIA = [Al_2O_3 / (Al_2O_3 + Na_2O + K_2O)] \times 100 \quad (1)$$

and the formula proposed by Harnois [58]:

$$CIW = [Al_2O_3 / (Al_2O_3 + Na_2O)] \times 100 \quad (2)$$

The CaO values are not considered in these calculations since, as will be shown later, they come from the small proportions of calcite present in the crusts and not from the phyllosilicates.

Additionally, the palaeoprecipitation regime was calculated using the climofunctions of Sheldon et al. [59].

$$P = 221\, e^{0.0197(\text{CIA-K})} \tag{3}$$

$$P = 14.265\,(\text{CIA-K}) - 37.632 \tag{4}$$

$$P = -259.34\,\ln(B) + 759.05 \tag{5}$$

where P is the palaeoprecipitation (mm/yr) and B in Equation (5) is the molecular ratio of $CaO + MgO + Na_2O$ to Al_2O_3 [59].

4. Results

4.1. Sedimentary Features

In the Villas section, the boundary between the Middle and Upper Jurassic is represented by a reddish terrigenous iron crust containing iron-coated grains (Figure 2). This crust is harder than the crust described in the other sections from the Central External Prebetic. According to Reolid and Abad [21], the iron crust is poorly preserved in the Prebetic. When preserved, it is a hard crust ranging from 2 to 10 cm in the Central Prebetic of Sierra de Cazorla (RGCH section). This crust is represented by a soft reddish to yellowish crust in the Eastern Prebetic (PC2 section), but commonly, it is not recorded or millimetric (PC 1 section). The iron crust related to the Middle–Upper Jurassic discontinuity has not been recorded in the Iberian Range.

Over the iron crust, and commonly, directly overlying the Middle Jurassic carbonates, there is the ferruginous oolithic limestone both in the External Prebetic and the Iberian Range [19,21,31,46]. Iron-coated grains are commonly recorded in these lithofacies together with abundant fossil remains such as ammonoids and belemnites that indicate the Middle Oxfordian Plicatilis Zone [31]. Over the ferruginous oolithic limestone, there are spongiolithic limestones with their lower part dated as Middle Oxfordian (Transversarium Zone) [31] (Figure 2).

In the Villas section, the iron crust is followed by a yellow ferruginous–carbonated breccia containing iron-coated grains (Figure 2) as well as a wide range of sedimentary rock fragments with different textures. This breccia does not occur in the other sections described in the External Prebetic [21,32] (Figure 2). There are no fossils of ammonoids and belemnites in the breccia of the Villas section. This lithofacies constituted a clear difference with respect to the rest of the studied outcrops. Over this lithofacies, there are spongiolithic limestones, the lower part of which dated to the Middle Oxfordian (Transversarium Zone). The age of the ferruginous–carbonated breccia is probably the same as the ferruginous oolithic limestone (Plicatilis Zone).

4.2. The Iron Crust Containing Iron-coated Grains

4.2.1. Mineralogical Composition

The XRD pattern of the whole rock and the calculated relative proportions indicate that the iron crust from the Villas section is mainly composed of clay minerals (44%), quartz (35%), and goethite (11%), together with minor proportions of calcite, dolomite, orthoclase and hematite (<5%) (Figure 3a).

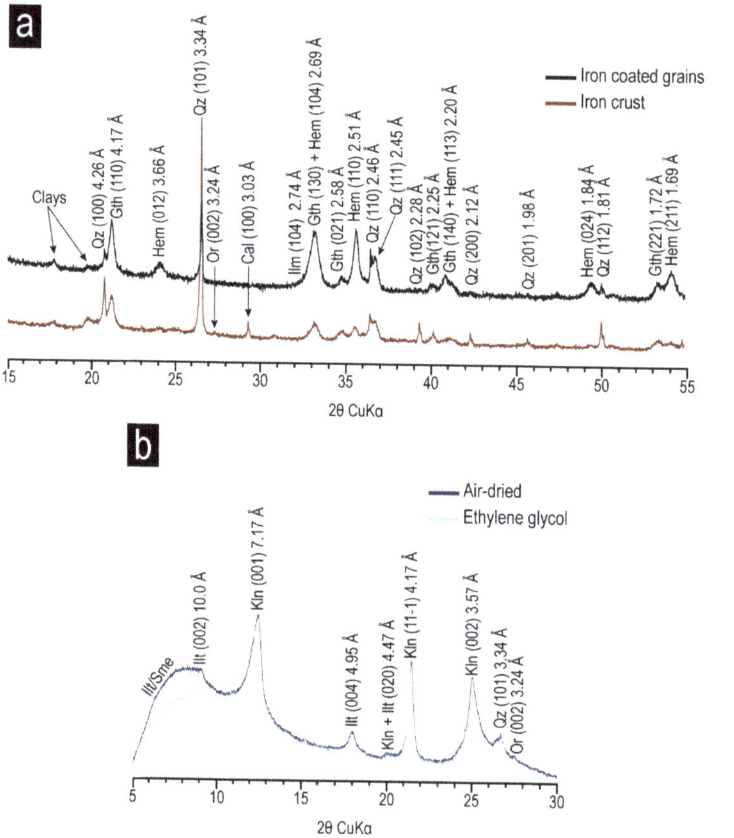

Figure 3. (**a**) XRD patterns obtained from the whole rock of the iron crust and the iron-coated grains from the Villas section; (**b**) XRD patterns obtained from the <2 μm fraction of the iron crust from the Villas section. *Clays* = clay minerals, *Qz* = quartz, *Gth* = goethite, *Hem* = hematite, *Or* = orthoclase, *Cal* = calcite, *Ilm* = ilmenite, *Ilt/Sme* = illite/smectite mixed layers, *Ilt* = illitic phases, *Kln* = kaolinite.

In the <2 μm fraction, abundant kaolinite (43%) and illitic phases (57%) are identified in the XRD pattern of the iron crust from the Villas section. Characteristic reflections of kaolinite at 7.15, 4.18, and 3.57Å are identified, and dickite and nacrite reflections are not detected. Illite and micas (quantified together) are included within the term "illitic phases", as well as random mixed-layer illite/smectite (Ilt/Sme), which were not quantified but were detected in the XRD pattern (Figure 3b).

The XRD patterns of the iron-coated grains included in the crust from the Villas section show that they are mainly composed of 34% of clay minerals (mainly kaolinite as deduced by SEM study), 23% of goethite, 23% of quartz, 10% of ilmenite, and 9% of hematite (Figure 3a).

According to the method proposed by Schulze [54] for determining the Al^{3+} content in the goethite structure, based on the position of de d_{110} and d_{111} peaks of this mineral phase, thereis an Al-substitution in goethite of 13% mol in the iron crust and of 7% in the iron-coated grains from the Villas section.

4.2.2. Textural Features

The optical microscopy images show that the iron crust from the Villas section is formed by a fine-grained matrix containing clay minerals and iron oxides whose small size did not allow their identification under optical microscopy resolution (Figure 4a). Abundant quartz grains (300 μm–1 mm) are differentiated within the matrix (Figure 4b).

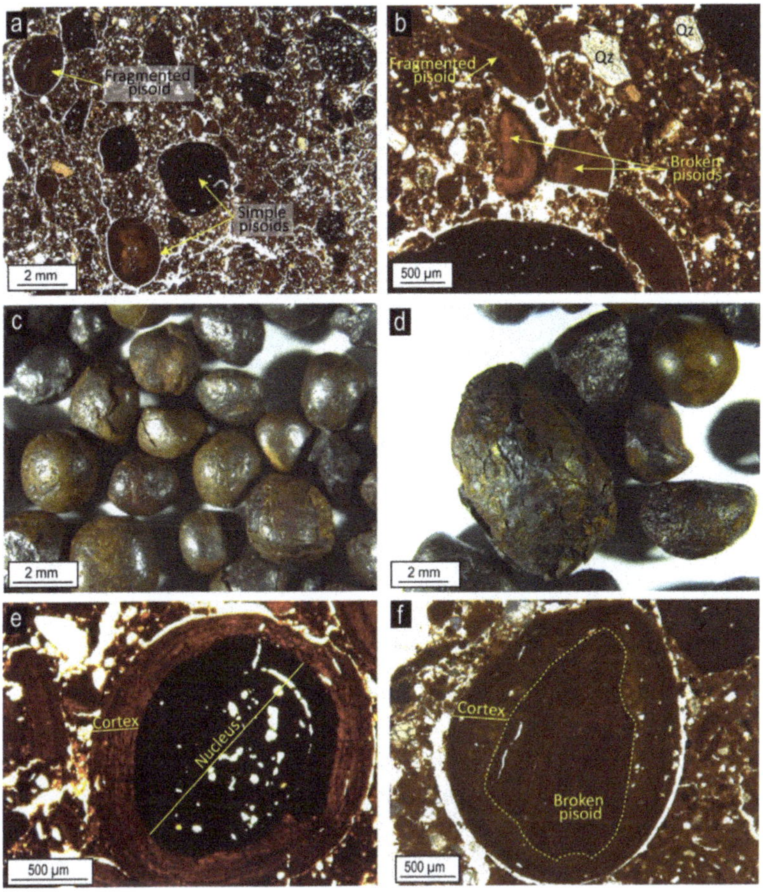

Figure 4. Optical microscopy (**a**,**b**,**e**,**f**) and binocular loupe (**c**,**d**) images of the iron crust containing iron-coated grains from the Villas section: (**a**) Simple and fragmented pisoids in the clay-rich matrix of the crust; (**b**) Fragmented pisoids, broken pisoids, and quartz clasts in the clay-rich matrix of the crust; (**c**,**d**) Spherical to ellipsoidal pisoids and macropisoids extracted from the crust; (**e**) Simple pisoid showing a nucleus composed by an indeterminate ferruginous lump and a cortex formed by several concentric layers; (**f**) fragmented pisoid whose nucleus is a broken pisoid. *Qz* = quartz.

The iron-coated grains included in the iron crust from the Villas section are spherical to ellipsoidal with sizes from 300 μm to 7 mm, encompassing macropisoids, pisoids, and ooids according to Bárdossy [55], being the pisoids the predominant ferruginous particles in the crust (Figure 4c). These ferruginous particles are included in the type A category of the iron-coated grains proposed by Reolid et al. [32] based on morphology and internal microstructure. According to these authors, type A iron-coated grains are constituted by a thin and regular lamination, forming concentric layers around a nucleus. The nuclei of the macropisoids, pisoids, and ooids from the Villas section are formed by an indeterminate ferruginous lump or by a fragment of a previously broken iron-coated grain, which allows them to be classified as simple and fragmented pisoids, according to Guerrak [60] classification (Figure 4a,b,e,f). The nucleus and the cortex (formed by concentric layers) of the simple pisoids are composed of a mixture of clay minerals, iron oxides, and occasional quartz grains. Some simple pisoids are broken and can act as a nucleus for the fragmented pisoids (Figure 4b,f). The fragmented pisoids also present several concentric layers with the same composition as those of the simple pisoids (Figure 4f).

Backscattered and secondary electron images of the iron crust from the Villas section show a clay-rich matrix formed by kaolinite and iron oxides (hematite and goethite as detected by XRD) (Figure 5a–c). Quartz grains with anhedral morphologies are also observed (Figure 5a). Kaolinite crystals appear as subhedral to euhedral nanometric plates (100 nm–1 μm), showing pseudohexagonal outlines with random orientation, as well as forming book-type aggregates (Figure 5a,b). The illitic phases are larger than kaolinite crystals (5–10 μm) and present subhedral morphologies with their sheets frequently separated (Figure 5c). Nanometric kaolinite crystals growing over illitic plates are also observed (Figure 5d).

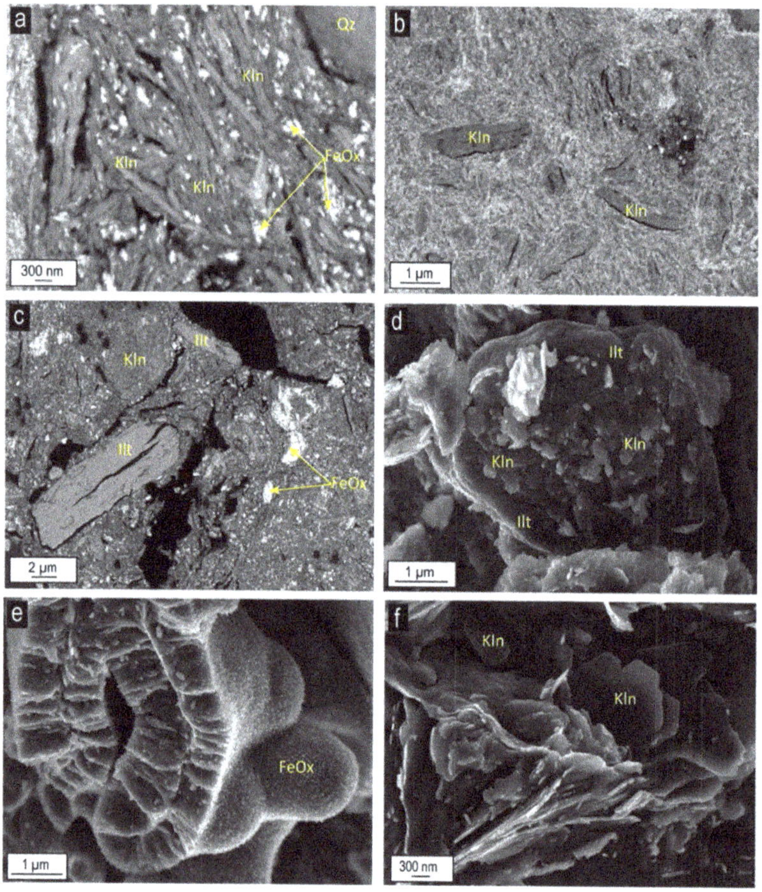

Figure 5. SEM images of the iron crust from the Villas section (**a**–**d**) and the iron crust from the PC2 section (**e**,**f**) showing: (**a**) a mixture of nanometric kaolinite plates and nanometric iron oxides forming the matrix of the crust; (**b**) nanometric to micrometric kaolinite booklets in the matrix of the crust; (**c**) micrometric subhedral illitic phases showing their sheets separated; (**d**) nanometric pseudohexagonal kaolinite plates over micrometric subhedral illitic phases; (**e**) tube-like filaments composed by nanometric acicular iron oxides; (**f**) nanometric pseudohexagonal kaolinite plates with random orientation. *FeOx* = iron oxides, *Kln* = kaolinite, *Qz* = quartz, *Ilt* = illitic phases.

The new analysis of the secondary electron images of the iron crust from the PC2 section shows areas with nanometric iron oxide crystals forming tube-like filaments (~1 μm in diameter) (Figure 5e), as well as areas with nanometric kaolinite crystals with subhedral shapes (Figure 5f).

The nucleus and the cortex of the iron-coated grains embedded in the clay-rich matrix of the iron crust from the Villas section are mainly composed of hematite and goethite with different proportions of kaolinite (Figure 6). Kaolinite presents the same morphological features as those observed in the clay-rich matrix of the iron crust, forming subhedral nanometric plates with random orientation, as well as book-type aggregates (Figure 6a,b). Some of the kaolinite book-type aggregates present iron oxide crystals between their sheets (Figure 6b). In addition, the concentric layers that form the cortex of the iron pisoids contain different amounts of kaolinite and iron oxides, allowing the different layers to be differentiated (Figure 6c,d).

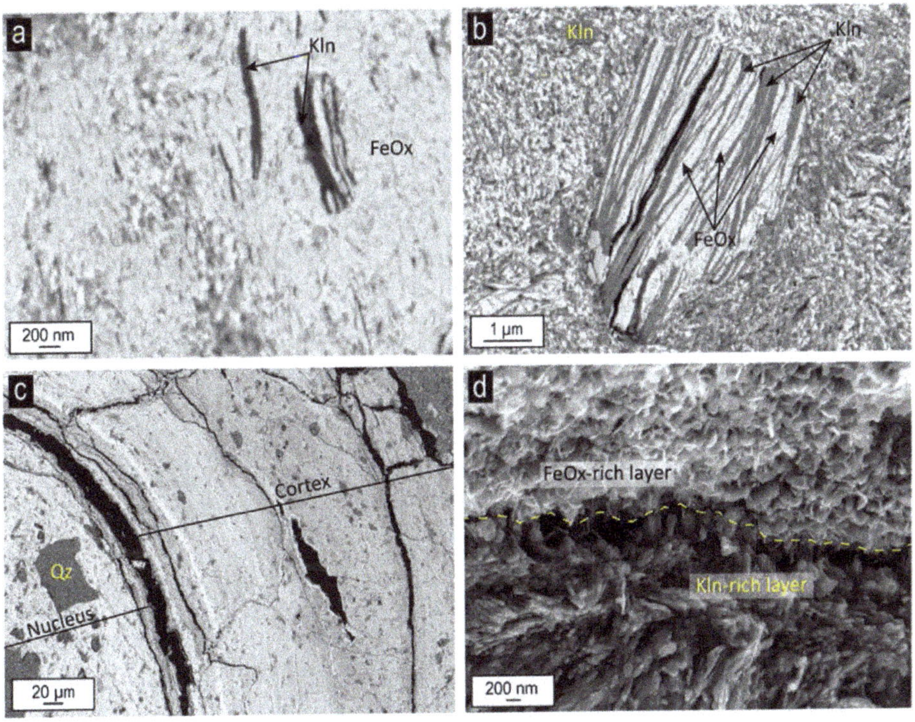

Figure 6. SEM images of the iron-coated grains in the iron crust from the Villas section showing: (**a**,**b**) nanometric kaolinite plates and iron oxides, with iron oxides occasionally growing between the kaolinite plates; (**c**) the nucleus and the cortex of a simple pisoid both composed by different proportions of iron oxides and kaolinite; (**d**) the cortex of a simple pisoid where two layers can be differentiated by their higher content in fibrous iron oxides crystals or kaolinite plates. *Kln* = kaolinite, *FeOx* = iron oxides, *Qz* = quartz.

4.2.3. Geochemical Composition

The geochemical composition of major and minor elements of the matrix of the iron crust and the iron-coated grains from the Villas section is represented in Tables 1 and 2. The geochemical composition of the previously analyzed iron crusts and iron-coated grains from the Prebetic reported by Reolid and Abad [21] and Reolid et al. [32] are also included in Tables 1 and 2 for comparison with the new data from the Villas section.

Table 1. Major element contents (wt.%) of the iron crust and the iron-coated grains obtained for the Villas section and those reported by Reolid and Abad [21] and Reolid et al. [32] for the iron crusts of the RGCH and PC2 sections and the iron-coated grains from the RGCH section. The obtained values for the CIA and CIW indexes calculated from all the iron crusts are also included, as well as the palaeoprecipitation regime (PPR) (mm/yr) obtained from the average of Formulae 3–5.

	Iron Crust			Iron-Coated Grains	
	Villas	RGCH	PC2	Villas	RGCH
SiO_2	39.33	3.84	11.95	11.70	5.39
Al_2O_3	12.91	2.52	5.41	7.32	6.47
Fe_2O_3	31.06	80.48	71.19	72.70	67.28
MnO	0.07	0.03	0.04	0.08	0.05
MgO	0.67	0.33	0.27	0.10	0.52
CaO	2.19	2.31	1.89	0.25	4.69
Na_2O	0.06	-	0.12	0.03	-
K_2O	0.51	0.10	0.31	0.02	0.24
TiO_2	0.88	0.14	0.56	0.62	0.53
P_2O_5	0.09	1.76	0.49	0.03	3.58
LOI	12.15	10.88	7.45	5.78	6.45
CIA	95.20	95.87	90.87	-	-
CIW	91.99	93.59	88.14	-	-
PPR	1380.20	1329.00	1325.06	-	-

Table 2. Trace element contents (ppm) of the iron crust and the iron-coated grains obtained for the Villas section and those reported by Reolid et al. [32] for the iron crust and the iron-coated grains for the RGCH section. The Ba/Sr, Sr/Cu, Rb/Sr, and Ga/Rb ratios calculated from these analyses are also included.

	Iron Crust		Iron-Coated Grains	
	Villas	RGCH	Villas	RGCH
Li	67.00	-	35.12	-
Rb	24.64	5.15	2.04	10.36
Cs	2.49	0.41	0.40	0.81
Be	5.44	-	6.79	-
Sr	22.43	44.53	14.90	65.86
Ba	80.4	27.29	34.50	51.69
Sc	23.05	14.23	25.60	15.10
V	424.02	1414.42	1067.87	1908.24
Cr	154.44	186.09	280.63	1058.94
Co	30.14	143.36	39.61	121.09
Ni	65.4	384.68	72.49	369.06
Cu	36.85	54.94	39.79	35.25
Zn	20.78	289.60	9.44	311.37
Ga	37.39	7.44	41.61	13.44
Y	88.14	66.28	80.87	107.45
Nb	14.87	9.57	11.72	40.42
Ta	1.33	0.31	0.87	1.02
Zr	168.66	62.52	125.25	147.71
Hf	4.76	1.39	3.23	4.31
Mo	2.69	56.07	8.18	10.18
Sn	3.8	0.64	2.58	3.01
Tl	0.18	0.13	0.06	0.56
Pb	59.67	150.74	139.22	120.89
U	2.87	13.17	4.74	8.02
Th	22.4	6.97	26.54	27.87
La	134.42	29.58	107.02	55.09
Ce	211.29	13.79	146.33	31.30

Table 2. *Cont.*

	Iron Crust		Iron-Coated Grains	
	Villas	RGCH	Villas	RGCH
Pr	32.58	4.42	25.92	10.38
Nd	128.71	19.03	104.25	43.83
Sm	28.15	4.09	24.51	10.35
Eu	6.49	0.95	5.98	2.30
Gd	20.62	5.13	18.20	11.09
Tb	2.88	0.87	2.52	1.81
Dy	12.67	5.76	14.16	12.07
Ho	2.86	1.41	2.84	2.67
Er	6.45	4.01	6.95	7.14
Tm	0.86	0.60	0.97	1.06
Yb	4.87	3.47	5.39	6.02
Lu	0.74	0.55	0.78	0.83
Ba/Sr	3.58	0.61	0.67	0.23
Sr/Cu	0.61	0.81	0.37	1.87
Rb/Sr	1.10	0.12	0.14	0.16
Ga/Rb	1.52	1.45	20.40	1.28

The content in major elements of the matrix of the iron crust from the Villas section (Table 1) fits well with the mineralogy detected by XRD, showing high contents in Fe_2O_3 (31.06%), SiO_2 (39.33%) and Al_2O_3 (12.92%), reflecting the high presence of clay minerals (kaolinite and illitic phases), as well as goethite. In the case of the iron crust from the RGCH and PC2 sections, their higher Fe_2O_3 content (80.4% and 71.19%, respectively), reflects the higher presence of hematite and goethite. The higher content of P_2O_5 (1.75%) in the RGCH iron crust (0.07%–0.48%) also reflects the presence of apatite reported by Reolid et al. [32]. In addition, the CaO content in all the crusts correlates with calcite content. The iron crusts from the Villas, RGCH, and PC2 sections contain minor proportions of calcite, which is reflected in their lower CaO content (2.19%, 2.31%, and 1.89%, respectively). For this reason, CaO content was not considered when calculating CIA and CIW indexes.

The iron-coated grains included in the iron crust from the Villas section show higher Fe_2O_3 (72.70%) and lower Al_2O_3 (7.32%) and SiO_2 (11.70%) contents than the iron crust, reflecting the higher presence of goethite with respect to clay minerals.According to the Fe_2O_3-Al_2O_3-SiO_2 diagram of Aleva [61], the iron crust from the RGCH and PC2 sections can be classified as ferrites, whereas the iron crust from the Villas section, plots on the kaolinite field but closer to the laterite field (Figure 7a).

Weathering trends are represented using the Fe_2O_3 + MgO-CaO + Na_2O + MgO-Al_2O_3 diagram of Mitchell and Sheldon [62] and the Al_2O_3-CaO + Na_2O-K_2O diagram of Nesbit and Young [63] (Figure 7b,c). The CaO + Na_2O + MgO content is low in all the iron crusts, whereas Fe_2O_3 + MgO content is higher in the RGCH and PC2 sections, and Al_2O_3 content is higher in the Villas section (Figure 7b). The bedrock of the iron crust from the Villas section has a high degree of disintegration to soil compared to that of the iron crust from the RGCH and PC2 sections (Figure 7c).

The calculated CIA and CIW values are higher in the iron crust from the RGCH (CIA = 95.87, CIW = 93.58) and Villas (CIA = 95.20, CIW = 91.99) sections than in the iron crust from the PC2 section (CIA = 90.89, CIW = 88.14). In the same way, the palaeoprecipitation regime is higher in the Villas (1380.20 mm/yr) and RGCH (1329.00 mm/yr) sections than in the PC2 section (1325.06 mm/yr).

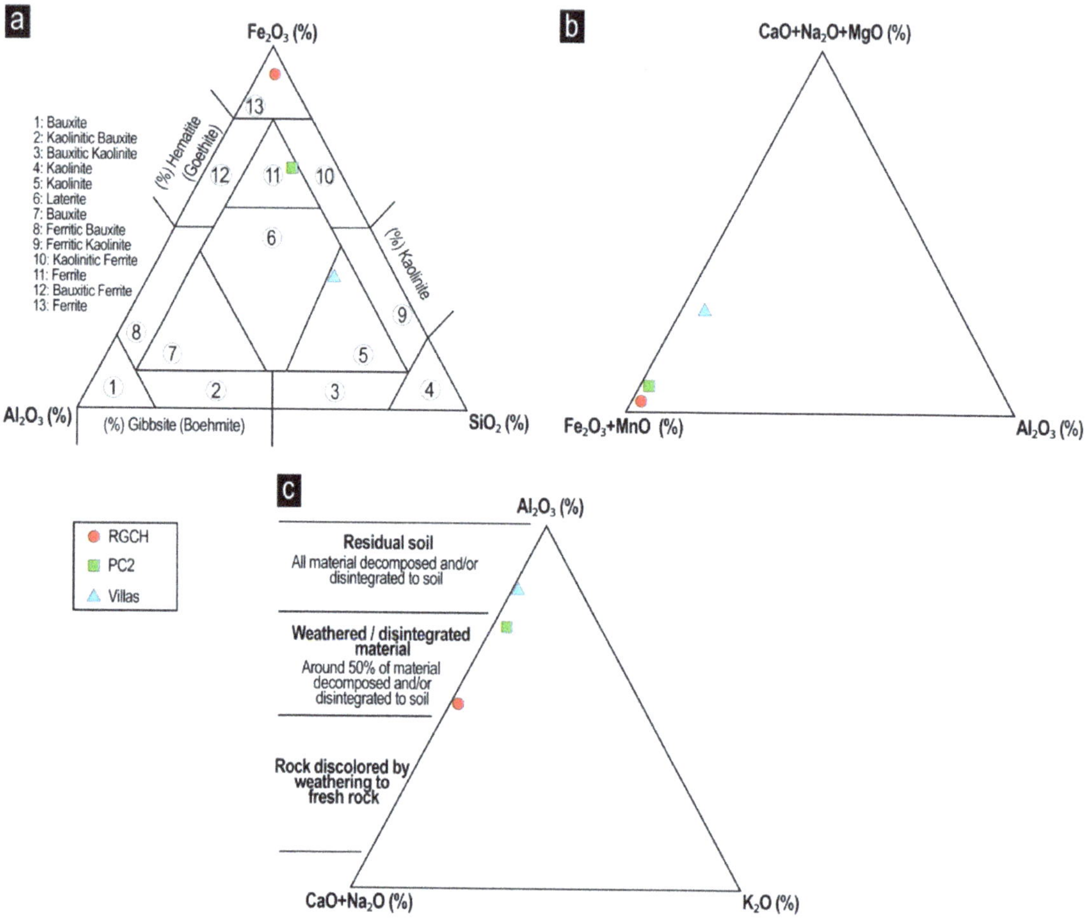

Figure 7. Ternary diagrams showing the geochemical composition of the iron crusts from the Villas, RGCH, and PC2 sections: (**a**) Fe_2O_3-SiO_2-Al_2O_3 diagram of bauxite chemical classification [61]; (**b**) Fe_2O_3 + MgO-CaO + Na_2O + MgO-Al_2O_3 diagram of Mitchell and Sheldon [62]; (**c**) Al_2O_3-CaO + Na_2O-K_2O diagram of Nesbit and Young [63]. The geochemical data of the RGCH and PC2 sections have been taken from Reolid et al. [32] and Reolid and Abad [21], respectively.

With respect to the trace element content (Table 2), the Ba/Sr, Sr/Cu and Rb/Sr ratios of the iron crust from the Villas section are higher (Ba/Sr_{crust} = 3.58, Sr/Cu_{crust} = 0.61 and Rb/Sr_{crust} = 1.10) than those calculated for the iron-coated grains included in the crust (Ba/Sr_{grains} = 0.67, Sr/Cu_{grains} = 0.37 and Rb/Sr_{grains} = 0.14). In the RGCH section, the Ba/Sr ratio is higher in the crust (0.61) than in the iron-coated grains (0.23), whereas the opposite occurred for the Sr/Cu and Rb/Sr ratios, which are higher in the iron-coated grains (Sr/Cu_{grains} = 1.87 and Rb/Sr_{grains} = 0.16) than in the crust (Sr/Cu_{crust} = 0.81 and Rb/Sr_{crust} = 0.12). The Ga/Rb ratios obtained for the Villas (Ga/Rb_{crust} = 1.52 and Ga/Rb_{grains} = 20.40) and the RGCH (Ga/Rb_{crust} = 1.45 and Ga/Rb_{grains} = 1.28) sections show similar values for the crusts but a clear much higher Ga/Rb ratio value in the iron-coated grains from the Villas section.Comparing the iron crusts and the iron-coated grains chondrite-normalized patterns to the bulk composition of the upper continental crust (Post-Archean Australian Shales, PAAS) of Taylor and McLennan [64], it can be observed that both the iron crust and the iron-coated grains from the Villas section are enriched in

LREE and HREE (Figure 8a). By contrast, a relative impoverishment in LREE but a slight enrichment in HREE occur in the crust from the RGCH section, whereas, except for Ce, the iron-coated grains are enriched in REE, according to the data reported by Reolid et al. [32] (Figure 8a). The REE fractionation is higher in the Villas section than in the RGCH section, this fractionation being higher in the crust than in the pisoids (Figure 8b,c). The iron crusts and the iron-coated grains of both sections show negative Eu and Ce anomalies, which are weaker in the Villas section (Figure 8).

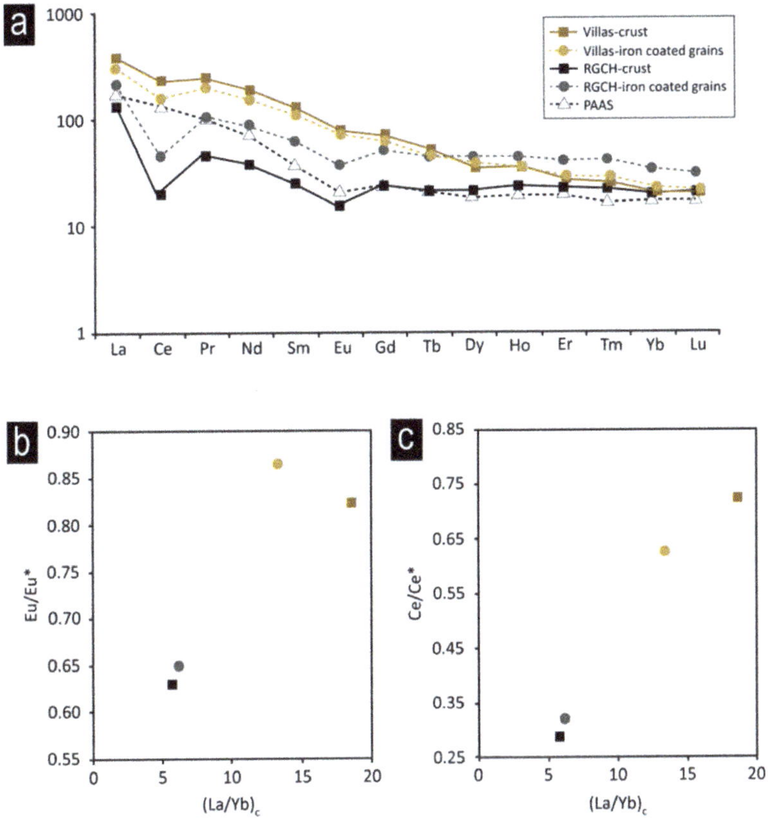

Figure 8. (**a**) Chondrite-normalized REE patterns of the crusts and the iron-coated grains from the Villas and RGCH sections; (**b**,**c**) plot of the Eu/Eu* and Ce/Ce* vs. $(La/Yb)_c$ values, respectively, in the crusts and the iron-coated grains from the Villas and RGCH sections. The subscript *c* refers to those values normalized to chondrite. The geochemical data of the RGCH section have been taken from Reolid et al. [32].

4.3. The Iron-Coated Grains of the Ferruginous–carbonated Breccia and the Ferruginous Oolithic Limestones

4.3.1. Textural Features

In the ferruginous–carbonated breccia of the Villas section, iron-coated grains and micritic clasts cemented by sparite are observed (Figure 9a). The iron-coated grains present the same textural features as those embedded in the clay-rich matrix of the iron crust, which has a range from 200 μm to 7 mm, encompassing iron macropisoids, pisoids, and ooids according to Bárdossy [55], being pisoids the most common ones. They are included in the type A category of Reolid et al. [32], differentiating simple and fragmented pisoids

(Figure 9b). The nucleus and the cortex are also composed of a mixture of clay minerals, iron oxides, and occasional quartz grains (Figure 9c,d).

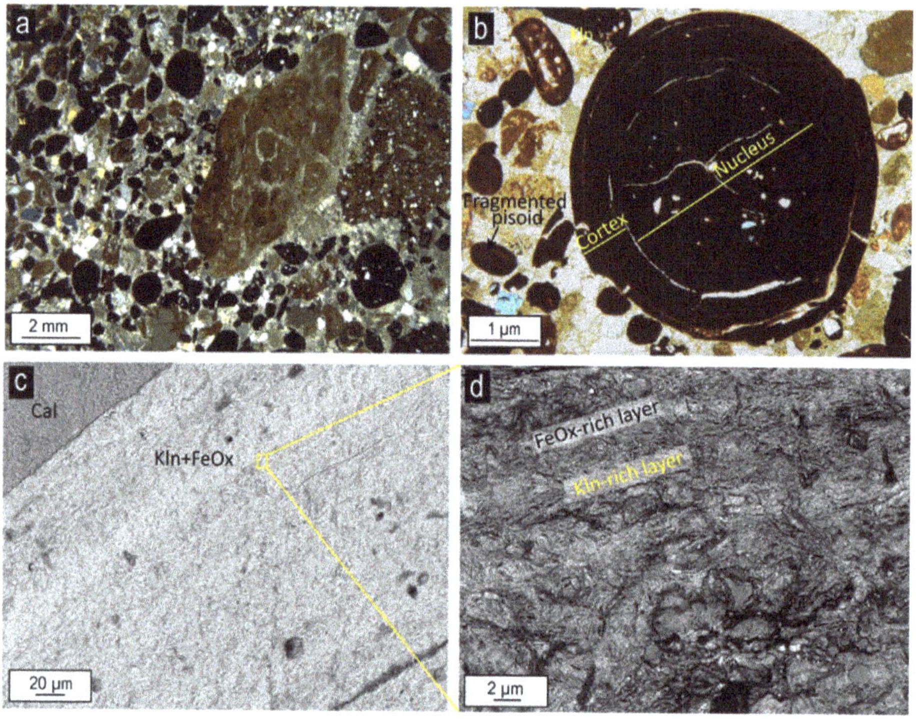

Figure 9. Optical microscopy (**a**,**b**) and SEM (**c**,**d**) images of the ferruginous–carbonated breccia from the Villas section showing: (**a**) iron-coated grains and micritic clasts in a sparite matrix; (**b**) simple and fragmented pisoids in the sparite matrix; (**c**,**d**) the cortex of a simple pisoid where the different layers can be differentiated by their higher content in iron oxides or kaolinite. *Cal* = calcite, *Kln* = kaolinite, *FeOx* = iron oxides.

In the ferruginous oolithic limestone from the RGCH, CHO and PC1 sections and the equivalent lithofacies of the AF section, iron-coated grains are found embedded in a micritic-sparitic matrix. Following the classifications given by Reolid et al. [32] and Guerrak [60], these iron-coated grains are included within type A and type B (Figure 10). Unlike the iron crust and the ferruginous–carbonated breccia from the Villas section, the four morphological types of type A iron-coated grains are found in the RGCH, CHO, and PC1 sections and in the AF section (Figure 10): simple, fragmented, complex and spastolith. Simple and fragmented pisoids are the most frequent particles and present the same morphological features as those found in the iron crust and the ferruginous–carbonated breccia from the Villas section (Figure 10a,d,e). Complex pisoids, although scarcer, can be observed as being formed by two or more nuclei composed of indeterminate ferruginous lumps evolved by several concentric layers (Figure 10b). Spastoliths, which are distorted and stretched iron-coated grains deformed by squashing [32], are more frequent than complex pisoids and present distorted and stretched grains (Figure 10c). The type B iron-coated grains can be classified as simple and complex, formed by one or two nuclei composed by a previous iron ooid, bioclast, foraminifera, or quartz grains (Figure 10f).

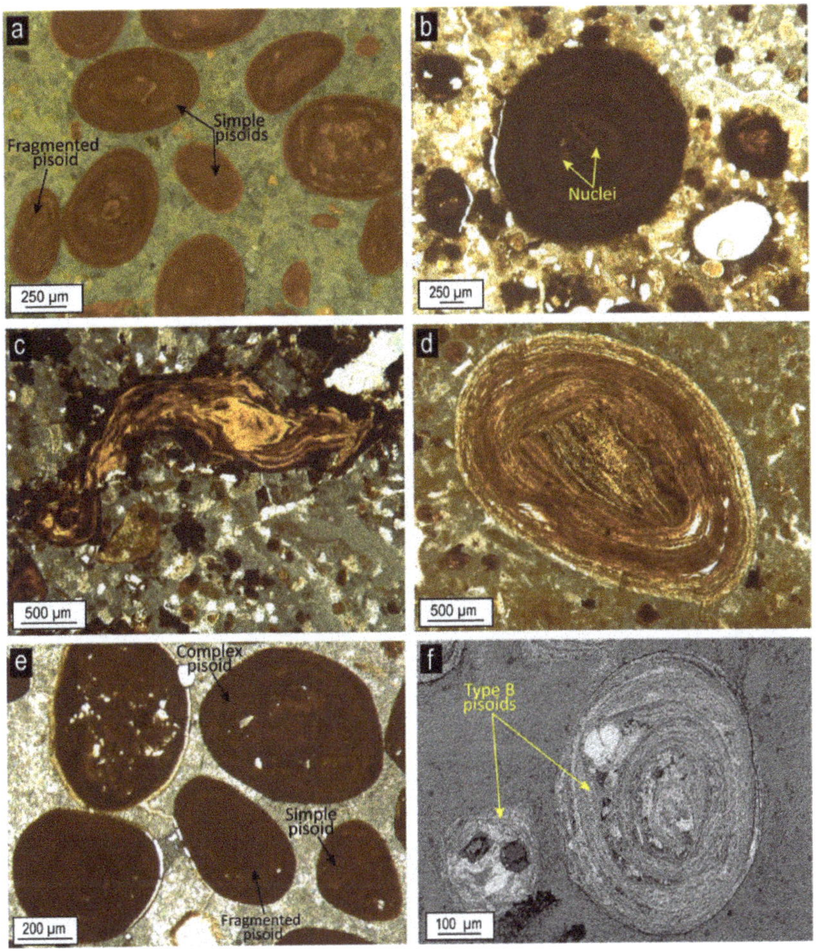

Figure 10. Optical microscopy (**a–e**) and SEM (**f**) images of the ferruginous oolithic limestones from the RGCH, CHO and PC1 sections (**a–d**) and the AF section (**e–f**) showing: (**a**,**d**,**e**) type A iron-coated grains including simple and fragmented pisoids; (**b**,**e**) complex type A iron-coated grains; (**c**) spastoliths; (**f**) simple (right) and complex (left) type B iron-coated grains.

4.3.2. Geochemical Composition

The EDS compositional maps of different iron-coated grains embedded in the iron crust and the ferruginous–carbonated breccia (Villas section) and in the ferruginous oolithic limestone (AF section) are represented in Figure 11. Those from the Villas section correspond to type A simple pisoids, composed of a nucleus and a more or less developed cortex, whereas the pisoids from the AF section are type B simple pisoids, which present two types of nuclei (a previous iron ooid and a foraminifera) and a cortex made up of several concentric layers. Regardless of the pisoid type, all the EDS elemental maps show that both the nucleus and the cortex are composed of a mixture of Al and Fe due to the presence of kaolinite, hematite, and goethite. In the more developed cortexes (Figure 11b,c), different content of Al and Fe are observed, allowing one layer to be differentiated from the other in both type A and type B simple pisoids. However, the compositional maps show clearly that the lamination of type A pisoids is thinner and less irregular than in type B.

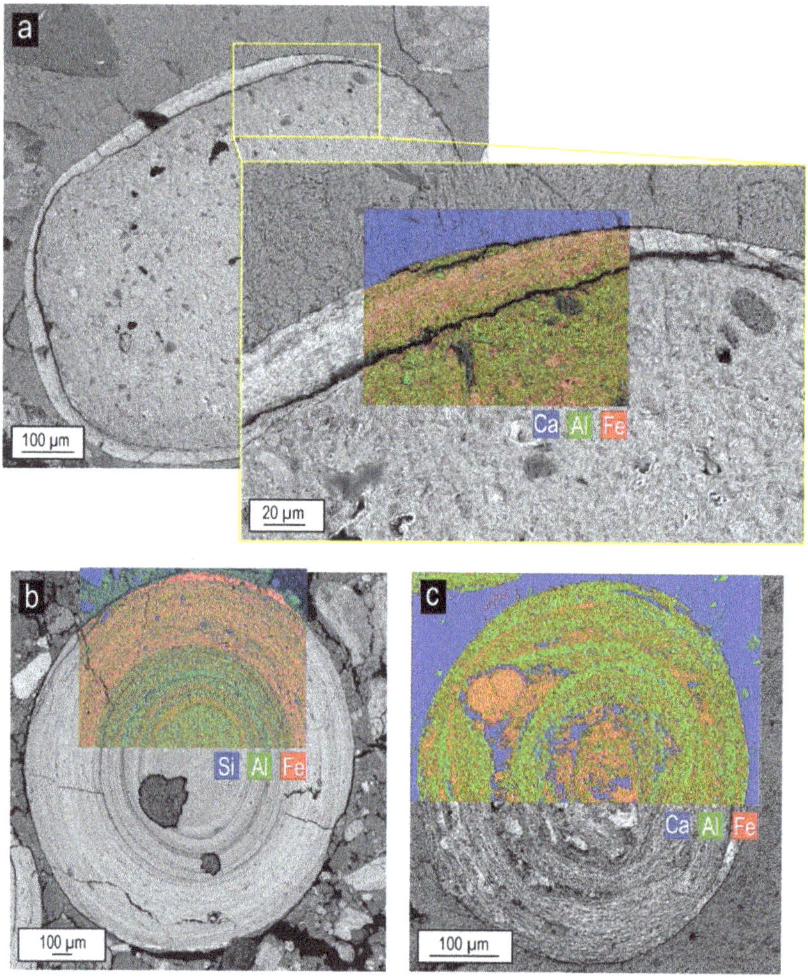

Figure 11. EDS compositional maps over SEM images of: (**a**) type A iron-coated grain (simple pisoid) in the ferruginous–carbonated breccia from the Villas section; (**b**) type A iron-coated grain (simple pisoid) in the iron crust from the Villas section; (**c**) type B iron-coated grain (simple pisoid) in the ferruginous oolithic limestone from the AF section.

XRF compositional maps of the ferruginous–carbonated breccia from the Villas also reveal that some metals, such as Ba, Co, Mn, Ti, and Cr, are concentrated in certain layers of the iron-coated grains (Figure 12). Ba and Ti are concentrated in the nucleus and some layers, coinciding with those areas where Si and Al contents are higher. By contrast, the Co, Cr, and Mn distribution within the iron-coated grains coincides with higher Fe contents (Figure 12).

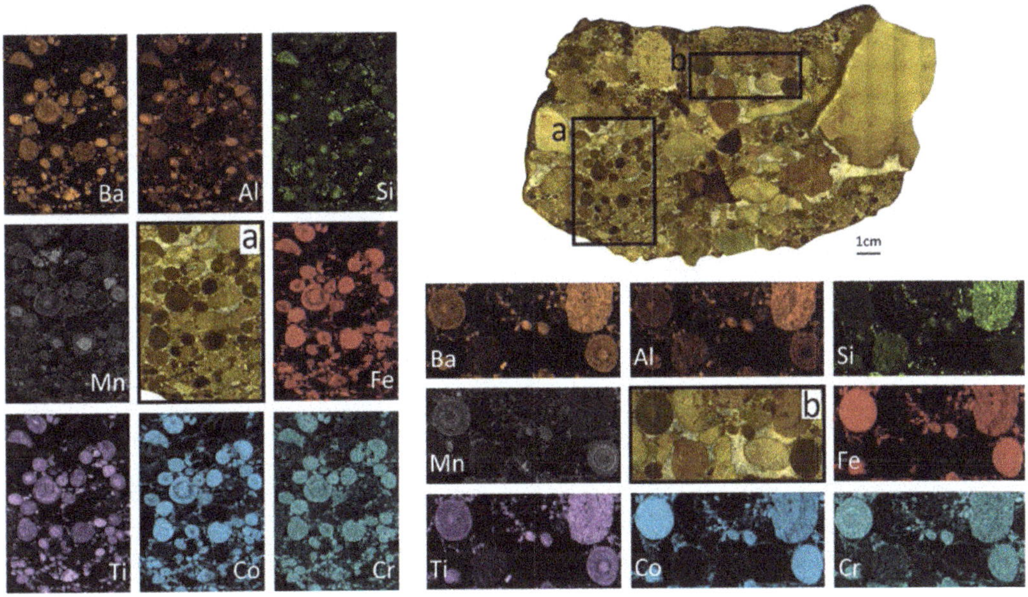

Figure 12. X-ray microfluorescence elemental maps of the ferruginous–carbonated breccia from the Villas section that allows us to observe differences in the distribution of elements in function of their affinities for Fe or Si-Al.

5. Discussion

5.1. Genesis of the Iron Crusts Containing Iron-Coated Grains

The electron microscopy images provide key textural and compositional information of the clay minerals for the interpretation of their origin. The euhedral to subhedral kaolinite plates showing pseudohexagonal outlines and the type-book aggregates, which are mixed with the iron oxides in the matrix of the iron crust from the Villas section, are indicative of an authigenic origin since these textures are too delicate to withstand erosion or transport processes [8,11,65]. Platy kaolinites and booklets have commonly been described as the result of chemical weathering processes during soil development [3,6–11]. In this way, kaolinite would have thus crystallized during edaphic processes after the dissolution of previous aluminosilicates present in the parent rock (e.g., orthoclase and illitic phases).

These genetical conditions are not compatible with the formation of illitic phases, which are commonly associated with diagenetic or hydrothermal environments [9]. In those environments, kaolinite would destabilize from 95–120 °C [66–68]. In addition, the genesis of Ilt/Sme from illite degradation due to K loss has also been described during chemical weathering processes [13,69–71].

Kaolinite has been observed over subhedral illitic phase plates, which may indicate that the illitic phases are earlier phases that acted as substrates for kaolinite crystallization, as described in other palaeosols and sediments [6,72]. Therefore, the illitic phases would probably have a detrital origin and may destabilize during the soil development, giving place to the genesis of kaolinite and Ilt/Sme mixed layers.

Goethite and hematite form the matrix of the iron crust together with the clay minerals, which also suggests an edaphic origin for these minerals [65,73]. On the other hand, quartz fragments showing anhedral morphologies reflect detrital phases that may represent the remains of the partial dissolution of the parent rock.

In the matrix of the iron crust from the PC2 section, iron oxide tube-like filaments are observed. The PC2 iron crust contains not only the iron oxide tube-like filaments but also subhedral kaolinite plates similar to those observed in the iron crust from the Villas section, pointing out an edaphic origin for this crust. The iron oxide tube-like filaments observed in the PC2 section would have been generated during the soil development, most probably due to microorganisms as has been described in other palaeosols [74,75].

Szamatek et al. [76] classify iron pisoids according to the kaolinite/hematite ratio as kaolinitic, kaolinitic–hematitic, hematitic–kaolinitic, or hematitic. Kaolinitic–hematitic and hematitic–kaolinitic pisoids have been previously described as pisoids composed of a nucleus of kaolinite or hematite surrounded by a cortex of alternating kaolinite and hematite layers [73]. The kaolinite and hematite content of the iron-coated grains from the iron crust of the Villas section allows them to be categorized as kaolinitic–hematitic pisoids, which is also supported by the different kaolinite and iron oxide contents observed in the cortex that allowed the differentiation of the different concentric layers.

The presence of kaolinite, goethite, and hematite in the iron-coated grains showing the same textural features as those observed in the matrix of the iron crust, along with their concentric structure, is typical of the edaphic process [6,9,55,65,77]. Therefore, the iron-coated grains included in the iron crust from the Villas section were formed in situ during the development of the soils. The secondary and backscattered electron images also showed Fe oxides growing between kaolinite sheets in the iron-coated grains. The Fe oxides cementing kaolinite have also been described in Lower Cretaceous lateritic palaeosols from the SE Iberian Range by Laita et al. [9], suggesting that these oxides generated at a late stage during the edaphic process.

The genesis of iron-coated grains requires hydrous conditions with permanent groundwater saturation [55,78] and is related to dissolution and re-precipitation processes that are controlled by different factors (e.g., climatic conditions, chemical composition of the groundwater, etc.) [79].

A possible later decrease in the water level may favoured the reworking and fragmentation of some of the iron-coated grains, giving place to the broken pisoids that acted as nuclei for new pisoids. Reworking processes affecting the iron-coated grains in different episodes of soil development have also been described in other palaeosols and sediments [6,9,80,81].

5.2. Provenance of the Iron-Coated Grains Included in the Ferruginous–Carbonated Breccia and the Ferruginous Oolithic Limestones

Eustatic and tectonic factors influenced the Middle–Upper Jurassic unconformity, which is linked to extensional tectonic activity related to an advanced rifting stage, followed by an oceanization and expansion phase recording in the Tethyan Domain during the Middle–Late Oxfordian [36,82]. Some authors report a sea-level fall during the Middle–Late Jurassic transition [25,32,83], whereas, for other authors, it was a global sea-level rise period [26,84,85]. The evidence of palaeosols development in the studied sections of the Prebetic agrees with a sea-level fall during the Late Callovian–Early Oxfordian.

The presence of type A iron-coated grains in the ferruginous oolithic limestones from the RGCH, CHO and PC1 sections is explained by reworking during the erosion of the iron crusts [21,32]. These ferruginous oolithic limestones represent Late Jurassic flooding events of the Prebetic, which developed in a sedimentation context dominated by storms [36,40,86]. The flooding events extended to the emerged areas, giving place to the erosion of the palaeosols previously developed in the coastal plain [21]. According to Olóriz et al. [31], the ferruginous oolithic limestones in the External Prebetic represent complex lithofacies with high energy features that combine the presence of components coming from proximal settings (iron-coated grains with edaphic origin, quartz grains coming from emerged lands, and benthic foraminifera of shallow environments) and others coming from distal environments (ammonoids, belemnites and planktic foraminifera).

Recent studies have pointed out that iron-coated grains included in Proterozoic oolithic ironstones formed by synsedimentary reworking processes in low-energy coastal environments [87]. According to these authors, negative Eu anomalies (as those reported by Reolid et al. [32] for the RGCH section) suggest a continental iron source. The iron-rich microenvironment generated due to the redistribution of the iron crust fragments and type A iron-coated grains allowed the development of new iron-coated grains with marine features (type B iron-coated grains) in the ferruginous oolithic limestones [32].

On the other hand, a ferruginous–carbonated breccia containing juts type A iron-coated grains is recorded, instead of the ferruginous oolithic limestone, in the Villas section. The textural features and chemical and mineralogical composition of type A iron-coated grains from this ferruginous–carbonated breccia is similar to that observed in the iron crust. Therefore, these iron-coated grains might have been reworked from the underlying iron crust. The absence of fossil remains, especially those typical of open marine conditions such as ammonoids and planktic foraminifera and type B coated grains, that occur in the ferruginous oolithic limestones, suggests that the Villas section was located inland, compared with the RGCH, CHO and PC1 sections where the breccia was not developed and marine remains are dominant.

In the equivalent lithofacies of the AF section (Iberian Range), the same type A and type B iron-coated grains are differentiated as those observed in the ferruginous oolithic limestones from the Prebetic. The mineralogy of these iron-coated grains is mainly composed of goethite, together with calcite, siderite, hematite, kaolinite and illite [19,88]. This mineralogy is quite similar to the reported in this study for the type A iron-coated grains. The electron microscopy images showed that the texture of the type B iron-coated grains from the AF section is similar to that of type B iron-coated grains reported by Reolid et al. [32] in the RGCH section. The environmental conditions for the genesis of the ferruginous oolithic limestone of the AF section are the same as in the Prebetic (PC, CHO, RGCH sections).

5.3. Palaeoclimatic Inferences from Clay Minerals and Geochemical Proxies

Reolid and Abad [21] and Reolid et al. [32] interpreted the iron crust containing iron-coated grains from the PC2 and RGCH sections as plinthitic palaeosols. These authors reported Al-substitution in goethite of 5 mol% in the iron-coated grains and of 10 mol% in the iron crusts, which are quite similar to those obtained for the Villas section (iron crust = 13 mol% and iron-coated grains = 7 mol%), these values being indicators of pedogenic conditions according to the values given by Fitzpatrick and Schwertmann [53].

The mineralogical assemblage reported for the iron crust from the PC2 section [21], is mainly formed by hematite, goethite, kaolinite, illite, calcite, and quartz. By contrast, the crust from the RGCH section does not contain clay minerals, and it is mainly formed by goethite and hematite, together with boehmite and apatite [32]. These differences in the mineralogical content of the iron crusts maybe related to a more or less evolved stage of the plinthitic palaeosols [21]. The iron crust from the PC2 section, which includes poorly developed iron-coated grains with incipient laminations and partial ferruginization, would correspond to a less evolved plinthite, whereas the iron crust from the RGCH section would correspond to a more evolved plinthite [21].

Plinthosols are well-drained lateritic soils with high Fe and/or Al content in relation to other components [89,90]. The term "plinthosol" is defined as an iron-rich mixture of clay minerals (mainly kaolinite) and silica, according to the Food and Agriculture Organization (FAO). This term is equivalent to that of ferrallitic soils categorized by Duchaufour [91] and to the oxisols described by the USDA classification [92].

According to the Fe_2O_3-Al_2O_3-SiO_2 diagram (Figure 7a), the iron crust from the RGCH and PC2 sections plot on the ferrite field due to their high Fe content, whereas the iron crust from the Villas section plots on the kaolinite field but close to that of laterite due to its higher Si and Al content. Therefore, the iron crust from the Villas section may correspond to a less drained plinthosol.

Plinthites develop in tropical environments where elevated temperatures and alternating wet and dry conditions favor the precipitation, mobilization, and accretion of weathered iron oxides [89]. The main process that consumes silicate minerals during chemical weathering is hydrolysis [62]:

$$\text{Hydrolysis} = \text{Al}/\text{Ca} + \text{Mg} + \text{K} + \text{Na} \quad (6)$$

Applying this equation to the studied crusts, it can be determined that the hydrolysis was higher in the iron crust from the Villas section (3.76) than in the iron crust from the PC2 (2.10) and RGCH (0.92) sections. These differences in the hydrolysis intensity can also be inferred from the Fe_2O_3 + MgO-CaO + Na_2O + MgO-Al_2O_3 diagram (Figure 7b), where the low CaO + Na_2O + MgO contents indicate that these elements were weathered and that the Al_2O_3 content is higher in the Villas section due to the immobility of Al during intense chemical weathering.

Sr is more soluble than Ba and Rb during chemical weathering, so higher Ba/Sr and Rb/Sr ratios reflect more weathered rocks [62]. The Ba/Sr and the Rb/Sr are higher in the iron crust from the Villas section (3.58 and 1.10, respectively) than in the iron crust from the RGCH section (0.61 and 0.12, respectively), indicating that the chemical weathering was somewhat more intense in the Villas section (Table 3), which also correlates with the trend observed from watered/disintegrated material (RGCH) to residual soil (Villas) in the Al_2O_3-CaO + Na_2O-K_2O diagram (Figure 7c). The higher fractionation of the REE in the Villas section compared to the RGCH also points to a higher chemical weathering intensity [93].

In soil systems, metal oxides can be scavenged by hematite and goethite [94,95] or being also adsorbed on clay minerals [93], which would explain the different distribution of Ba, Ti, Co, Cr, and Mn in the Fe-rich or kaolinite-rich layers of the iron-coated grains from the Villas section.

The CIA and CIW values lower than 65 reflect compositions of illite, smectite, and feldspar relating to a low degree of chemical weathering, whereas the genesis of kaolinite though intense chemical weathering results in CIA and CIW values close to 100 [66,96–98]. The plinthites studied here show high CIA (from 90.87 to 95.87) and CIW (from 88.14 to 95.59), with the lowest values corresponding to PC2 section (Tables 1 and 3). Sr/Cu ratios higher than 10 indicate hot and dry climates, Sr/Cu ratios between 5 and 10 reflect semi-arid and semi-humid conditions, while Sr/Cu ratios lower than 5 point to a warm and humid climate [99]. Ga/Rb ratios lower than 0.25 agree with cold and dry environments, whereas Ga/Rb ratios higher than 0.25 indicate warm and humid conditions [100]. In the Villas and RGCH sections, both the crusts and the iron-coated grains show Sr/Cu ratios lower than 5 and Ga/Rb higher than 0.25, reflecting warm and humid climatic conditions (Tables 2 and 3).

The higher chemical weathering and palaeoprecipitation regime affecting the Villas sections indicates that the palaeoclimatic conditions were somewhat humid inland in the studied area.

Table 3. CIA, CIW, Ba/Sr, Rb/Sr, Sr/Cu, and Ga/Rb values obtained for the iron crusts from the Villas, RGCH, and PC2 sections and their paleoclimatic significance [66,96,98–100].

CIA			CIW			Ba/Sr		Rb/Sr		Sr/Cu		Ga/Rb	
Villas	RGCH	PC2	Villas	RGCH	PC2	Villas	RGCH	Villas	RGCH	Villas	RGCH	Villas	RGCH
95.20	95.87	90.87	91.99	95.59	88.14	3.58	0.61	1.10	0.12	0.61	0.81	1.52	1.45
CIA or CIW value → Weathering intensity						Higher Ba/Sr and Rb/Sr ratios				<5	Warm/humid	<0.25	Cold/dry
<65			Low-moderate			=				5–10	Semi-arid/ Semi-humid	>0.25	Warm/humid
>75			High			Greater weathering				>10	Hot/dry		

Plate tectonic reconstructions suggest that during the Middle to Upper Jurassic, Iberia was situated at tropical to subtropical latitudes as part of the African plate [101]. This would have provided the necessary palaeoclimatic and paleoenvironmental conditions for the development of the plinthites in the South Iberian Palaeomargin.

With respect to the global palaeoclimatic context, a short warming episode was recorded worldwide in the Middle Callovian [102–105]. However, the end of the Callovian to early Oxfordian was a time of climatic cooling [106,107] as well as a sea-level fall [108]. Many carbonate platforms were exposed, and stratigraphic successions are characterized by the presence of hardgrounds, palaeokarst, and hiatuses [26,109,110].

Similar palaeoclimatic conditions are described in west Iberia. Azerêdo et al. [109] describe a sea-level fall and a change from humid to seasonal semi-arid climatic conditions in the Lusitanian Basin (west-central Portugal) during the Middle–Late Jurassic. According to these authors, a palaeokarstic surface is recorded over the Middle Jurassic carbonates, followed by ferruginous clays (composed of kaolinite and iron oxides) and crust/pisoids, which are consistent with subaerial exposure, as occurs with the plinthitic palaeosols reported in this study.

During the Middle Oxfordian, the decline in $\delta^{18}O$ values is related to rising sea water temperatures [111], accompanied by a secular sea-level rise [106]. In many Tethyan palaeomargins, including the Iberian Palaeomargins, this sea-level rise is recorded as a flooding of the Middle Jurassic carbonate platforms with the recovery of sedimentation and the development of hemipelagic facies [112–114].

The data reported in this study reflect a tropical to subtropical climate during the development of the iron crusts and iron-coated grains at any moment during the Middle Callovian–Early Oxfordian transition in the South Iberian Palaeomargin.

Marine iron ooids with similar textural features as those studied here (type B iron-coated grains) have also been described in Callovian–Oxfordian sections of nearby basins of the Western Tethys (e.g., Paris Basin) [115]. Previous studies indicate that the iron-coated grains in the ferruginous oolithic limestones from the correlatable AF section (Iberian Range, NE Iberia) are a marine rework of iron-coated grains from continental laterites [29,88]. This is congruent with our observations. Therefore, the tropical to subtropical palaeoclimatic conditions occurring during the Middle–Upper Jurassic transition in the South Iberian Palaeomargin may also have taken place in the East Iberian Palaeomargin.

6. Conclusions

The combination of XRD, geochemical, and optical and electron microscopy analysis has allowed the mineralogical and textural characterization of the iron crusts and iron-coated grains from the Central and Eastern External Prebetic (SE Iberia) and the Iberian Range (NE Iberia), drawing the following conclusions:

1. The mineralogical composition of the iron crust is mainly composed of clay minerals (kaolinite, illitic phases, and Ilt/Sme) and/or iron oxides (hematite and goethite), along with the presence of iron-coated grains, this indicates that they are plinthitic palaeosols, characteristic of tropical environments.
2. The platy morphologies of the kaolinite crystals and the presence of book-type aggregates in the matrix of the iron crusts point to an authigenic origin for the kaolinite through chemical weathering processes during the palaeosols development. The kaolinite crystals growing on subhedral illitic phases plates indicate that the illitic phases are previous phases, probably detrital.
3. The mineralogy and texture of the iron-coated grains included in the iron crusts (type A) reflect an in situ origin during the plinthites formation. Subsequent reworking processes caused the break of some of these iron-coated grains that eventually acted as a nucleus of new iron-coated grains. The reworking processes also led to the incorporation of these iron-coated grains into the above ferruginous oolithic limestones (RGCH, CHO, PC, and AF sections) and the ferruginous–carbonated breccia (Villas section). New iron-coated grains (type B) were later formed in the ferruginous oolithic limestones in a marine context.
4. The CIA and CIW values and the Ba/Sr, Rb/Sr, Sr/Cu, and Ga/Rb ratios agree with an intense chemical weathering related to warm and humid conditions during the development of the palaeosols. The higher Ba/Sr ratio value of the iron crust from

the Villas section comparted to that of the RGCH section reflects a higher chemical weathering intensity in the former.

5. The similarity of the iron-coated grains from the equivalent ferruginous oolithic ironstone of the AF section in the Iberian Range suggests the same origin as those from the Central and Eastern External Prebetic. Therefore, the warm and humid conditions related to a tropical climate recorded in the South Iberian Palaeomargin during the Callovian–Oxfordian could have also taken place in the East Iberian Palaeomargin.

Author Contributions: Conceptualization, E.L., M.R. and I.A.; methodology, E.L. and M.R.; formal analysis, E.L.; investigation, E.L., M.R. and I.A.; resources, M.R. and I.A.; data curation, E.L.; writing—original draft preparation, E.L.; writing—review and editing, M.R. and I.A.; visualization, E.L.; supervision, M.R. and I.A.; project administration, M.R. and I.A.; funding acquisition, M.R. and I.A. All authors have read and agreed to the published version of the manuscript.

Funding: This research was funded by the Research groups RNM-200 and RNM-325 of the Junta de Andalucía. This publication is part of the grant JDC2022-048348-I funded by the MCIN/AEI/10.13039/501100011033 and by the European Union "NextGenerationEU"/PRTR.

Data Availability Statement: The original contributions presented in the study are included in the article, further inquiries can be directed to the corresponding author.

Acknowledgments: Technical and human support provided by Centro de Instrumentación Científico-Técnica (CICT)-Servicios Centrales de Apoyo a la Investigación (SCAI)-Universidad de Jaén (UJA, MICINN, Junta de Andalucía, FEDER) is gratefully acknowledged. The authors would like to acknowledge Antonio Piedra (technician of the Department of Geology) for preparing thin sections and polished slabs.

Conflicts of Interest: The authors declare no conflicts of interest.

References

1. Chamley, H. *Clay Sedimentology*; Springer: Berlin, Germany, 1989; 623p, ISBN 3-540-50889-9.
2. Velde, B. *Origin and Mineralogy of Clays: Clays and the Environment*; Springer: Berlin, Germany, 1995; 334p, ISBN 978-3-540-58012-6.
3. Mack, G.; Calvin, W.; Curtis, H. Classification of paleosols. *Geol. Soc. Am. Bull.* **1993**, *105*, 129–136. [CrossRef]
4. Thiry, M. Palaeoclimatic interpretation of clay minerals in marine deposits: An outlook from the continental origin. *Earth-Sci. Rev.* **2000**, *49*, 201–221. [CrossRef]
5. Tabor, N.J.; Myers, T.S.; Michel, L.A. Sedimentologist's Guide for Recognition, Description, and Classification of Paleosols. In *Terrestrial Depositional Systems: Deciphering Complexities through Multiple Stratigraphic Methods*; Zeigler, K.E., Parker, W.G., Eds.; Elsevier: Amsterdam, The Netherlands, 2017; pp. 165–208, ISBN 9780128032435.
6. Laita, E.; Bauluz, B.; Aurell, M.; Bádenas, B.; Yuste, A. Weathering events recorded in uppermost Hauterivian–lower Barremian clay-dominated continental successions from the NW Iberian Range: Climatic vs. tectonic controls. *J. Iber. Geol.* **2022**, *48*, 45–63. [CrossRef]
7. Raucsik, B.; Varga, A. Climato-environmental controls on clay mineralogy of the Hettangian–Bajocian successions of the Mecsek Mountains, Hungary: An evidence for extreme continental weathering during the early Toarcian oceanic anoxic event. *Palaeogeogr. Palaeoclim. Palaeoecol.* **2008**, *265*, 1–13. [CrossRef]
8. Bauluz, B.; Yuste, A.; Mayayo, M.J.; Canudo, J.I. Early kaolinization of detrital Weald facies in the Galve Sub-basin (Central Iberian Chain, north-east Spain) and its relationship to palaeoclimate. *Cretac. Res.* **2014**, *50*, 214–227. [CrossRef]
9. Laita, E.; Bauluz, B.; Aurell, M.; Bádenas, B.; Canudo, J.I.; Yuste, A. A change from warm/humid to cold/dry climate conditions recorded in lower Barremian clay-dominated continental successions from the SE Iberian Chain (NE Spain). *Sediment. Geol.* **2020**, *403*, 105673. [CrossRef]
10. Righi, D.; Meunier, A. Origin of Clays by Rick Weathering and Soil Formation. In *Origin and Mineralogy of Clays. Clays and the Environment*; Velde, B., Ed.; Springer: Berlin/Heidelberg, Germany; New York, NY, USA, 1995; pp. 43–161, ISBN 978-3-540-75633-0.
11. Campo, M.D.; Bauluz, B.; del Papa, C.; White, T.; Yuste, A.; Mayayo, M.J. Evidence of cyclic climatic changes recorded in clay mineral assemblages from a continental Paleocene-Eocene sequence, northwestern Argentina. *Sediment. Geol.* **2018**, *368*, 44–57. [CrossRef]
12. Wilson, M.J. The origin and formation of clay minerals in soils: Past, present and future perspectives. *Clay Miner.* **1999**, *34*, 7–25. [CrossRef]
13. Galán, E. Genesis of clay minerals. In *Developments in Clay Science*; Bergaya, F., Theng, B.K.G., Lagaly, G., Eds.; Elsevier: Amsterdam, The Netherlands, 2006; pp. 1129–1162.

14. Dera, G.; Pellenard, P.; Neige, P.; Deconinck, J.-F.; Pucéat, E.; Dommergues, J.-L. Distribution of clay minerals in Early Jurassic Peritethyan seas: Palaeoclimatic significance inferred from multiproxy comparisons. *Palaeogeogr. Palaeoclim. Palaeoecol.* **2009**, *271*, 39–51. [CrossRef]
15. Cecca, F.; Garin, B.M.; Marchand, D.; Lathuiliere, B.; Bartolini, A. Paleoclimatic control of biogeographic and sedimentary events in Tethyan and peri-Tethyan areas during the Oxfordian (Late Jurassic). *Palaeogeogr. Palaeoclim. Palaeoecol.* **2005**, *222*, 10–32. [CrossRef]
16. Bucur, I.I.; Reolid, M. Incidence of the early Toarcian global change on Dasycladales (Chlorophyta) and the subsequent recovery: Comparison with end-Triassic Mass Extinction. *Earth-Sci. Rev.* **2024**, *249*, 104666. [CrossRef]
17. Aurell, M.; Bádenas, B.; Bello, J.; Delvene, G.; Meléndez, G.; Pérez-Urresti, I.; Ramajo, J. El Calloviense y el Jurásico Superior en la Cordillera Ibérica Nororiental y la zona de enlace con la Cordillera Costero Catalana, en los sectores de Sierra de Arcos, Calanda y Xerta-Paüls. *Cuad. Geol. Ibér.* **1999**, *25*, 73–110.
18. Kadiri, K. Jurassic ferruginous hardgrounds of the "Dorsale Calcaire" and the Jbel Moussa Group (Internal Rif, Morocco): Stratigraphical context and paleoceanographic consequences of mineralization processes. *Geol. Romana* **2002**, *36*, 33–69.
19. Ramajo, J.; Aurell, M.; Cepría, J. Análisis de facies de la Capa de oolitos ferruginosos de Arroyofrío en la Sierra de Arcos (Jurásico, Cordillera Ibérica septentrional). *J. Iber. Geol.* **2002**, *28*, 45–64.
20. Reolid, M.; El Kadiri, K.; Abad, I.; Olóriz, F.; Jiménez-Millán, J. Jurassic microbial communities in hydrothermal manganese crust of the Rifian Calcareous Chain, Northern Morocco. *Sediment. Geol.* **2011**, *233*, 159–172. [CrossRef]
21. Reolid, M.; Abad, I. The Middle-Upper Jurassic unconformity in the South Iberian Palaeomargin (Western Tethys): A history of carbonate platform fragmentation, emersion and subsequent drowning. *J. Iber. Geol.* **2019**, *45*, 87–110. [CrossRef]
22. Picard, S.; Garcia, J.P.; Le'cuyer, C.; Sheppard, S.; Capetta, H.; Emig, C. δ18O values of co- existing brachiopods and fish temperature di¡erences and estimates of paleodepths. *Geology* **1998**, *26*, 975–978. [CrossRef]
23. Podlaha, O.G.; Mutterlose, J.; Veizer, J. Preservation of $\delta^{18}O$ and $\delta^{13}C$ in belemnite rostra from the Jurassic/early Cretaceous successions. *Am. J. Sci.* **1998**, *298*, 324–347. [CrossRef]
24. Price, G.D. The evidence and implications of polar ice during the Mesozoic. *Earth-Sci. Rev.* **1999**, *48*, 183–210. [CrossRef]
25. Dromart, G.; Garcia, J.-P.; Picard, S.; Atrops, F.; Lécuyer, C.; Sheppard, S. Ice age at the Middle–Late Jurassic transition? *Earth Planet. Sci. Lett.* **2003**, *213*, 205–220. [CrossRef]
26. Norris, M.S.; Hallam, A. Facies variations across the Middle-Upper Jurassic boundary in Western Europe and the relationship to sea-level changes. *Palaeogeogr. Palaeoclim. Palaeoecol.* **1995**, *116*, 189–245. [CrossRef]
27. Gygi, R.A.; Coe, A.L.; Vail, P.R. Sequence stratigraphy of the Oxfordian and Kimmeridgian stages (Late Jurassic) in northern Switzerland. Mesozoic and Cenozoic Sequence Stratigraphy of European Basins. *SEPM Spec. Publ.* **1998**, *60*, 527–544.
28. Leinfelder, R.R.; Wilson, R.C. Third-Order Sequences in an Upper Jurassic Rift-Related Second-Order Sequence, Central Lusitanian Basin, Portugal. Mesozoic and Cenozoic Sequence Stratigraphy of European Basins. *SEPM Spec. Publ.* **1998**, *60*, 507–525.
29. Aurell, M.; Fernandez-Lopez, S.; Melendez, G. The Middle-Upper Jurassic oolitic ironstone level in the Iberian range (Spain). Eustatic implications. *Geobios* **1994**, *27*, 549–561. [CrossRef]
30. Courville, P.; Collin, P.Y. La série du Callovien et de l'Oxfordien de Veuxhaulles (Châtillonnais, Côte d'Or): Problèmes de datation, de géométrie et de paléoenvironnements dans une série "condensée". *Bull. Sci. Bourgogne* **1997**, *49*, 29–43.
31. Olóriz, F.; Reolid, M.; Rodríguez-Tovar, F.J. Taphonomy of fossil macro-invertebrate assemblages as a tool for ecostratigraphic interpretation in Upper Jurassic shelf deposits (Prebetic Zone, southern Spain). *Geobios* **2008**, *41*, 31–42. [CrossRef]
32. Reolid, M.; Abad, I.; Martín-García, J.M. Palaeoenvironmental implications of ferruginous deposits related to a Middle–Upper Jurassic discontinuity (Prebetic Zone, Betic Cordillera, Southern Spain). *Sediment. Geol.* **2008**, *203*, 1–16. [CrossRef]
33. García-Hernández, M.; López-Garrido, A.C.; Rivas, P.; Sanz de Galdeano, C.; Vera, J.A. Mesozoic paleogeographic evolution in the external zones of the Betic Cordillera (Spain). *Geol. Mijnbouw.* **1980**, *59*, 155–168.
34. Martín-Algarra, A.; Vera, J.A. La Cordillera Bética y las Baleares en el contexto del 996 Mediterráneo Occidental. In *Geología de España;* Vera, J.A., Ed.; Instituto Geológico y Minero de España: Madrid, Spain, 2004; 884p, ISBN 978-8478405466.
35. Jerez-Mir, L. Geología de la Zona Prebética en la Transversal de Elche de la Sierra y Sectores Adyacentes (Provincias de Albacete y Murcia). Ph.D. Thesis, Universidad de Granada, Granada, Spain, 1973.
36. García-Hernández, M.; López-Garrido, A.C.; Martín-Algarra, A.; Molina, J.M.; Ruiz-Ortiz, P.A.; Vera, J.A. Las discontinuidades mayores del Jurásico de las Zonas Externas de las Cordilleras Béticas: Análisis e interpretación de los ciclos sedimentarios. *Cuad. Geol. Ibérica.* **1989**, *13*, 35–52.
37. Reolid, M.; Molina, J.M.; Nieto, L.M.; Rodríguez-Tovar, F.J. *The Toarcian Oceanic Anoxic Event in the South Iberian Palaeomargin;* Springer Nature: Berlin, Germany, 2018.
38. López-Garrido, A.C.; García-Hernández, M. Ciclos sedimentarios mayores en la primera fase carbonatada de la plataforma prebética (Lías-Valanginiense inferior). *II Congr. Geol. España* **1988**, *SGE 1*, 107–110.
39. Foucault, A. Etude Géologique des Environs des Sources du Guadalquivir (Provinces de Jaen et de Grenade, Espagne Méridionale). Ph.D. Thesis, University of Paris, Paris, France, 1971.
40. Reolid, M. Dinámica Eco-Sedimentaria Durante el Oxfordiense Medio-Kimmeridgiense Temprano en la Zona Prebética: Interpretación Ecoestratigráfica y Secuencial. Ph.D. Thesis, Universidad de Granada, Granada, Spain, 2003.
41. Pendas, F. Definición morfológica de los embalses subterráneos del alto sureste español. *I Congr. Hisp. Luso-Am. Geol. Económica Sección Hidrogeol.* **1971**, *2*, 529–550.

42. Rodríguez-Estrella, T. Geología e Hidrogeología del Sector de Alcaraz-Lietor-Yeste (Prov. de Albacete). Síntesis Geológica de la Zona Prebética. Ph.D. Thesis, Universidad de Granada, Granada, Spain, 1978.
43. Rodríguez-Tovar, F.J. Evolución Sedimentaria y Ecoestratigráfica en Plataformas Epicontinentales del Margen Subibérico Durante el Kimmeridgiense Inferior. Ph.D. Thesis, Universidad de Granada, Granada, Spain, 1993.
44. Olóriz, F.; Reolid, M.; Rodríguez-Tovar, F. Fossil assemblages, lithofacies, taphofacies and interpreting depositional dynamics in the epicontinental Oxfordian of the Prebetic Zone, Betic Cordillera, southern Spain. *Palaeogeogr. Palaeoclim. Palaeoecol.* **2002**, *185*, 53–75. [CrossRef]
45. Olóriz, F.; Reolid, M.; Rodríguez-Tovar, F.J. A Late Jurassic carbonate ramp colonized by sponges and benthic microbial communities (External Prebetic, southern Spain). *Palaios* **2003**, *18*, 528–545. [CrossRef]
46. Gómez, J.J.; Goy, A. Las unidades litoestratigráficas del Jurásico medio y superior, en facies carbonatadas del sector levantino de la Cordillera Ibérica. *Estud. Geol.* **1979**, *35*, 569–598.
47. Linares-Girela, L. Datos sobre las series Jurásico-Cretácicas en el sector de Peñas de S. Pedro-Chinchilla de Montearagón (prov. de Albacete). *Bol. Geol. Min.* **1976**, *XXXVII-IV*, 355–364.
48. Martin-Islán, J.D. Using XPowder: A Software Package for Powder X-ray Diffraction Analysis. Available online: www.xpowder.com (accessed on 25 June 2024).
49. Biscaye, P.E. Mineralogy and sedimentation of recent deep-sea clay in the Atlantic Ocean and adjacent seas and oceans. *Geol. Soc. Am. Bull.* **1965**, *76*, 803–832. [CrossRef]
50. Schultz, L.G. *Quantitative Interpretation of Mineralogical Composition From X-ray and Chemical Data for the Pierre Shale*; USGS Professional Paper 391-C; U.S. Government Publishing Office: Washington, DC, USA, 1964; pp. 1–30.
51. Smith, D.K.; Johnson, G.G., Jr. Digitized database quantification, DDBQ, analysis of complex mixtures using fully digitized patterns. *Adv. X-ray Anal.* **2000**, *42*, 276–286.
52. Hillier, S. Quantitative analysis of clay and other minerals in sandstones by X-ray powder diffraction (XRPD). In *Clay Mineral Cements in Sandstones*; Worden, R.H., Morad, E., Eds.; Blackwell Publishing: Malden, MA, USA, 2003; pp. 213–251, ISBN 1-40510-587-9.
53. Fitzpatrick, R.; Schwertmann, U. Al-substituted goethite—An indicator of pedogenic and other weathering environments in South Africa. *Geoderma* **1982**, *27*, 335–347. [CrossRef]
54. Schulze, D.G. The Influence of Aluminum on Iron Oxides. VIII. Unit-Cell Dimensions of Al-Substituted Goethites and Estimation of Al from them. *Clays Clay Miner.* **1984**, *32*, 36–44. [CrossRef]
55. Bárdossy, G. Karst Bauxites. In *Developments in Economic Geology*; Elsevier: Amsterdam, The Netherlands, 1982; 441p.
56. Whitney, D.L.; Evans, B.W. Abbreviations for names of rock-forming minerals. *Am. Mineral.* **2010**, *95*, 185–187. [CrossRef]
57. Nesbitt, H.W.; Young, G.M. Early Proterozoic climates and plate motions inferred from major element chemistry of lutites. *Nature* **1982**, *299*, 715–717. [CrossRef]
58. Harnois, L. The CIW index: A new chemical index of weathering. *Sediment. Geol.* **1988**, *55*, 319–322. [CrossRef]
59. Sheldon, N.D.; Retallack, G.J.; Tanaka, S. Geochemical Climofunctions from North American Soils and Application to Paleosols across the Eocene-Oligocene Boundary in Oregon. *J. Geol.* **2002**, *110*, 687–696. [CrossRef]
60. Guerrak, S. Metallogenesis of cratonic oolitic ironstone deposits in the Bled el Mass, Azzel Matti, Ahnet and Mouydir basins, Central Sahara, Algeria. *Int. J. Earth Sci.* **1987**, *76*, 903–922. [CrossRef]
61. Aleva, G.J.J. *Laterites: Concepts, Geology, Morphology and Chemistry*; International Soil Reference and Information Centre (ISRIC): Wageningen, The Netherlands, 1994; 169p, ISBN 90.6672.053.0.
62. Mitchell, R.L.; Sheldon, N.D. Sedimentary provenance and weathering processes in the 1.1 Ga Midcontinental Rift of the Keweenaw Peninsula, Michigan, USA. *Precambrian Res.* **2016**, *275*, 225–240. [CrossRef]
63. Nesbitt, H.; Young, G. Prediction of some weathering trends of plutonic and volcanic rocks based on thermodynamic and kinetic considerations. *Geochim. Cosmochim. Acta* **1984**, *48*, 1523–1534. [CrossRef]
64. Taylor, S.R.; McLennan, S.M. *The Continental Crust: Its Composition and Evolution*; Blackwell: Oxford, UK, 1985; ISBN 0632011483.
65. Yuste, A.; Camacho, I.; Bauluz, B.; Mayayo, M.J.; Laita, E. Palaeoweathering events recorded on siliciclastic continental deposits (Albian, Lower Cretaceous) in NE Spain. *Appl. Clay Sci.* **2020**, *190*, 105598. [CrossRef]
66. Bauluz, B.; Mayayo, M.J.; Fernandez-Nieto, C.; Lopez, J.M.G. Geochemistry of Precambrian and Paleozoic siliciclastic rocks from the Iberian Range NE Spain: Implications for source-area weathering, sorting, provenance, and tectonic setting. *Chem. Geol.* **2000**, *168*, 135–150. [CrossRef]
67. Lanson, B.; Beaufort, D.; Berger, G.; Bauer, A.; Cassagnabère, A.; Meunier, A. Authigenic kaolin and illitic minerals during burial diagenesis of sandstones: A review. *Clay Miner.* **2002**, *37*, 1–22. [CrossRef]
68. Brosse, E.; Margueron, T.; Cassou, C.; Sanjuan, B.; Canham, A.; Girad, J.P.; Lacharpahne, J.C.; Sommer, F. The formation and stability of kaolinite in Brent sandstone reservoirs: A modelling approach. In *Clay Mineral Cements in Sandstones*; Worden, R.H., Morad, S., Eds.; Blackwell Publishing: Malden, MA, USA, 2003; pp. 383–408, ISBN 1-40510-587-9.
69. Jeong, G.Y.; Hillier, S.; Kemp, R.A. Changes in mineralogy of loess–paleosol sections across the Chinese Loess Plateau. *Quat. Res.* **2011**, *75*, 245–255. [CrossRef]
70. Schaetzl, R.; Thompson, M.L. *Soils: Genesis and Geomorphology*; Cambridge University Press: Cambridge, UK, 2015; 779p, ISBN 978-1-107-01693-4.

71. Meenakshi; Shrivastava, J.; Chandra, R. Pedogenically degenerated illite and chlorite lattices aid to palaeoclimatic reconstruction for chronologically constrained (8–130 ka) loess-palaeosols of Dilpur Formation, Kashmir, India. *Geosci. Front.* **2020**, *11*, 1353–1367. [CrossRef]
72. Entrena, A.; Fornós, J.J.; Auqué, L.F.; Gràcia, F.; Laita, E. Mineralogical and Sedimentological Characterization of the Clay-Rich Sediments from Ases Cave (Cova Dets Ases, Mallorca, Spain): Origin and Classification. *Minerals* **2022**, *12*, 1473. [CrossRef]
73. Youssef, E.S.A. Sedimentological studies on the central Wadi Kalabsha kaolin deposits, Southwest of Aswan, Egypt. *J. Miner. Pet. Econ. Geol.* **1996**, *91*, 353–363. [CrossRef]
74. Kenny, R.; Krinsley, D.H. Silicified Mississippian Paleosol Microstructures: Evidence for Ancient Microbial-Soil Associations. *Scanning Microsc.* **1992**, *6*, 359–366.
75. Mahaney, W.; Krinsley, D.; Allen, C.; Ditto, J.; Langworthy, K.; Batchelor, A.; Lecompte, M.; Milner, M.; Hart, K.; O'Reilly, S.S.; et al. Reassessment of the Microbial Role in Mn-Fe Nodule Genesis in Andean Paleosols. *Geomicrobiol. J.* **2015**, *32*, 27–41. [CrossRef]
76. Szamatek, K.; Barczuk, A.; Youssed, E.A.A. Genesis and mineralogy of lateritic kaolin at Aswan area (Sw Egypt). *Arch. Mineral. (Wars.)* **1993**, *2*, 81–97.
77. Taylor, G.; Eggleton, R.A.; Foster, L.D.; Tilley, D.B.; Le Gleuher, M.; Morgan, C.M. Nature of the Weipa Bauxite deposit, northern Australia. *Aust. J. Earth Sci.* **2008**, *55*, S45–S70. [CrossRef]
78. Retallack, G.J. *Soils of the Past*; Blackwell Science: Malden, MA, USA, 2001; 30p.
79. Bhattacharyya, D.P.; Kakimoto, P.K. Origin of ferriferous ooids: An SEM study of ironstone ooids and bauxite pisoids. *J. Sediment. Petrol.* **1982**, *53*, 849–857.
80. Mücke, A.; Badejoko, T.A.; Akande, S.O. Petrographic-microchemical studies and origin of the Agbaja Phanerozoic Ironstone Formation, Nupe Basin, Nigeria: A product of a ferruginized ooidal kaolin precursor not identical to the Minette-type. *Miner. Depos.* **1999**, *34*, 284–296. [CrossRef]
81. Aurell, M.; Bádenas, B.; Castanera, D.; Gasca, J.; Canudo, J.; Laita, E.; Liesa, C. Latest Jurassic–Early Cretaceous synrift evolution of the Torrelapaja Subbasin (Cameros Basin): Implications for Northeast Iberia palaeogeography. *Cretac. Res.* **2021**, *128*, 104997. [CrossRef]
82. Marques, B.; Olóriz, F.; Rodríguez-Tovar, F.J. Interactions between tectonics and eustasy during the Upper Jurassic and lowermost Cretaceous. Examples from the south ofIberia. *Bull. Soc. Géol. Fr.* **1991**, *162*, 1109–1124.
83. Aurell, M.; Robles, S.; Bádenas, B.; Rosales, I.; Quesada, S.; Meléndez, G.; García-Ramos, J. Transgressive–regressive cycles and Jurassic palaeogeography of northeast Iberia. *Sediment. Geol.* **2003**, *162*, 239–271. [CrossRef]
84. Legarreta, L. Evolution of a Callovian-Oxfordian carbonate margin in the Neuquén Basin of west-central Argentina: Facies, architecture, depositional sequences and global sea-level changes. *Sediment. Geol.* **1991**, *70*, 209–240. [CrossRef]
85. Hallam, A. A review of the broad pattern of Jurassic sea-level changes and their possible causes in the light of current knowledge. *Palaeogeogr. Palaeoclim. Palaeoecol.* **2001**, *167*, 23–37. [CrossRef]
86. Olóriz, F.; Reolid, M.; Rodríguez-Tovar, F.J. Taphonomy of fossil macroinvertebrates for improving ecostratigraphy: Shifting eco-sedimentary conditions from carbonate to carbonate-siliciclastic shelf (Upper Jurassic, Prebetic Zone). *Geobios* **2005**, *2*, 123–124.
87. Lechte, M.; Halverson, G.; Wallace, M.; Gibson, T.; van Smeerdijk Hood, A.; Wang, C.; Bui, T.H.; Maloney, K.; Millikin, A. Oolitic ironstones, continental iron flux and reverse weathering in the Proterozoic Eon: Insights from the Tonian Katherine Group, Yukon. *Earth-Sci. Rev.* **2024**, *253*. [CrossRef]
88. Aurell, M. El Jurásico Superior de la Cordillera Ibérica Central (Provincias de Zaragoza y Teruel). Análisis de Cuenca. Ph.D. Thesis, Universidad de Zaragoza, Zaragoza, Spain, 1990.
89. Eze, P.; Udeigwe, T.; Meadows, M. Plinthite and Its Associated Evolutionary Forms in Soils and Landscapes: A Review. *Pedosphere* **2014**, *24*, 153–166. [CrossRef]
90. Luko-Sulato, K.; Mounier, S.; Furlan, L.M.; Govone, J.S.; Bueno, G.T.; Rosolen, V. Spatial variation of soil organic matter and metal mobility in wetland soils: Implications for biogeochemical processes in lateritic landscape. *Catena* **2024**, *237*. [CrossRef]
91. Duchaufour, P. *Pedology. Pedogénesis and Classification*; Springer: New York, NY, USA, 1982; 462p, ISBN 978-9401160056.
92. Fookes, P.G. A Review: Genesis and classification of tropical residual soils for engineers. In *Geotechnics in the African Environment*; Blight, G.E., Fourie, A.B., Luker, I., Mouton, D.J., Scheurenberg, R.J., Eds.; CRC Press: Rotterdam, The Netherlands, 1994; pp. 423–442, ISBN 978-90-5410-007-2.
93. Laita, E.; Bauluz, B.; Yuste, A. The role of clay minerals in the concentration and distribution of critical metals in lateritic palaeosols from NE Iberia. *Appl. Clay Sci.* **2024**, *249*, 107264. [CrossRef]
94. McLaren, R.G.; Lawson, D.M.; Swift, R.S. Sorption and desorption of cobalt by soils and soil components. *Eur. J. Soil Sci.* **1986**, *37*, 413–426. [CrossRef]
95. Bradl, H.B. Adsorption of heavy metal ions on soils and soils constituents. *J. Colloid Interface Sci.* **2004**, *277*, 1–18. [CrossRef] [PubMed]
96. Price, J.R.; Velbel, M.A. Chemical weathering indices applied to weathering profiles developed on heterogeneous felsic metamorphic parent rocks. *Chem. Geol.* **2003**, *202*, 397–416. [CrossRef]
97. Bahlburg, H.; Dobrzinski, N. A Review of the Chemical Index of Alteration (CIA) and Its Application to the Study of Neoproterozoic Glacial Deposits and Climate Transitions. In *The Geological Record of Neoproterozoic Glaciations*; Memoir 36; Arnaud, E., Halverson, G.P., ShieldsZhou, G.A., Eds.; Geological Society: London, UK, 2011; pp. 81–92.

98. Fathy, D.; Abart, R.; Wagreich, M.; Gier, S.; Ahmed, M.S.; Sami, M. Late Campanian Climatic-Continental Weathering Assessment and Its Influence on Source Rocks Deposition in Southern Tethys, Egypt. *Minerals* **2023**, *13*, 160. [CrossRef]
99. Lerman, A.; Imboden, D.M.; Gat, J.R. *Physics and Chemistry of Lakes*; Springer: Berlin, Germany, 1995; 334p, ISBN 978-3-642-85134-6.
100. Xu, J.; Liu, Z.; Bechtel, A.; Meng, Q.; Sun, P.; Jia, J.; Cheng, L.; Song, Y. Basin evolution and oil shale deposition during Upper Cretaceous in the Songliao Basin (NE China): Implications from sequence stratigraphy and geochemistry. *Int. J. Coal Geol.* **2015**, *149*, 9–23. [CrossRef]
101. Stampfli, G.; Borel, G. A plate tectonic model for the Paleozoic and Mesozoic constrained by dynamic plate boundaries and restored synthetic oceanic isochrons. *Earth Planet. Sci. Lett.* **2002**, *196*, 17–33. [CrossRef]
102. Dromart, G.; Garcia, J.-P.; Gaumet, F.; Picard, S.; Rousseau, M.; Atrops, F.; Lecuyer, C.; Sheppard, S.M.F. Perturbation of the carbon cycle at the Middle/Late Jurassic transition: Geological and geochemical evidence. *Am. J. Sci.* **2003**, *303*, 667–707. [CrossRef]
103. Pellenard, P.; Fortwengler, D.; Marchand, D.; Thierry, J.; Bartolini, A.; Boulila, S.; Pierre-Yves Collin, P.-Y.; Enay, R.; Galbrun, B.; Gardin, S.; et al. Integrated stratigraphy of the Oxfordian global stratotype section and point (GSSP) candidate in the Subalpine Basin (SE France). *Vol. Jurass.* **2014**, *12*, 1–44.
104. Carmeille, M.; Bourillot, R.; Pellenard, P.; Dupias, V.; Schnyder, J.; Riquier, L.; Mathieu, O.; Brunet, M.-F.; Enay, R.; Grossi, V.; et al. Formation of microbial organic carbonates during the Late Jurassic from the Northern Tethys (Amu Darya Basin, Uzbekistan): Implications for Jurassic anoxic events. *Glob. Planet. Chang.* **2020**, *186*, 103127. [CrossRef]
105. Price, G.D.; Heszler, B.; Charlton, L.-M.T.; Cox, J. A climate perturbation at the Middle–Late Jurassic Transition? Evaluating the isotopic evidence from south-central England. *Palaeogeogr. Palaeoclim. Palaeoecol.* **2023**, *628*, 111755. [CrossRef]
106. Lécuyer, C.; Picard, S.; Garcia, J.; Sheppard, S.M.F.; Grandjean, P.; Dromart, G. Thermal evolution of Tethyan surface waters during the Middle-Late Jurassic: Evidence from $\delta^{18}O$ values of marine fish teeth. *Paleoceanography* **2003**, *18*. [CrossRef]
107. Wierzbowski, H. Seawater temperatures and carbon isotope variations in central European basins at the Middle–Late Jurassic transition (Late Callovian–Early Kimmeridgian). *Palaeogeogr. Palaeoclim. Palaeoecol.* **2015**, *440*, 506–523. [CrossRef]
108. Haq, B.U. Jurassic sea-level variations: A reappraisal. *GSA Today* **2018**, *28*, 4–10. [CrossRef]
109. Azerêdo, A.C.; Wright, V.P.; Ramalho, M.M. The Middle–Late Jurassic forced regression and disconformity in central Portugal: Eustatic, tectonic and climatic effects on a carbonate ramp system. *Sedimentology* **2002**, *49*, 1339–1370. [CrossRef]
110. Lorin, S.; Courville, P.; Collin, P.Y.; Thierry, J.; Tort, A. Modalités de réinstallation d'une plate-forme carbonatée après une crise sédimentaire: Exemple de la limite Oxfordien moyen—Oxfordien supérieur dans le Sud-Est du Bassin de Paris. *Bull. Soc. Géol. Fr.* **2004**, *175*, 289–302. [CrossRef]
111. Dera, G.; Brigaud, B.; Monna, F.; Laffont, R.; Pucéat, E.; Deconinck, J.-F.; Pellenard, P.; Joachimski, M.M.; Durlet, C. Climatic ups and downs in a disturbed Jurassic world. *Geology* **2011**, *39*, 215–218. [CrossRef]
112. Martire, L. Stratigraphy, facies and synsedimentary tectonics in the Jurassic Rosso Ammonitico Veronese (Altopiano di Asiago, NE Italy). *Facies* **1996**, *35*, 209–236. [CrossRef]
113. Olóriz, F.; Marques, B.; Caracuel, J.E. The Middle-Upper Oxfordian of Central Sierra Norte (Mallorca, Spain), and progressing ecostratigraphic approach in Western Tethys. *Geobios* **1998**, *31*, 319–336. [CrossRef]
114. Coimbra, R.; Immenhauser, A.; Olóriz, F.; Rodríguez-Galiano, V.; Chica-Olmo, M. New insights into geochemical behaviour in ancient marine carbonates (Upper Jurassic Ammonitico Rosso): Novel proxies for interpreting sea-level dynamics and palaeoceanography. *Sedimentology* **2015**, *62*, 266–302. [CrossRef]
115. Collin, P.Y.; Loreau, J.P.; Courville, P. Depositional environments and iron ooid formation in condensed sections (Callovian–Oxfordian, south-eastern Paris basin, France). *Sedimentology* **2005**, *52*, 969–985. [CrossRef]

minerals

MDPI

Article

The Miocene Source-to-Sink Evolution of Fibrous Clay Minerals in Hyperalkaline Playa-Lakes, Duero Basin (Central Spain)

Pablo del Buey [1,2,3,*], María Esther Sanz-Montero [3], Juan Pablo Rodríguez-Aranda [3,4], Mónica Sánchez-Román [1] and Fernando Nieto [2]

1 Geobiology Laboratory, Department of Earth Sciences, Faculty of Science, Vrije Universiteit Amsterdam, De Boelelaan 1100, 1081 HZ Amsterdam, The Netherlands; m.sanchezroman@vu.nl
2 Department of Mineralogy and Petrology, Faculty of Science, University of Granada, Avda. de Fuente Nueva, s/n, 18071 Granada, Spain; nieto@ugr.es
3 Department of Mineralogy and Petrology, Faculty of Geological Sciences, Complutense University of Madrid, C/Profesor José Antonio Novais 12, 28040 Madrid, Spain; mesanz@ucm.es (M.E.S.-M.); juanparo@ucm.es (J.P.R.-A.)
4 IES Camilo José Cela, Av. Monte, 16, Pozuelo de Alarcón, 28223 Madrid, Spain
* Correspondence: pablodelbuey@ucm.es

Abstract: Palygorskite is assumed to be the predominant clay mineral in playa-lakes, where it may be detrital or authigenic in origin. Discriminating between detrital and authigenic clays is crucial to elucidate paleoenvironmental conditions in lacustrine deposits. This study provides insight into the sedimentary evolution of clay minerals from source, lacustrine Miocene marlstones and mudstones, to sink, represented by three recent hyperalkaline playa-lakes in Central Spain. XRD, TEM, and AEM analyses show concomitant detrital and authigenic palygorskites in the three playa-lakes. The inherited palygorskites exhibit degradation features, larger widths, and common and ideal compositions, in contrast to neoformed particles. The latter are narrower. Depending on the hydrochemical composition of each playa-lake, neoformed palygorskites are enriched in a different octahedral cation (Al_2O_3, MgO, and Fe_2O_3). Iron-rich palygorskites are only formed in association with authigenic saponites in one of the playa-lakes. The same effect of magnesium competition between smectite and palygorskite is observed in Miocene mudstones, where palygorskite is relatively enriched in iron. In hyperalkaline, seasonal playa-lakes lying in the vicinity, slight physicochemical differences play a crucial role in the crystallochemical composition of authigenic palygorskites, highlighting the use of this mineral as a geochemical proxy.

Keywords: TEM-AEM; palygorskite; chlorite; mudstones; marlstones; sedimentation

1. Introduction

The occurrence of the fibrous clay minerals, palygorskite, and sepiolite in lacustrine rocks has been widely reported in the literature [1–3]. Fibrous clay minerals can precipitate from multiple mechanisms. Sepiolite is a common neoformed mineral precipitated from a colloidal silicate precursor in magnesium-rich environments via desiccation [4]. Palygorskite can precipitate from aluminic precursors [5,6] and directly from dissolution [7,8]. Ref. [9] revealed that the desiccation of gel precursors is responsible for the textural features of palygorskite and sepiolite in Miocene palustrine facies at Nullarbor Plain (Australia). More recently, microbially induced palygorskite neoformation has been recognized in hypersaline playa-lakes in relation to desiccated extracellular polymeric substances within microbial mats [10].

The transformation of pre-existing clay minerals (e.g., chlorite, illite, stevensite, and kerolite) is another common mechanism of palygorskite and sepiolite formation [11–13]. Detrital clay minerals in lacustrine environments usually exhibit textural features related to degradation (i.e., weathering and loss of elements) due to reworking under new hydrochemical conditions [14]. Detrital clay minerals in playa-lake systems consist of mixtures of clay minerals, mainly illite, kaolinite, and dioctahedral smectite [2], which contribute to the precipitation of magnesium smectite. In environments where detrital clay minerals are present in low proportions, the purer magnesium phyllosilicates, sepiolite and kerolite, can precipitate [15].

In the case of neoformation, supersaturation is the fundamental control of authigenic clay mineral precipitation [16]. Ref. [5] summarized the physicochemical parameters of sepiolite and palygorskite neoformation. These environmental parameters were high alkalinity (pH around 8–9.5), high Mg+Fe/Si ratios, and low P_{CO2} between water and sediment. In addition, the stability of palygorskite and sepiolite is a function of pH, which the latter is unstable at pH $\geq$ 12. In waters that are not in equilibrium with atmospheric CO_2, the effect of P_{CO2} is less obvious due to its effect on pH value [17]. Finally, a significant increment in the CO_2 concentration, biotic or abiotically induced, could result in the dissolution of magnesium clay minerals [18].

Among others, two Cenozoic continental basins of Central Spain, the Madrid and Duero Basins, are well-known to bear palygorskite-rich lacustrine and palustrine facies associated with alluvial fan systems in transition to lacustrine environments [11,19–21]. In some intervals of the intermediate Miocene unit of the Madrid Basin [22,23], sepiolite is abundant enough to be economically exploited in open-pit mines. The extra-basin source of magnesium required for the precipitation of neoformed sepiolite is still a subject of debate [24]. In the Duero Basin, the deposition of palygorskite is commonly linked to the transformation of detrital phyllosilicates in relation to alluvial fans [11,25].

The aim of this investigation is to analyze the depositional processes of authigenic and clastic fibrous clay minerals and the associated clays in modern hyperalkaline playa-lakes under variable hydrochemical conditions. The transport of clays from Miocene lacustrine sources into the new environment involves physicochemical changes. The results show that there are two distinct populations of fibrous clay minerals in playa-lakes, one of detrital origin and the other of neoformed palygorskites. In each playa-lake, neoformed palygorskites and sepiolites are enriched in an octahedral cation (MgO, Al_2O_3, and Fe_2O_3) in comparison to Miocene fibrous clay minerals. The specific enrichment in one or another of the cations is controlled by the interplay of dissolution, precipitation, and transformation processes.

2. Geological Setting

The Caballo Alba, Las Eras, and Bodón Blanco lakes are located in the Coca–Olmedo endorheic wetland in the southwestern sector of the Duero Basin (Central Spain) (Figure 1 up). The three playa-lakes overlie horizontally bedded Miocene and Quaternary deposits in the Duero Basin (Figure 1 down). The Miocene sediments in the vicinity of Caballo Alba lake (Figure 1 down) consist of gray-greenish marlstones, dolo-marlstone, and mudstones [26]. Miocene facies are deposited in the marginal subenvironments of saline lakes belonging to the Cuestas Unit [27]. Quaternary conglomerates, sandstones, and siltstones were deposited in the fluvial systems running through the study area (Figure 1 down) [28].

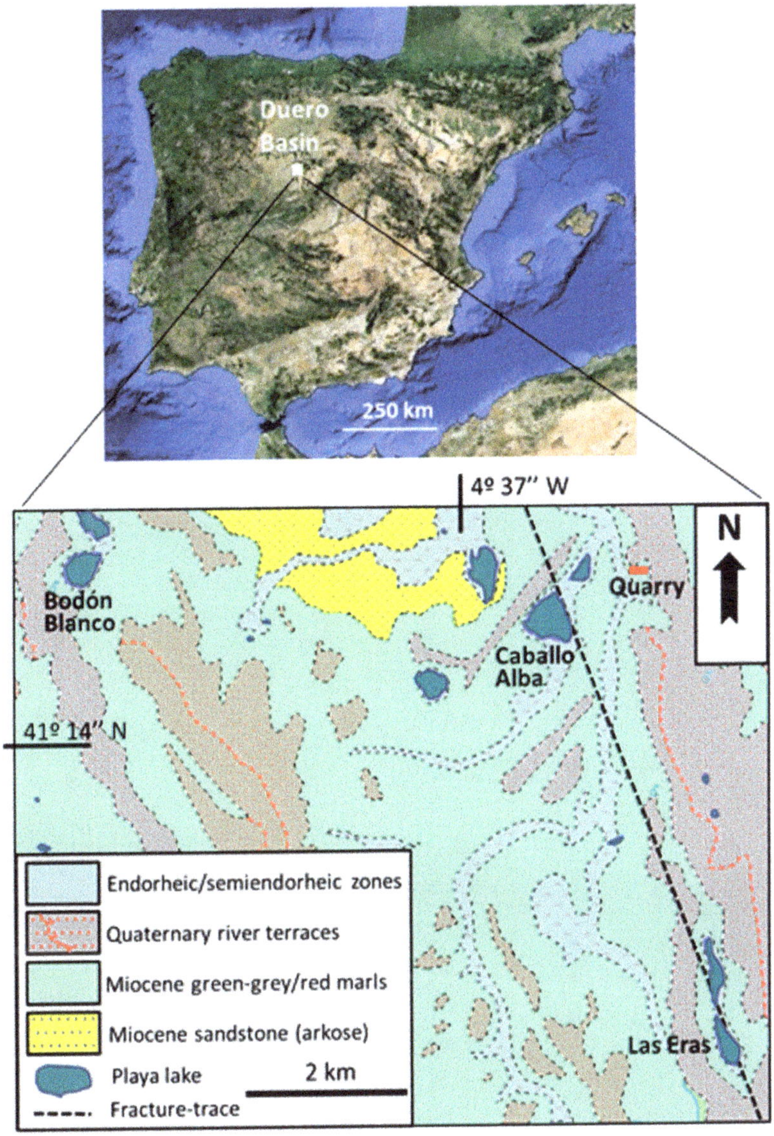

Figure 1. Location of the study area: (**up**) satellite image (source of the image Google Earth); (**down**) geological map (based on MAGNA 1:50.000, n° 428) showing the location of the Caballo Alba, Las Eras, and Bodón Blanco playa-lakes.

Caballo Alba, with a maximum area of 0.17 km^2, is a shallow basin (up to 0.5 m deep) fed by groundwater [26]. In addition, low water inputs were derived from a small, extremely ephemeral stream. Caballo Alba presents high pH values ranging from 9.0 to 10.4, thus presenting brackish-to-saline water (1–16 $g \cdot L^{-1}$), whereas Las Eras reached the highest pH value recorded, "up to 11.3". Overall, the hydrochemical composition of Caballo Alba, Las Eras, and Bodón Blanco is characterized by high carbonate concentrations, with chloride dominant over sulfate. Nevertheless, the hydrochemical composition is very variable among the playa-lakes: the analysis of most of the data mainly shows Na^+-Cl^- composition, but some values correspond to Na^+-Cl^--$SO_4{}^{2-}$-$HCO_3{}^-$ for Caballo Alba and Las Eras. Bodón Blanco hydrochemistry is classified into three different areas of the Piper–Hill–Langelier diagram, including Na^{+}-HCO_3^{-} [29]. Finally, the concentration of dissolved potassium is an order of magnitude higher in Las Eras than in Caballo Alba and Bodón Blanco [30].

The region is characterized by a continental Mediterranean climate, with low mean annual rainfall of 400–500 mm·year^{-1} and average temperatures of 13 °C [29]. Mostly from March to May, the Caballo Alba playa-lake hosts benthic and floating microbial mats that expand over the lake (Figure 2A,B). In spring, macroscopic filamentous green algae of the genus *Spirogyra* become entangled with *Cyanobacteria* (*Oscillatorya* genus) and Diatoms (Figure 2B) [31]. The lake is also colonized by charophyte and the associated ostracod biota. Eventually, the photosynthetic activity of some microbes produces oxygen bubbles that detach benthic microbial mats, which float and rise to the surface [32]. Ref. [31] described the intracellular precipitation of calcite and dolomite during the conjugation cycle of *Spirogyra*. Calcite and dolomite are the most abundant carbonates in these hyperalkaline playa-lakes. From June to October, coinciding with periods of low rainfall, the lake floor dries out and Caballo Alba, Las Eras, and Bodón Blanco become playas covered by a white mineral crust (Figure 2A,C). The crust contains an assemblage of authigenic carbonates (e.g., calcite, dolomite, nesquehonite, hydromagnesite, natron, and trona), sulfates (e.g., hexahydrite, starkeyite, gypsum, thenardite, and bloedite), chlorides (halite), and magnesium clays (Figure 2C) [26,29]. Some magnesium carbonates like hydromagnesite, magnesite, and huntite can precipitate due to the heterotrophic degradation of microbial mass [33]. Underneath the desiccated microbial mat mass, some quartzite pebbles, cobbles, and boulders were bound and trapped (Figure 2D).

Figure 2. Sedimentary evolution in the Caballo Alba playa-lake: (**A**) white mineral precipitates (e.g., hydromagnesite, northupite, natron, and trona, among others) on the shore and in ponds during desiccation; (**B**) floating microbial mats during the wet season (December 2018); (**C**) desiccation cracks in the sediment (July 2020); (**D**) quartzite pebbles, cobbles, and boulders trapped beneath a desiccated microbial mat (black arrows).

2.1. *Playa-Lake Sediments*

Twenty-six surface sediment samples comprising mineral and organic matter from the three playa-lakes were collected at different sites (i.e., located within each playa-lake) and periods (i.e., seasons).

2.2. *Miocene Deposits*

Well-exposed outcrops of Miocene rocks surrounding these playa-lakes are scarce. To study the Miocene deposits that constitute the watershed of these playa-lakes, a stratigraphic column was made in a quarry. This small quarry with a height of 10 m (Figures 1 and 3), where common clays are extracted for the ceramic industry (IGME and SIEMCALSA, 2016), is located 1 km NE of Caballo Alba.

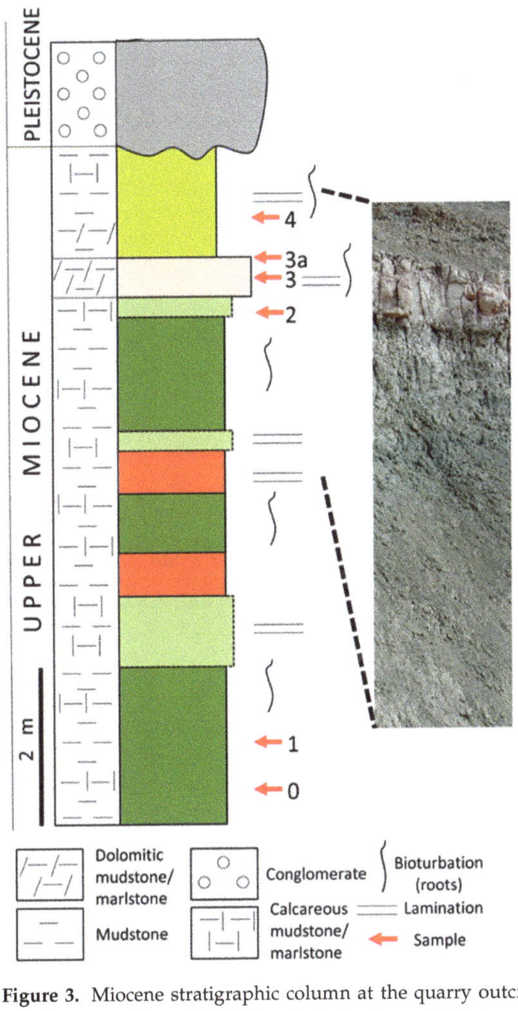

Figure 3. Miocene stratigraphic column at the quarry outcrop where marlstone and mudstone samples were collected (0–4). Sample 3a: laminar whitish transition between sample 3 and sample 4.

The stratigraphic section of Miocene sedimentary rocks is 9.5 m thick, and it is situated at 41°14′58.50″ N//4°35′38.33″ W, lying between 773 m and 783 m.a.s.l. This section consists of marlstones, dolo-marlstones, and mudstones. The different colors in the logged section reflect the colors of these rocks in the outcrop, regardless of their lithology (Figure 3).

The lower part of the section is composed of green marlstone up to 2–3 m thick. From 3 to 6.5 m, an alternation of greenish to reddish mudstones is partially covered. From 6.5 to 7 m, there is a whitish marlstone level before a reddish dolo-marlstone level with bioturbation. There is a laminar whitish transition from the reddish dolo-marlstone to the upper 2 m of the brownish-green level of mixed calcitic/dolo-mudstone. At the topmost of the succession is a fluvial channel filled with surrounded quartzite pebbles, cobbles, and boulders [26]. Six marlstone and mudstone samples along the stratigraphic column were collected in July 2020 (Table S1, Figure 3). Three samples were used for clay mineralogy, TEM, and EDX analyses (Samples 0, 3, and 4).

3. Materials and Methods

3.1. Bulk and Clay Mineral Assemblages (XRD)

Bulk mineralogy was analyzed by random powder X-ray diffraction (XRD) analysis using a Bruker D8 Advance diffractometer with Cu K_{α} radiation (λ = 1.54060 Å) and equipped with a Sol-X detector. The samples were scanned from 2 to 65° 2θ with a step size of 0.02° 2θ and a counting time of 0.5 s per step. XRD interpretation was based on the method described by [34] using EVA Bruker software (Version 7).

For clay mineralogy analysis, carbonates, sulfates, and organic matter were removed from samples using 200 mL of 0.2 M HCl dissolution and continuously agitated for 20 min [35,36]. Briefly, a <2 µm fraction of the treated sediment was separated using a centrifuge technique [37] with a Thermo Scientific Sorvall ST 16 Centrifuge (extraction cycles of 1000× *g* during 3 min). Then, the <2 µm fraction was analyzed as oriented mounts prepared by smearing the slurries on glass slides. Three XRD diagrams were obtained for each fraction: after air drying (AD), ethylene glycol solvation (EG), and heating at 550 °C for at least 2 h. These samples were scanned from 2 to 35° 2θ with a step size of 0.02° 2θ and a counting time of 1 s per step. The semiquantification of clay mineral assemblages was performed by using the mineral intensity factor method (MIF) as described in [10]. The names of clay mineral phases were abbreviated according to [38] as follows: mica (Mca), chlorite (Chl), kaolinite (Kln), sepiolite (Sep), palygorskite (Plg), and smectite (Sme).

3.2. Textural Features and Crystallochemical Compositions of Fibrous Clay Minerals (TEM-AEM)

Digital analyses of TEM images are the most effective method for measuring the form of anisotropic particles and distinguishing particle populations within a sample [39]. TEM observations and EDX analyses of 13 samples, 10 of which correspond to playa-lake sediments and 3 of Miocene marlstones (Samples 0, 3, and 4), were obtained from powdered portions prepared using C-coated Cu grids. The TEM data were obtained using two electron microscopes for semiquantitative chemical analyses at the CNME (Complutense University, Madrid, Spain), namely a JEOL 3000F microscope operating at 300 kV and a JEOL JEM 2100 microscope operating at 200 kV. Semiquantitative analyses of particles were performed in the STEM mode using an Oxford Instruments EDX microanalysis system. In order to ensure the reproducibility of the data for quantitative chemical analyses (AEM) of Plg particles from Caballo Alba, data were obtained in the STEM mode using the Thermo Fisher Scientific Talos F200X, operated at 200 kV at the CIC (University of Granada, Granada, Spain). For this task, individual spectra corresponding to each pixel of homogeneous areas were summed up to produce the average spectrum of the entire analyzed area. Albite, biotite, muscovite, spessartine, olivine, and titanite standards were used to obtain K-factors for the transformation of intensity ratios to concentration ratios according to [40,41]. The structural formulae of palygorskite were calculated on the basis of 42 negative charges, i.e., $O_{20}\,(OH)_2{\cdot}(OH_2)_4$. All the Fe was considered as trivalent.

4. Results

4.1. Miocene Clay Minerals

The mineralogical clay assemblages of Miocene beds (Figure 4) are composed of Mca, Chl, Kln, Sep, Plg, and Sme. The assemblage comprising Mca-Kln-Plg is ubiquitous across the stratigraphic section, whereas the Chl-Sep-Sme assemblage is irregularly distributed. From bottom to top, the Mca-Chl-Kln-Sep-Plg assemblage is recognized in samples of marlstone and dolo-marlstone. Only the brownish-greenish mixed calcitic/dolo-mudstone has a representative amount of Sme content, as evidenced by the shift in the 001 reflection under ethylene glycol solvation (EG). In this sample, Chl and Sep contents are absent or below the detection limit of the X-ray diffraction technique (<5% wt) (Table S2).

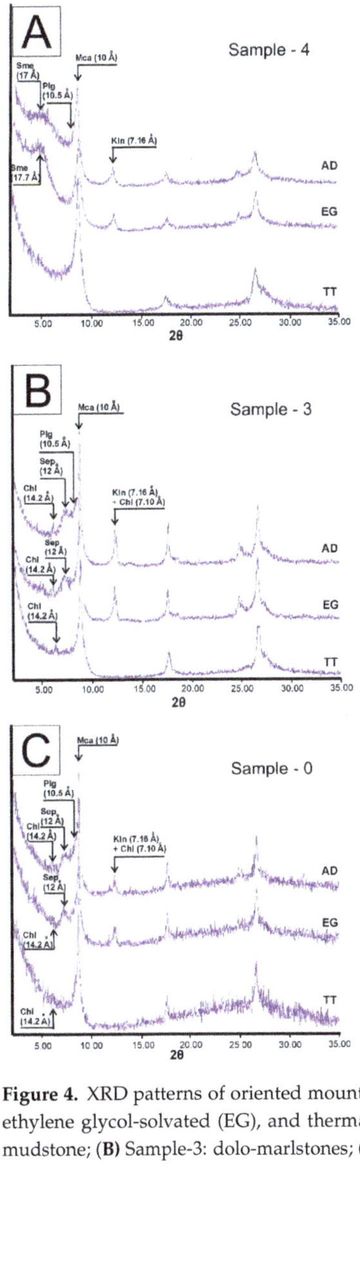

Figure 4. XRD patterns of oriented mounts of marlstone and mudstone samples: air-dried (AD), ethylene glycol-solvated (EG), and thermally treated (TT) (550 °C). **(A)** Sample-4: calcitic/dolo-mudstone; **(B)** Sample-3: dolo-marlstones; **(C)** Sample-0: green marlstones.

Textural and Crystallochemical Characterization of Miocen Fibrous Clay Minerals

Textural characterization is based on the width of the Plg fibers and other textural features like length, width/length ratio, fiber curliness, and fiber arrangement [42]. The length and width/length ratios are not described in all images due to the impossibility of completely measuring the length of most fibers in TEM images (Figure 5). Fiber curliness and fiber arrangement were the other key features analyzed. Due to the impossibility of distinguishing between Sep and Plg only by their morphological features [42], crystallochemical analyses of Miocene fibrous clay minerals were performed (Figure 6). X-ray diffraction showed that the most abundant fibrous clay mineral in samples of marlstone and dolo-marlstone is Sep, whereas in calcitic/dolo-mudstone samples, Plg is only present but scarce. In these terms, the fibers from dolo-marlstone are mostly Sep and only Plg for marlstone and calcitic/dolo-mudstone.

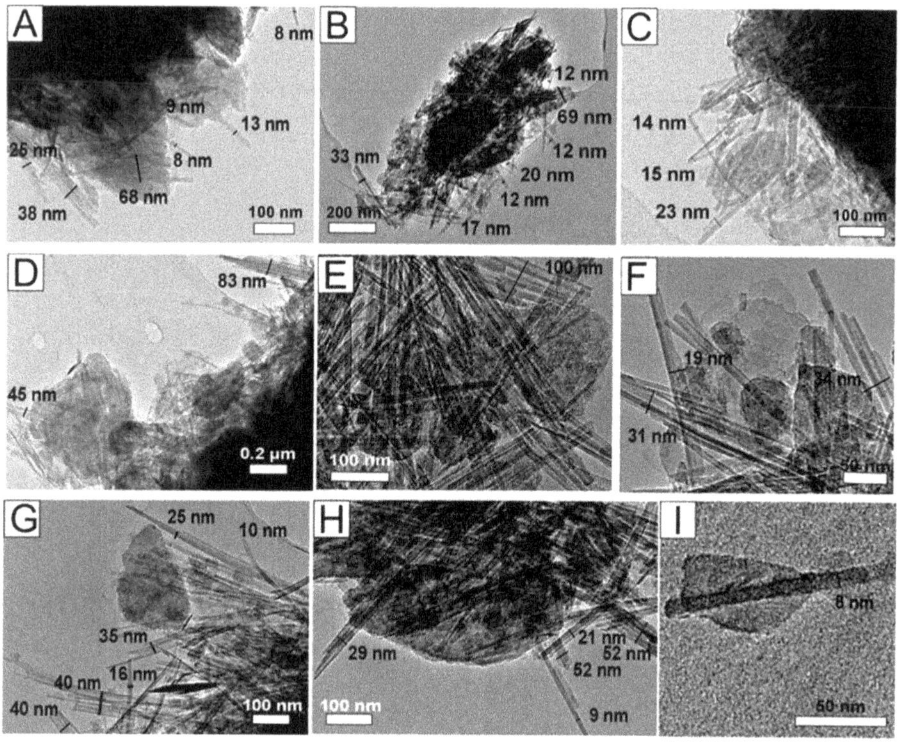

Figure 5. TEM micrographs of fibrous clay minerals from Miocene marlstones and mudstone. Sample 0–marlstone (**A–C**): nanometer-sized Sep and minor Plg laths, rods, and bundles from marlstone samples at the base of the sedimentary succession. Sample 3—dolo-marlstone (**D–H**): rods and bundles of Sep and minor Plg. Sample 4—calcitic/dolo-mudstone (**I**): laths of Plg from the topmost sample from the Miocene sedimentary succession.

The basal unit of the stratigraphic section (Sample 0, Figure 5A–C), which is composed of green marlstones, presents fibers in three different aggregation states (Figure 5): laths (8–17 nm width), rods (12–38 nm width), and bundles (68–69 nm width). These fibers with independent states of aggregation are curved to straight. The basal marlstone bed consists of ideal Plg (Sample 0, Figure 6). The dolo-marlstones in the intermediate interval contain Sep–Plg fibers (Sample 3, Figure 5D–H) in three different aggregation states: laths (9–16 nm), rods (19–45 nm width), and bundles (40–100 nm width). Most fibers in this

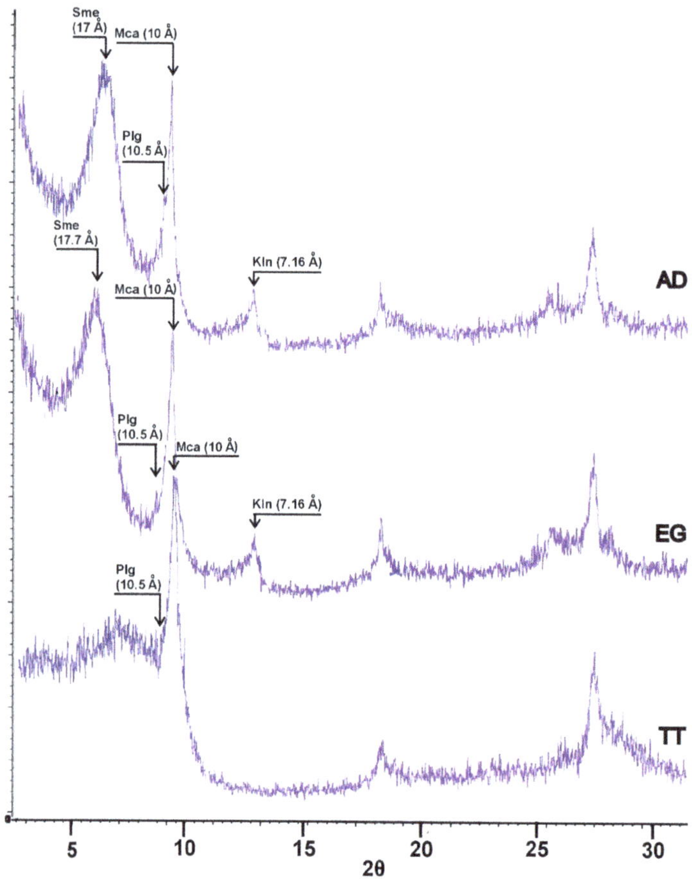

Figure 7. XRD patterns of oriented mounts of a desiccated microbial mat from the Caballo Alba playa-lake: air-dried (AD), ethylene glycol-solvated (EG), and thermally treated (TT) (500 °C).

4.2.1. Textural and Crystallochemical Characterization of Fibrous Clay Minerals from Playa-Lakes

Fibrous clay minerals (mostly Plg) were classified according to crystallochemical classification by [43]. In this sense, there are three different groups of Plg particles according to their enrichment with some octahedral cations: Al_2O_3 (Figure 8), Fe_2O_3 (Figure 9), and MgO (Figure 10). Plg with intermediate compositions, which are the majority of these minerals, are shown in Figure 11. In the corners of the ternary diagram, the Plg particles from the three playa-lakes enriched in one of the octahedral cations Al_2O_3, MgO, or Fe_2O_3 are shown (Figure 6). Notably, a group of Plg from Caballo Alba are enriched in Fe_2O_3 $[Si_8O_{20}\ (Mg_2\ Fe_2)_{0\cdot6}\ (Mg_2\ Al_2)_{0\cdot4}\ (OH)_2 \cdot (OH_2)_4]\cdot nH_2O$, while some minor particles corresponding to Las Eras and Bodón Blanco can be classified as magnesium-rich Plg/aluminum-rich Sep $[Si_{12}O_{30}Mg_8(OH)_4(OH_2)_2]_{0\cdot5}[Si_8O_{20}\ (Mg_2\ Fe_2)_{0\cdot1}\ (Mg_2\ Al_2)_{0\cdot9}\ (OH)_2 \cdot (OH_2)_4]_{0\cdot5}\cdot nH_2O$. Most aluminum-rich Plg $[Si_8O_{20}\ (Mg_2\ Fe_2)_0\ (Mg_2\ Al_2)_1\ (OH)_2 \cdot (OH_2)_4]_1\cdot nH_2O$ minerals correspond to the particles of Bodón Blanco. The four varieties of Plg particles are described next in terms of width, curliness, and arrangement of the fibers (Table S2).

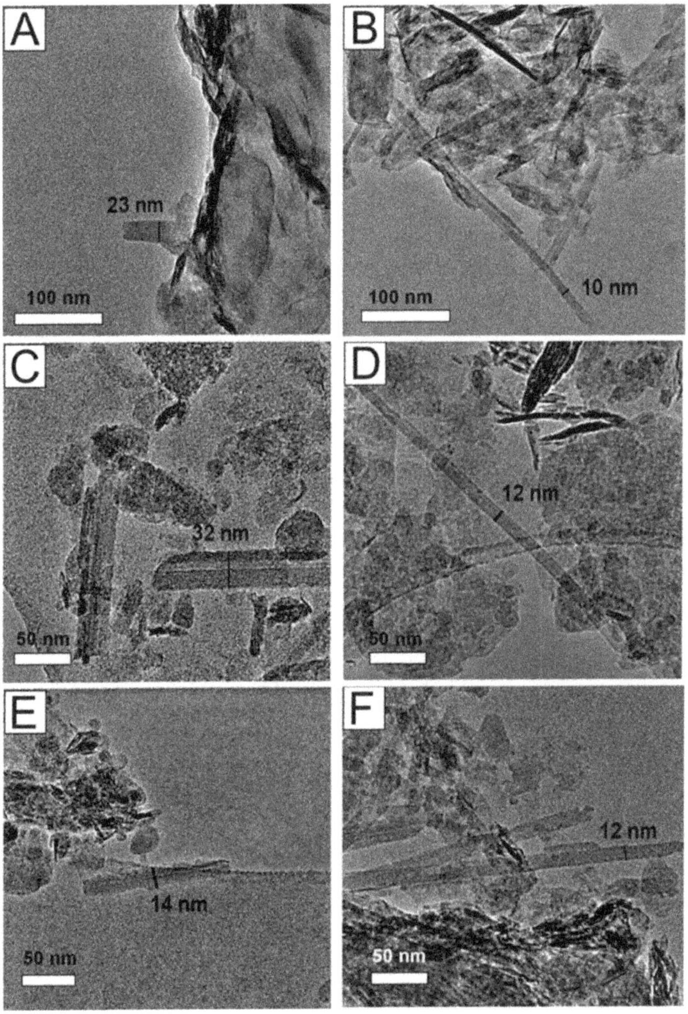

Figure 8. Aluminum-rich Plg fibers deposited in the playa-lakes: Bodón Blanco (**A–C**), Caballo Alba (**D**,**E**), and Las Eras (**F**): (**A**) Plg lath associated with particles of Sme; (**B**) Plg lath; (**C**) Plg rods; (**D–F**) Plg lath.

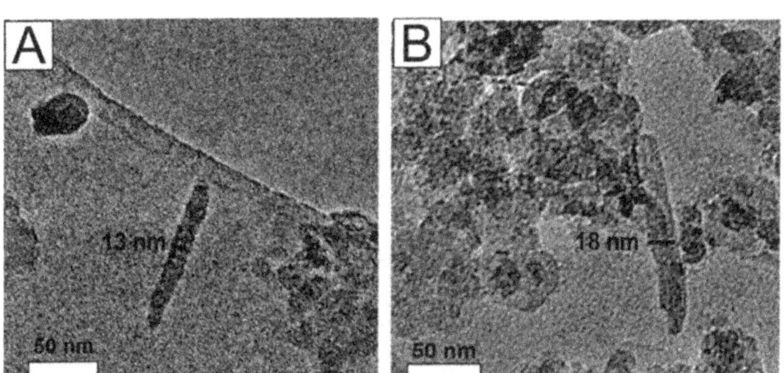

Figure 9. *Cont.*

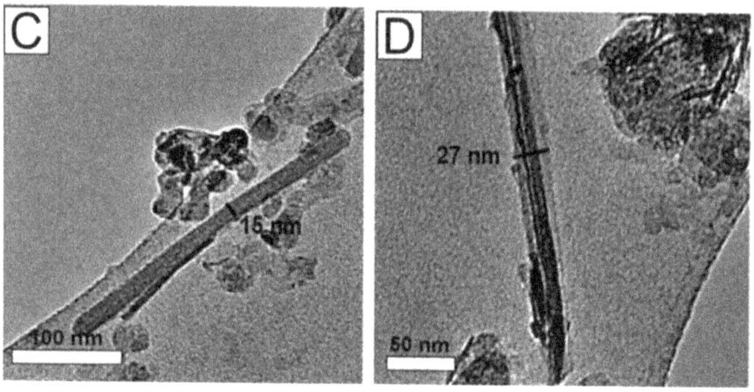

Figure 9. Iron-rich Plg from the Caballo Alba playa-lake: (**A**) nanometer-sized Plg particle; (**B**) Plg lath shorter than 100 nm; (**C**) Plg lath longer than 100 nm; (**D**) Plg rod.

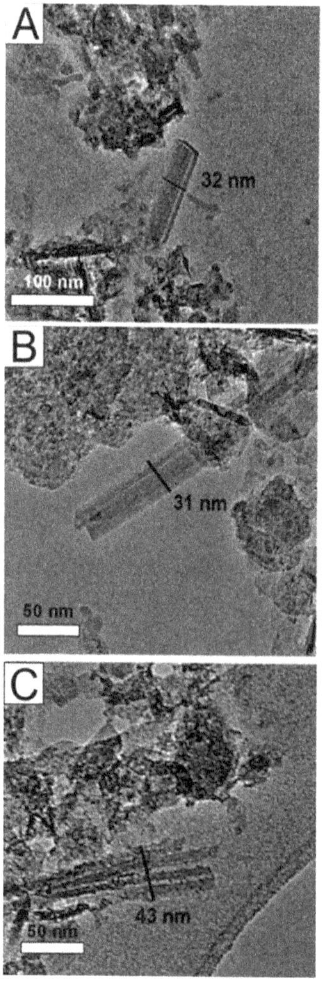

Figure 10. Magnesium Plg and aluminum Sep from the Las Eras playa-lake: (**A**) rod of aluminum Sep, 100 nm in length; (**B**) rod of aluminum Sep; (**C**) bundle of magensium Plg.

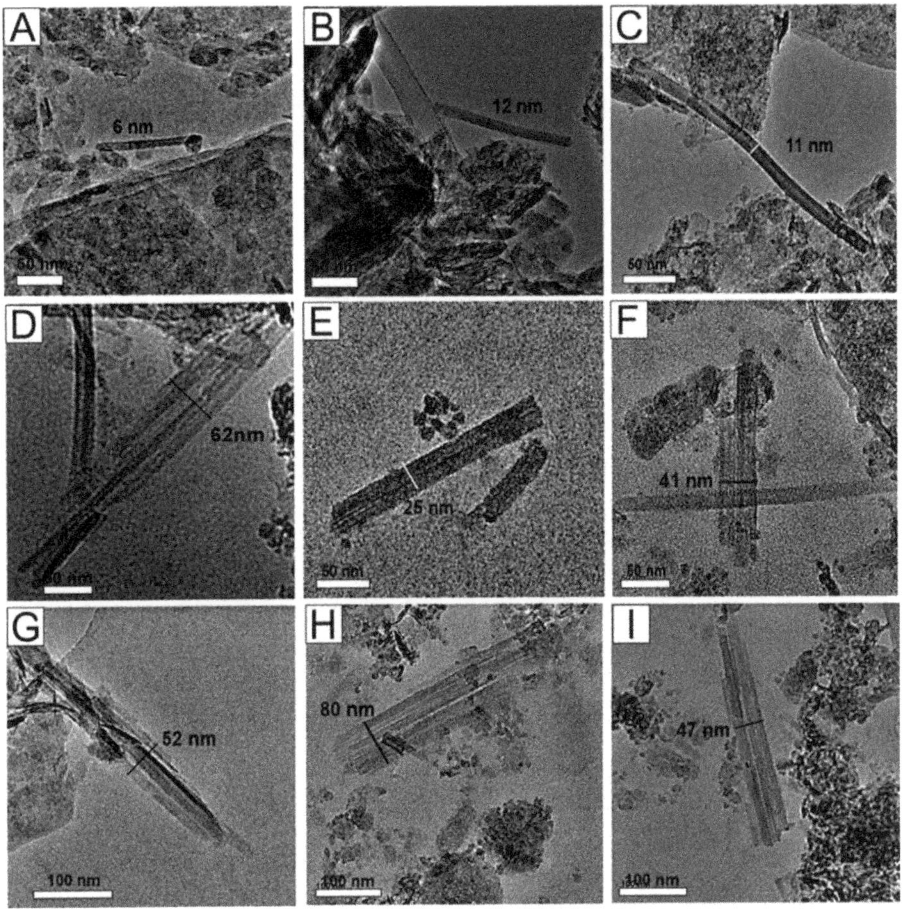

Figure 11. Intermediate composition Plg deposited in the playa-lakes: Caballo Alba (**A**), Las Eras (**B–I**), and Bodón Blanco (**G**): (**A**) nanometer-sized Plg lath; (**B**,**C**) Plg laths; (**D–I**) partially degraded bundles of Plg.

Aluminum-Rich Plg

Aluminum-rich Plg particles are recognized in the three studied hyperalkaline playa-lakes. In Bodón Blanco (Figure 8A–C), Plg consists of lath and rods enriched in Al_2O_3 (20.17%–27.95% wt). The width of Plg particles varies from 10 nm of laths to 23–32 nm of rods. The laths of Plg (Figure 8D,E) in Caballo Alba range from 21.06 to 25.54% wt of Al_2O_3. The width of the laths of Plg particles varies from 12 to 14 nm. Las Eras aluminum-rich Plg laths (21.57% wt) are scarce and present widths of 12 nm (Figure 8F). Many of these particles are straight and rigid (Figure 8A,C,E,F), and only a few of them are curved (Figure 8B,D).

Iron-Rich Plg

Iron-rich Plg particles are only found in the Caballo Alba playa-lake (Figure 9). In relation to their length, two populations of fibers can be distinguished. The smallest laths are up to 100 nm in length (Figure 9A,B). By contrast, the larger fibers, laths (13–18 nm width), and rods (27 nm width) are more than 100 nm in length. All these Plg fibers are straight and rigid. The content of Fe_2O_3 of these Plg is particularly high (10.40%–21.71% wt).

The results of quantitative TEM-AEM analyses of Plg from Caballo Alba and their structural formulae were obtained from one sample (Table 1). According to the Plg crystallochemical classification established by [44], three types of Plg can be distinguished in terms of the number of Mg atoms per half unit cell (p.h.u.c) vs. the number of octahedral positions occupied (p.h.u.c): types II, III and IV. The results of the structural formulae for iron-rich Plg are highlighted in green. Type IV represents aluminum-rich Plg. Maximum, minimum, and mean values for each tetrahedral, octahedral, interlayer, and the sum of octahedral cations are presented in Table 1.

Table 1. Structural formulae proposed for the studied Plg from the Caballo Alba playa-lake.

Sample	Si	^{IV}Al	^{VI}Al	Mg	Fe^{3+}	∑OC	Ca	K	Na	Type
CA I/01	7.4	0.6	1.41	2.19	0.47	4.07	0.1	0.23	0.14	Type II
CA I/02	7.45	0.55	1.09	2.55	0.49	4.13	0.09	0.32	0.17	Type III
CA I/03	7.10	0.9	0.88	2.88	0.64	4.4	0.07	0.29	0.16	Type III
CA I/04	7.36	0.64	0.53	3.37	0.38	4.28	0.21	0.11	0.64	Type III
CA I/05	7	1	0.57	2.78	0.8	4.15	0.21	0.19	0.72	Type III
CA I/06	6.94	1.06	0.74	3.54	0.42	4.7	0.05	0.25	0.15	Type III
CA I/07	7.21	0.79	0.82	3.57	0.31	4.7	0.09	0.07	0.03	Type III
CA I/08	7.23	0.77	1.6	1.46	0.51	3.57	0.23	0.51	0.55	Type IV
CA I/09	6.68	1.32	1.29	1.35	0.74	3.38	0.52	0.61	0.86	Type IV
CA I/10	7.47	0.53	2.18	0.7	0.43	3.31	0.13	0.82	0.24	Type IV
CA I/11	7.35	0.65	2.23	0.85	0.34	3.42	0.12	0.71	0.3	Type IV
MAX	7.47	1.32	2.23	3.57	0.8	4.7	0.52	0.82	0.86	
MIN	6.68	0.53	0.53	0.7	0.31	3.31	0.07	0.07	0.03	
MEAN	7.20	0.80	1.21	2.29	0.50	4.01	0.17	0.37	0.36	

Magnesium Plg and Aluminum Sep

Magnesium-rich Plg and aluminum Sep particles are found in the Las Eras playa-lake (Figure 10). The state of the fiber arrangement varies from rods (31–32 nm width) to bundles (43 nm width), from straight and rigid to curved. The MgO content ranges from 20.23% to 27.79% wt.

Common and Ideal Plg

This group of minerals is the most representative in terms of the number of particles analyzed (Figure 11). The compositional limits can be established in the ternary diagram by the "y" and "x" values. The "y" value corresponds to the proportion of Sep and Plg polysomes [43], whereas "x" indicates the Fe_2O_3 content in Plg polysomes. In this line, "y" values vary from 0.1 beyond to 0, and "x" varies from 0.1 to 0.4. This implies that all Plg particles in this field can be mostly classified as common and ideal Plg particles [43].

There are two populations of fibrous clay minerals among the three playa-lakes: laths of Plg (6 nm to 12 nm widths) (Figure 11A–C) and partially degraded rods and bundles of Plg (25 nm to 62 nm widths) (Figure 11D–I). Another textural difference between these two populations is the fiber curliness: laths are curved (Figure 11A–C), whereas partially degraded rods and bundles are straight and rigid (Figure 11D–I).

5. Discussion

The mineralogical, textural, and chemical analyses provide supporting evidence indicating that a mixed assemblage of detrital and authigenic Plg is deposited in the three hyperalkaine playa-lakes studied. The crystallochemical compositions of the detrital Plg particles analyzed are equivalent to Miocene marlstones and mudstones. Alternatively, authigenic Plg particles, and to a minor extent aluminum Sep, are characterized by greater proportions of octahedral cations (Al_2O_3, MgO, and Fe_2O_3) in the playa-lakes. High pH and minor differences in the geochemical compositions of the studied playa-lakes influence the precipitation of different types of Plg as well as the precipitation of Sme and the dissolution of Chl. The authigenic processes of iron-rich Plg precipitation are closely associated with the Caballo Alba playa-lake and Miocene calcitic/dolo-mudstone, where Plg is present along with Sme content.

Plg deposited in recent lacustrine environments is commonly detrital [1]. A similar interpretation applies to most of the Plg deposited in the three studied playa-lakes. Due to the shallow nature of these playa-lakes and their surface area, Plg was homogeneously distributed throughout the lakes. The small particle size also favors the transport of clay minerals in suspension from one area to another [45]. Plg particles migrate and move preferentially over Sme and Kln [46]. Rods and bundles of detrital Plg are straight and rigid and close to the ideal composition (Figure 10A–C). The widths measured in the detrital bundles of Plg in these playa-lakes, ranging from 25 to 80 nm, have been also reported in Plg fibers deposited in other worldwide geological settings [47–49].

In relation to mineralogy, the crystallochemical composition and textural features, including width, fiber curliness, and the arrangement of Plg fibers deposited in modern lakes, fit well with the features of fibrous clay minerals forming the Miocene strata in the watershed [26]. There are many examples of detrital Plg deposits and their parent rocks in different geological formations worldwide [50–52].

Our results show that these alterations of detrital clay minerals and the precipitation of authigenic ones are related to the high pH values that these playa-lakes consistently record. As the pH increases, the separation and dispersion degrees of Plg fibers increase, whereas the contents of SiO_2, Al_2O_3, MgO, and Fe_2O_3 in Plg remain constant [53]. On the edges of these fibers, degradation is evident. Plg is stable only at high pH values [7]. There is no significant variation with the increase in pH value (i.e., up to 13), neither at the d_{110} peak intensity nor at the full width at half maximum [53]. In addition, Sep is more abundant than Plg in the studied Miocene marlstones, whereas Plg is more abundant in the hyperalkaline lakes studied. Although Sep is a mineral with low solubility, its long-term interaction with hyperalkaline waters (i.e., $8 > pH < 13$) results in a greater decrease in the length of Sep fibers and subsequent dissolution of Si-OH groups than those reported by [54].

A second population of fibrous clay minerals is considered authigenic because of their notable differences in composition and texture, with regard to the detrital, concerning width, and particle aggregation. The measured widths of these neoformed Plg fibers range from 6 to 32 nm. Similar ranges of widths are common in neoformed Plg [6,55,56]. Plg and Sep are clay minerals that are typically formed in lacustrine evaporitic environments [1], where Plg predominates over other clay minerals, especially in mudflats and playa-lakes [2]. Consequently, Plg formation in sedimentary environments can be controversial, as it is either inherited from source areas or authigenic [57]. Those authigenic Plg and Sep fibers distinctly consist of laths and rods enriched with octahedral cations Al_2O_3, MgO, or Fe_2O_3 (Figures 8–10). The composition of a minority of authigenic Plg is closer to common and ideal terms (Figure 11A–C). Overall, neoformed Plg are laths from 50 nm up to 300 nm in length. Although authigenic aluminum-rich Sep deposits are detected in Las Eras, the greater abundance of Plg compared to Sep could be linked to an increase in aluminum activity [58]. Unlike the well-preserved bundles of authigenic aluminium-rich Sep (Figure 10A,B) in Las Eras, the inherited magnesium Plg bundles (Figure 10C) exhibit degraded edges. Aluminum-rich Plg deposits are also of authigenic origin because they are laths and rods, straight, and rigid, with no signs of erosion and, in some cases, associated with Sme (Figure 8A). These Plg deposits are frequent in nature due to the transformation of Sme [43], and they could be the result of an epitaxial intergrowth with Sme [59]. Other authors have theorized about the transformation of Plg into Sme under alkaline pH and environmental conditions [60]. Clay mineral intergrowth between Sme and Plg has also been observed in Paleogene calcretes in Central Spain, where the transformation in the solid state is unlikely due to their structural differences [61].

Chl clay minerals are abundant in the Miocene layers but scarce or absent in the studied playa-lakes sediments (Figures 5, S1 and S2). The dissolution of Chl may explain this notable decrease. The dissolved Chl appears to be the most probable source of iron for iron-rich Plg exclusively detected in Caballo Alba. Thus, iron-rich Plg is of neoformed origin. The incongruent dissolution of Chl occurs at alkaline pH [62,63], whereas Plg precipitation is favored by high pH levels (8–9.5) and high Mg+Si/Al ratios [5]. The iron present in the form of Fe^{2+} in the structure of Chl may oxidize to Fe^{3+} in solution as a function of pH [64,65]. Thus, Plg neoformed in the Caballo Alba playa-lake should preferentially incorporate the available Fe^{3+} in the solution. The authors of [66] carried out a Mössbauer characterization and concluded that iron in Plg is ferric and occupies M1 and M2 sites. Ref. [67] involving XRF studies revealed a very similar iron concentration between bulk samples of these Miocene rocks (av. 16.26%) and these playa-lake sediments (av. 16.7%). In addition, Fe^{3+} is strongly correlated with other characteristic elements of clay minerals (e.g., Si^{4+}, Al^{3+}, Ti^{4+}, and K^{+}) [67]. The preferential incorporation of Fe^{3+} into the structure of Plg is favored by the incorporation of Mg^{2+} into authigenic saponite (Sap), which is abundant in Caballo Alba. The mean structural formula of these Sap is $[(Si_{3\cdot35–4}Al_{00\cdot0–0\cdot65})\,O_{10}\,(Al_{0\cdot18–1\cdot70}\,Fe_{0\cdot11–0\cdot50}\,Mg_{0\cdot27–2\cdot12}\,Mn_{0\cdot00–0\cdot04}\,Ti_{0\cdot00–0\cdot08})\,(K_{0\cdot00–0\cdot61}\,Na_{0\cdot00–0\cdot27}\,Ca_{0\cdot03–0\cdot20})\,(OH)_2\cdot nH_2O]$ [30]. Beyond the scope of this study, Sap authigenesis is active in this lake and could involve also some Chl transformation into iron-rich Sme, in contrast to Las Eras and Bodón Blanco, where Sap is absent (Figures S1 and S2). This has a significant impact on the geochemical Mg^{2+} correlation with clay-forming elements (e.g., Si^{4+}, Al^{3+}, Ti^{4+}, and K^{+}) in Caballo Alba in contrast to Las Eras, where it is absent [67].

Miocene calcitic/dolo-mudstone contains an assemblage of Sme and iron-rich Plg (Figure 4A) that is analog to the Caballo Alba clay mineral assemblage (Figure 7). As in modern deposits, the relative enrichment in Fe_2O_3 of Miocene Plg coincides with the Sme content of the montmorillonitic (Mnt) composition $[(Si_{3\cdot54–4}\,Al_{0\cdot00–0\cdot46})\,O_{10}\,(Al_{0\cdot69–1\cdot59}\,Fe_{0\cdot20–0\cdot73}\,Mg_{0\cdot21–0\cdot73}\,Mn_{0\cdot00–0\cdot17}\,Ti_{0\cdot00–0\cdot07})\,(K_{0\cdot12}\,Na_{0\cdot00–0\cdot18}\,Ca_{0\cdot01–0\cdot19})\,(OH)_2\cdot nH_2O]$ [30]. Our results suggest that Sap enables higher Fe^{3+} uptake by Plg than Mnt (Figure 6). An association of iron-rich Plg and magnesium-rich Sap has been reported in the Neogene geological record of alkaline lakes [3]. The Plg fibers described by [3] comprised up to 1.05 Fe atoms p.h.u.c, whereas Caballo Alba iron-rich Plg reached 0.8 Fe atoms p.h.u.c. The proposed structural formulae $[(Si_7\,Al_1)\,O_{20}\,(Al_{0\cdot57}\,Mg_{2\cdot78}\,Fe_{0\cdot8})\,(Ca_{0\cdot21}\,K_{0\cdot19}\,Na_{0\cdot72})\,(OH)_2\cdot(OH_2)_4\cdot 4H_2O]$ fits with an iron-rich Plg (Table 1). Iron-rich Plg deposited in Caballo Alba provides insight into iron-rich Plg deposits that were analyzed elsewhere in [68] considering the Maderuelo deposit (i.e., up to 1.31 Fe atoms p.h.u.c) (Duero Basin, Spain); in [69] considering the Pefkaki deposit (i.e., up to 0.99 Fe atoms p.h.u.c) (Greece); and in [10] considering the Altillo Grande playa-lake (i.e., $x \geq 0.5$) (Guadiana Basin, Spain).

Table 1 presents the classification of studied Plg in Caballo Alba, according to [44]. These results are very similar to Plg reported in southern Iranian soils of Hormozgan province, where types III (i.e., trioctahedral Plg) and IV (i.e., aluminum-rich Plg) are the dominant Plg types and Type II is the minority [70], as in Caballo Alba. Nevertheless, the Iranian Plg deposits are mostly neoformed [70], whereas in the studied playa-lakes, they are mostly inherited. Ref. [71] compared clay mineral assemblages in the Persian Gulf and the Gulf of Oman, and they found that Plg is more abundant in cores from the Iranian side of the Persian Gulf, reflecting the influence of Zagros hinterland. In the strait of Hormuz, Plg and Sme almost disappear, and in the Gulf of Oman, Mca and Chl are the dominant clay minerals [71]. In the NW side of the Persian Gulf, bordering Kuwait and Iraq coasts, Plg with no trend of distribution it is most probably derived from ancient Mesopotamian flood plain and submerged estuarine sediments of the Shatt-al-Àrab (i.e., Tigris and Euphrates rivers) transported by fluvial influxes and dust storms associated with NW winds (Shamal) [72,73].

The Plg present in calcretes and other sedimentary rocks associated with lacustrine or playa environments in continental settings can be of authigenic [74–76] or inherited origins [77]. The authigenic Plg of late Miocene–Pliocene lacustrine deposits of the Kızılırmak Formation (Kırsehir Region, Turkey) exhibit some similarities with aluminum-rich Plg of Caballo Alba and Bodón Blanco. The transformation of detrital Sme into fiber and fiber bundles of Plg represents a more evolved stage [74] than that observed for the studied playa-lakes (Figure 8A). This feature is evident in the aggregation stage of the fibers but not in the width of laths; there are some similarities in the widths ranging from 10 to 50 nm with laths of sharp edges [72], as shown in Figure 8. The dissolution–precipitation mechanism is proposed as the underlying process rather than the pure transformation of detrital Sme into Plg, as stated previously by [61,74]. In other sedimentary successions of Central Asia (i.e., Kazakhstan), the authigenic origin of Plg and its association with Sme in the Chu-Surasu Basin (late Cretaceous) [75] or Chl in the Bastau Formation (mid-Miocene) [76] points to the weathering process of igneous rocks under dry conditions. The detrital Plg deposits of calcretes of Thar Dessert (Rajasthan state, India) are inherited from marine sedimentary rocks of the area belonging to the Jaisalmer Formation (Cretaceous to Eocene age) related to aeolian flux during increased aridity over the last 30 ka in relation to fluctuating SW monsoonal strength [77]. In this same line, the abundance of detrital Plg in the studied hyperalkaline playa-lakes opens a window to potential paleoclimatic studies (i.e., aridity) along their sedimentary record.

6. Conclusions

The Caballo Alba, Las Eras, and Bodón Blanco hyperalkaline playa-lakes host a mixture of authigenic and detrital clay minerals that vary from lake to lake and differ from Miocene source rocks, with their precipitation dependent on slight differences in physicochemical conditions. Miocene marlstones and mudstones comprise an assemblage of Mca-Chl-Kln-Sep-Plg with Sme only present in the topmost mudstone. In contrast, in the three studied playa-lakes, the relative abundances of Sep and Chl are very low due to the dissolution effect by high pH waters. Alternatively, mineralogical, textural, and crystallochemical features indicate that Plg particles of different octahedral compositions precipitate in the three lakes. The concurrent authigenesis of Sme occur in Caballo Alba, in contrast to Las Eras and Bodón Blanco. Authigenic Plg occurs as the laths and rods with lengths of 50 nm and distinctively are enriched in one octahedral cation (MgO, Al_2O_3, and Fe_2O_3) depending on sedimentary conditions. Aluminum-rich Plg particles from the three playa-lakes could be intergrowth with Sme favored by high pH. Iron-rich Plg precipitates only in association with authigenic Sme (Sap) in the Caballo Alba playa-lake. In this case, Sme of Sap composition preferentially uptakes magnesium from the environment, and the available dissolved iron is mainly derived from Chl dissolution, due to its scarce presence in these playa-lakes. The same effect of magnesium competition between Sm and Plg is observed in a Miocene mudstone, where Plg is relatively enriched in iron. The exact conditions of Sm precipitation in some Miocene rocks should be constrained in future paleoenvironmental studies. Our findings in spatially close and hydrochemical equivalent playa-lakes provide evidence that authigenic Plg-bearing mineral assemblages can be used as geochemical proxies. Future perspectives include unraveling the authigenic mechanisms of Sap in Caballo Alba and seeking to define more in-depth iron-rich and aluminum-rich Plg as well as Sep precipitation. An investigation of the relationship between arid conditions and detrital Plg deposited in playa-lakes during the Holocene is also pending.

Supplementary Materials: The following supporting information can be downloaded at: https://www.mdpi.com/article/10.3390/min15010050/s1, Table S1: Bulk mineralogy; semiquantification results obtained by XRD powder method of marlstones sampled from the Vallesian stratigraphic section. In yellow: Selected samples for clay mineralogy.; Table S2: Clay mineralogy. Semi-quantification results obtained by XRD oriented mounts method of Miocene marlstones and mudstones.; Table S3: Clay mineralogy. Semi-quantification results obtained by XRD oriented mounts method of Caballo Alba sediments.; Figure S1: XRD patterns of oriented mounts of a representative sample from Las Eras: air dried (AD), ethylene glycol solvated (EG) and thermal treated (TT) (550 °C). The clay mineral assemblage comprise: mica (Mca), palygorskite (Plg) and kaolinite (Kln).; Table S4: Clay mineralogy. Semi-quantification results obtained by XRD oriented mounts method of Las Eras sediments.; Figure S2: XRD patterns of oriented mounts of a representative sample from Bodón Blanco: air dried (AD), ethylene glycol solvated (EG) and thermal treated (TT) (550 °C). The clay mineral assemblage comprise: sepiolite (Sep), palygorskite (Plg), mica (Mca) and kaolinite (Kln).; Table S5: Clay mineralogy. Semi-quantification results obtained by XRD oriented mounts method of Bodón Blanco sediments.

Author Contributions: Conceptualization, P.d.B., M.E.S.-M. and J.P.R.-A.; methodology, J.P.R.-A., M.S.-R. and F.N.; software, F.N.; investigation, P.d.B., M.E.S.-M., J.P.R.-A. and M.S.-R.; writing—original draft preparation, P.d.B.; writing—review and editing, P.d.B., M.E.S.-M., J.P.R.-A., M.S.-R. and F.N.; visualization, J.P.R.-A.; supervision, M.E.S.-M., J.P.R.-A., M.S.-R. and F.N.; project administration, M.E.S.-M. and M.S.-R.; funding acquisition, M.E.S.-M. All authors have read and agreed to the published version of the manuscript.

Funding: This research was funded by the Spanish Ministry of Science and Innovation through the National Research Project PID2021-123735OB-C22 along with financial support from the Dutch Research Council project OCENW.KLEIN.037.

Data Availability Statement: Dataset available on request from the authors.

Acknowledgments: This project is part of the scientific activities of the Research Group UCM-910404 and Geobiology group at Vrije Universiteit Amsterdam. Pablo del Buey acknowledges support from a Predoctoral Grant (CT27/16-CT28/16 UCM) and a Postdoctoral Grant (Margarita Salas CT18/22) at Department of Earth Sciences at Vrije Universiteit Amsterdam and Department of Mineralogy and Petrology at University of Granada. Óscar Cabestrero Aranda is acknowledged for his assistance in the field. Finally, Esteban Urones Garrote (CNME-UCM) and Josefina Pedrajas Jurado (CIC-UGR) are acknowledged for their kind assistance with TEM-AEM analyses and images of Plg and Sep particles.

Conflicts of Interest: The authors declare no conflict of interest.

References

1. Singer, A. Palygorskite in sediments: Detrital, diagenetic or neoformed—A critical review. *Int. J. Earth Sci.* **1979**, *68*, 996–1008. [CrossRef]
2. Calvo, J.P.; Blanc-Valleron, M.M.; Rodríguez-Aranda, J.P.; Rouchy, J.M.; Sanz-Montero, M.E. Authigenic clay minerals in continental evaporitic environments. In *Palaeoweathering Palaeosurfaces and Related Continental Deposits*; Wiley: Hoboken, NJ, USA, 1999; Volume 27, pp. 129–151.
3. Akbulut, A.; Kadir, S. The geology and origin of sepiolite, palygorskite and saponite in Neogene lacustrine sediments of the Serinhisar-Acipayam Basin, Denizli, SW Turkey. *Clays Clay Miner.* **2003**, *51*, 279–292. [CrossRef]
4. Leguey, S.; Pozo, M.; Medina, J.A. Polygenesis of sepiolite and palygorskite in a fluvio-lacustrine environment in the Neogene basin of Madrid. *Mineral. Petrogr. Acta* **1985**, *29*, 287–301.
5. Jones, B.F.; Galán, E. Sepiolite and palygorskite. In *Hydrous Phyllosilicates*; Bailey, S.W., Ed.; Reviews in Mineralogy; Mineralogical Society of America: Washington, DC, USA, 1988; Volume 19, pp. 631–674.

6. Chen, T.; Xu, H.; Lu, A.; Xu, X.; Peng, S.; Yue, S. Direct evidence of transformation from smectite to palygorskite: TEM investigation. *Sci. China D Earth Sci.* **2004**, *47*, 985–994. [CrossRef]
7. Singer, A.; Norrish, K. Pedogenic palygorskite occurrences in Australia. *Am. Mineral.* **1974**, *59*, 508–517.
8. Chahi, A.; Fritz, B.; Duplay, J.; Weber, F.; Lucas, J. Textural transition and genetic relationship between precursor stevensite and sepiolite in lacustrine sediments (Jbel Rhassoul, Morocco). *Clays Clay Miner.* **1997**, *45*, 378–389. [CrossRef]
9. Miller, C.R.; James, N.P. Autogenic microbial genesis of middle Miocene palustrine ooids; Nullarbor Plain, Australia. *J. Sediment. Res.* **2012**, *82*, 633–647. [CrossRef]
10. Del Buey, P.; Cabestrero, Ó.; Arroyo, X.; Sanz-Montero, M.E. Microbially induced palygorskite-sepiolite authigenesis in modern hypersaline lakes (Central Spain). *Appl. Clay Sci.* **2018**, *160*, 9–21. [CrossRef]
11. Suárez, M.; Robert, M.; Elsass, F.; Martín-Pozas, J.M. Evidence of *a* precursor in the neoformation of palygorskite—New data by analytical electron microscopy. *Clay Miner.* **1994**, *29*, 255–264. [CrossRef]
12. Galán, E.; Ferrero, A. Palygorskite-sepiolite clays of Lebrija, Southern Spain. *Clays Clay Miner.* **1982**, *30*, 191–199. [CrossRef]
13. Pozo, M.; Casas, J.C. Origin of kerolite and associated Mg clays in palustrine-lacustrine environments. The Esquivias deposit (Neogene Madrid Basin, Spain). *Clay Miner.* **1999**, *34*, 395–418. [CrossRef]
14. Millot, G. *Geology of Clays*; Springer: London, UK, 1970; 429p.
15. Deocampo, D.M.; Cuadros, J.; Wing-Dudek, T.; Olives, J.; Amouric, M. Saline lake diagenesis as revealed by coupled mineralogy and geochemistry of multiple ultrafine clay phases: Pliocene Olduvai Gorge, Tanzania. *Am. J. Sci.* **2009**, *309*, 834–868. [CrossRef]
16. Stoessell, R.K. 25 °C and 1 atm. Dissolution experiments of sepiolite and kerolite. *Geochim. Cosmochim. Acta* **1988**, *52*, 365–373. [CrossRef]
17. Deocampo, D.M. Evaporative evolution of surface waters and the role of aqueous CO_2 in magnesium silicate precipitation: Lake Eyasi and Ngorongoro crater, northern Tanzania. *S. Afr. J. Geol.* **2005**, *108*, 493–504. [CrossRef]
18. Deocampo, D.M.; Blumenschine, R.J.; Ashley, G.M. Wetland diagensis and traces of early hominids, Olduvai Gorge, Tanzania. *Quat. Res.* **2002**, *57*, 271–281. [CrossRef]
19. Cuevas, J.; Vigil de la Villa, R.; Ramírez, S.; Petit, S.; Meunier, A.; Leguey, S. Chemistry of Mg smectites in lacustrine sediments from the Vicálvaro sepiolite deposit, Madrid Neogene Basin (Spain). *Clays Clay Miner.* **2003**, *51*, 457–472. [CrossRef]
20. Galán, E.; Pozo, M. Palygorskite and Sepiolite Deposits in Continental Environments. Description, Genetic Patterns and Sedimentary Settings. In *Developments in Palygorskite-Sepiolite Research. A New Outlook on these Nanomaterials*; Galán, E., Singer, A., Eds.; Elsevier: New York, NY, USA, 2011; Volume 3, pp. 125–166.
21. Fernández-Macarro, B.; Armenteros, L.; Blanco, J.A. Procesos de alteración y paleosuelos ligados a la sedimentación miocena del noreste de Segovia, Depresión del Duero. *Acta Geol. Hispánica* **1988**, *23*, 269–281.
22. Doval, M.; Domínguez, M.C.; Brell, J.M.; García, E. Mineralogía y sedimentología de las Facies distales del borde norte de la Cuenca del Tajo. *Bol. Soc. Esp. Mineral.* **1985**, *8*, 257–269.
23. Ordóñez, S.; Calvo, J.P.; García del Cura, M.Á.; Alonso-Zarza, A.M.; Hoyos, M. Sedimentology of sodium sulphate deposits and special clays from the Tertiary Madrid Basin (Spain). In *Lacustrine Facies Analysis*; Anadón, P., Cabrera, L., Kelts, K., Eds.; Special Publication International Association of Sedimentologists; Blackwell Scientific Publications: Oxford, UK, 1991; Volume 13, pp. 39–55.
24. García-Romero, E. Génesis de arcillas magnésicas en la cuenca de Madrid: Interrogantes planteados. *Bol. Geol. Miner.* **2004**, *115*, 629–640.
25. Murray, H.H.; Pozo, M.; Galán, E. An introduction to palygorskite and sepiolite deposits. Location, geology and uses. In *Developments in Palygorskite and Sepiolite Research. A New Outlook on These Nanomaterials*; Galán, E., Singer, A., Eds.; Elsevier: New York, NY, USA, 2011; Volume 3, pp. 85–100.
26. Sanz-Montero, M.E.; Rodríguez-Aranda, J.P.; Del Buey, P. Influencia del sustrato cenozoico en el origen y sedimentación de la laguna hiperalcalina de Caballo Alba (Segovia). *Geogaceta* **2021**, *70*, 31–34.
27. Alonso-Gavilán, G.; Armenteros, I.; Carballeira, J.; Corrochano, A.; Huerta, P.; Rodríguez, J.M. Cuenca del Duero. In *Geología de España*; Vera, J.A., Ed.; SGE-IGME: Madrid, Spain, 2004; pp. 550–556.
28. Portero, J.M.; Carreras, F.; Olivé, A.; Del Olmo, P. *Mapa Geológico de España Escala 1:50.000, Hoja 428, Olmedo*; IGME: Madrid, Spain, 1982.
29. Cabestrero, Ó.; Sanz-Montero, M.E. Brine evolution in two inland evaporative environments: Influence of microbial mats in mineral precipitation. *J. Paleolimnol.* **2018**, *59*, 139–157. [CrossRef]
30. Del Buey, P. Interacciones Bióticas en la Neoformación de Minerales en Lagunas Salinas y Alcalinas, con Énfasis en los Minerales de la Arcilla. Ph.D. Thesis, Universidad Complutense de Madrid, Madrid, Spain, 2022.
31. Del Buey, P.; Sanz-Montero, M.E. Biomineralization of ordered dolomite and magnesian calcite by the green alga *Spirogyra*. *Sedimentology* **2022**, *70*, 685–704. [CrossRef]
32. Sanz-Montero, M.E.; del Buey, P.; Cabestrero, Ó.; Sánchez-Román, M. Isotopic signatures of microbial Mg-carbonates deposited in an ephemeral hyperalkaline lake (Central, Spain): Paleoenvironmental implications. *Minerals* **2023**, *13*, 617. [CrossRef]

33. Sanz-Montero, M.E.; Cabestrero, Ó.; Sánchez-Román, M. Microbial Mg-Rich Carbonates in an Extreme Alkaline Lake (Las Eras, Central Spain). *Front. Microbiol.* **2019**, *10*, 148. [CrossRef] [PubMed]
34. Chung, F.H. Quantitative interpretation of X-ray diffraction patterns. I. Matrix flushing method for quantitative multicomponent analysis. *J. Appl. Crystallogr.* **1974**, *7*, 519–931. [CrossRef]
35. Kübler, K. *Cristallinité de L'illite, Méthodes Normalisées de Préparations, Méthodes Normalisées de Mesure*; Série ADX; Cahiers Institut de Geologie: Neuchâtel, Switzerland, 1987; 13p.
36. Nieto, F.; Arroyo, X.; Aróstegui, J. XRD-TEM-AEM comparative study of n-alkilamonnium smectites and interestratified minerals in shallow-diagenetic carbonate sediments of the Basque-Cantabrian Basin. *Am. Mineral.* **2016**, *101*, 385–398. [CrossRef]
37. Jackson, M.L. *Soil Chemical Analysis Advanced Course*; CABI Digital Library: Madison, WI, USA, 1969; 895p.
38. Warr, L.N. IMA-CNMNC approved mineral symbols. *Mineral. Mag.* **2021**, *85*, 291–320. [CrossRef]
39. Yebra, Á. Influencia de la Mineralogía, Quimismo y Textura en las Aplicaciones Básicas Industriales de la Sepiolita. Ph.D. Thesis, Universidad de Granada, Granada, Spain, 2000.
40. Cliff, G.; Lorimer, G.W. The quantitative analysis of thin specimens. *J. Microsc.* **1975**, *103*, 203–207. [CrossRef]
41. Elert, K.; Nieto, F.; Azañon, J.M. Effects of lime treatments on marls. *Appl. Clay Sci.* **2017**, *135*, 611–619. [CrossRef]
42. García-Romero, E.; Suárez, M. Sepiolite-palygorskite: Textural study and genetic considerations. *Appl. Clay Sci.* **2013**, *86*, 129–144. [CrossRef]
43. Suárez, M.; García-Romero, E. Sepiolite-palygorskite: A continous polysomatic series. *Clays Clay Miner.* **2013**, *61*, 461–472. [CrossRef]
44. García-Romero, E.; Súarez, M. On the chemical composition of sepiolite and palygorskite. *Clays Clay Miner.* **2010**, *58*, 1–20. [CrossRef]
45. Velde, B. Composition and mineralogy of clay minerals. In *Origin and Mineralogy of Clays*; Velde, B., Ed.; Springer: New York, NY, USA, 1995; pp. 8–42.
46. Neaman, A.; Singer, A. The effects of palygorskite on chemical and physico-chemical properties of soils: A review. *Geoderma* **2004**, *123*, 297–303. [CrossRef]
47. Muir, A. Notes on the soils of Syria. *Eur. J. Soil. Sci.* **1951**, *2*, 163–182. [CrossRef]
48. Rogers, L.E.R.; Quirk, J.P.; Norrish, K. Ocurrence of an aluminum sepiolite in a soil having an unusual water relationships. *J. Soil. Sci.* **1956**, *7*, 177–185. [CrossRef]
49. Singer, A. The texture of palygorskite from the Rift Valley, southern Israel. *Clay Miner.* **1981**, *16*, 415–419. [CrossRef]
50. Ravikovitch, S.; Pines, F.; Ben-Yair, M. Composition of colloids in soils of Israel. *J. Soil. Sci.* **1960**, *11*, 82–91. [CrossRef]
51. McLean, S.A.; Allen, B.L.; Craig, J.R. The occurrence of sepiolite and attapulgite on the southern High Plains. *Clays Clay Miner.* **1972**, *20*, 143–149. [CrossRef]
52. Shadfan, H.; Mashhady, A.S. Distribution of palygorskite in sediments and soils of Eastern Saudi Arabia. *Soil Sci. Soc. Am. J.* **1985**, *49*, 243–250. [CrossRef]
53. Cui, J.; Zhang, Z.; Han, F. Effects of pH on the gel properties of montmorillonite, palygorskite and montmorillonite-palygorskite composite clay. *Appl. Clay Sci.* **2020**, *190*, 105543. [CrossRef]
54. Martínez-Ramirez, S.; Puertas, F.; Blanco-Varela, M.T. Stability of sepiolite in neutral and alkaline media at room temperature. *Clay Mineral.* **1996**, *31*, 225–232. [CrossRef]
55. Grim, R.E. *Clay Mineralogy*; Mc Graw-Hill: New York, NY, USA, 1968.
56. Hodge, T.; Turchenek, L.W.; Oades, J.M. Ocurrence of palygorskite in ground water rendzinas (petrocalcic Calciaquolls) in Southeast South Australia. In *Palygorskite-Sepiolite: Occurrences, Genesis and Uses*; Developments in Sedimentology; Elsevier: Amsterdam, The Netherlands, 1984; Volume 37, pp. 199–210.
57. Verrechia, E.P.; Le Coustumer, M.N. Occurrence and genesis of palygorskite and associated clay minerals in a Pleistocene calcrete complex, Sde Boquer, Negev desert, Israel. *Clay Mineral.* **1996**, *31*, 183–202. [CrossRef]
58. Birsoy, R. Formation of sepiolite-palygorskite and related minerals from solution. *Clays Clay Miner.* **2002**, *50*, 736–745. [CrossRef]
59. Krekeler, M.P.S.; Hammerley, E.; Rakovan, J.; Guggenheim, S. Microscopy studies of the palygorskite to smectite transformation. *Clays Clay Miner.* **2005**, *53*, 92–99. [CrossRef]
60. Golden, D.C.; Dixon, J.B.; Shadfan, H.; Kippenberger, L.A. Palygorskite and sepiolite alteration to smectite under alkaline conditions. *Clays Clay Miner.* **1985**, *33*, 44–50. [CrossRef]
61. Rodas, M.; Luque, F.J.; Mas, R.; Garzón, M.G. Calcretes, palycretes and silcretes in the Paleogene detrital sediments of the Duero and Tajo Basins, Central Spain. *Clay Mineral.* **1994**, *29*, 273–285. [CrossRef]
62. Malmström, M.; Banwart, S.; Lewenhagen, J.; Duro, L.; Bruno, J. The dissolution of biotite and chlorite at 25 °C in the near-neutral pH region. *J. Contam. Hydrol.* **1996**, *21*, 201–213. [CrossRef]
63. Rochelle, C.A.; Bateman, K.; MacGregor, R.; Pearce, J.M.; Savage, D.; Wetton, P.D. Experimental determination of chlorite dissolution rates. *MRS Online Proc. Lib.* **1996**, *353*, 149–156. [CrossRef]
64. Stumm, W.; Morgan, J.J. *Aquatic Chemistry*; Wiley-Interscience: New York, NY, USA, 1996.

sample are straight and rigid, whereas rods and bundles are curved (Figure 5G). These clay minerals are classified as common to magnesium Plg (Sample 3, Figure 6). Finally, the brownish-green mixed calcitic/dolo-mudstones at the topmost part of the stratigraphic column present only a few straight Plg laths (8 nm width) (Sample 4, Figure 5I), which are classified as ideal Plg, with higher content of Fe_2O_3 (x = 0.4–0.5) than Plg particles comprising the basal marlstones (Sample 4, Figure 6).

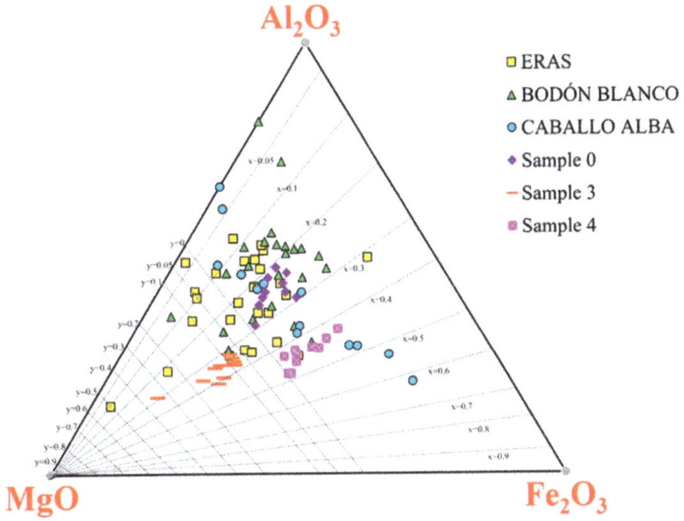

Figure 6. AEM analyses of Plg in the Miocene and playa-lakes deposits, plotted in a ternary graph, modified from [43].

The weight percentage results of semiquantitative analyses using TEM-AEM of Plg fibers from the Miocene mudstones and marlstones and the three playa-lakes are plotted in Figure 6. The graph shows that there is no significant variability in the chemical composition. Most of them are plotted in composition fields corresponding to aluminum-rich Plg to common ideal ones, according to [43]. The composition of Miocene fibrous clay minerals is characteristic of each marlstone and mudstone sample. The mean structural formula for Plg marlstone (Sample 0) is $[Si_8O_{20}\ (Mg_2\ Fe_2)_{0.25}\ (Mg_2\ Al_2)_{0.75}\ (OH)_2 \cdot (OH_2)_4] \cdot nH_2O$. The dolo-marlstone sample (Sample 3) is characterized by having the most magnesium-rich composition of Plg of the Miocene deposits $[Si_{12}O_{30}Mg_8(OH)_4(OH_2)_2]_{0.1}[Si_8O_{20}\ (Mg_2\ Fe_2)_{0.35}\ (Mg_2\ Al_2)_{0.65}\ (OH)_2\ (OH_2)_4]_{0.9} \cdot n\ H_2O$. The uppermost beds (Sample 4) consist of calcitic/dolo-mudstone with Plg relative enrichment in Fe_2O_3 content with the mean structural formula $[Si_8O_{20}\ (Mg_2\ Fe_2)_{0.45}\ (Mg_2\ Al_2)_{0.55}\ (OH)_2 \cdot (OH_2)_4] \cdot nH_2O$.

4.2. Clay Mineralogy of the Playa-Lakes

The surface sediment samples and microbial mats that proliferate in these three playa-lakes were characterized in terms of clay mineralogy. The clay mineralogy assemblage of Caballo Alba comprises Mca, Plg, Sme, and Kln. In addition, Sep is irregularly distributed and is always below 10% in relative abundance. Sme is always present, with remarkably higher relative abundances (up to 45%) than those of the Miocene strata (Figure 7, Table S3), as well as the Las Eras and Bodón Blanco playa-lake deposits. The sediments of the Las Eras playa-lake lack Sme (Figure S1, Table S4), whereas in the Bodón Blanco playa-lake, Sme represents up to 10% of the clay mineral assemblage (Figure S2, Table S5).

65. Lowson, R.T.; Josick-Comarmond, M.C.; Rajaratnam, G.; Brown, P.L. The kinetics of dissolution of the chlorite as a function of pH and at 25 °C. *Geochim. Cosmochim. Acta* **2005**, *69*, 1687–1699. [CrossRef]
66. Huang, Y.J.; Li, Z.; Li, S.Z.; Shi, Z.L.; Yin, L.; Hsia, Y.F. Mössbauer investigations of palygorskite from Xuyi, China. *Nucl. Instrum. Methods Phys. Res. Sect. B Beam Interact. Mater. Atoms* **2007**, *260*, 657–662. [CrossRef]
67. Grunwald, C. Estudio geoquímico de depósitos lagunares en la Cuenca del Duero con énfasis en la distribución del arsénico. Master's Thesis, Universidad Complutense de Madrid, Madrid, Spain, 2024.
68. Torres-Ruíz, J.; López-Galindo, A.; González-López, J.M.; Delgado, A. Geochemistry of Spanish sepiolite-palygorskite deposits: Genetic considerations based on trace elements and isotopes. *Chem. Geol.* **1994**, *112*, 221–245. [CrossRef]
69. Chryssikos, G.D.; Gionis, V.; Kacandes, G.H.; Statopoulou, E.T.; Suárez, M.; García-romero, E.; Sánchez del Río, M. Octahedral cation distribution in palygorskite. *Am. Mineral.* **2009**, *94*, 200–203. [CrossRef]
70. Abbaslou, H.; Abtahi, A.; Martin Peinado, F.J.; Owliaie, H.; Khormali, F. Mineralogy and characteristic of soils developed on Persian Gulf and Oman Sea basin, southern Iran: Implications for soil evolution in relation to sedimentary parent material. *Soil Sci.* **2013**, *178*, 568–584. [CrossRef]
71. Stoffers, P.; Ross, D.A. Late Pleistocene and Holocene sedimentation in the Persian Gulf-Gulf of Oman. *Sediment. Geol.* **1979**, *23*, 181–208. [CrossRef]
72. Al-Bakri, D.; Khalaf, F.; Al-Ghadban, A. Mineralogy, genesis, and sources of surficial sediments in the Kuwait marine environment, northern Arabian Gulf. *J. Sed. Petrol.* **1984**, *54*, 1266–1279.
73. Aqrawi, A.A.M. Petrography and mineral content of sea-floor sediments of the Tigris-Euphrates Delta, North-West Arabian Gulf, Iraq. *Estuar. Coast. Shelf. Sci.* **1994**, *38*, 569–582. [CrossRef]
74. Kadir, S.; Eren, M.; Külah, T.; Önalgil, N.; Cesur, M.; Gürel, A. Genesis of Late Miocene-Pliocene lacustrine palygorskite and calcretes from Kirsehir, central Anatolia, Turkey. *Clay Mineral.* **2014**, *49*, 473–494. [CrossRef]
75. Munara, A.; Cathelineau, M.; Carpentier, C.; Abylay, N. Clays as indicator of paleoclimate and source rocks in The Chu-Sarysu Basin (Kazakhstan). *Kazakhstan J. Oil Gas Ind.* **2023**, *5*, 21–35. [CrossRef]
76. Voigt, S.; Weber, Y.; Frisch, K.; Barteinstein, A.; Hellwing, A.; Petschick, R.; Bahr, A.; Pross, J.; Koutsodendris, A.; Voigt, T. Climatically forced moisture supply, sediment flux and pedogenesis in Miocene mudflat deposits of south-east Kazakhstan, Central Asia. *Depos. Rec.* **2017**, *3*, 209–232. [CrossRef]
77. Hameed, A.; Raja, P.; Ali, M.; Upreti, N.; Kumar, N.; Tripathi, J.K.; Srivastava, P. Micromorphology, clay mineralogy, and geochemistry of calcic-soils from western Thar Desert: Implications from origin of palygorskite and southwestern monsoonal fluctuations over the last 30 ka. *Catena* **2018**, *163*, 378–398. [CrossRef]

minerals

MDPI

Article

Effect of Interaction Between Expandable Minerals and Glycerin-Based Fluids on the Occurrence of Accretion

Ana P. O. Sousa [1], Mário C. S. Lima [1], Waleska R. P. Costa [1], Renalle C. A. M. Nascimento [2], João M. P. Q. Delgado [3,*], Antonio G. B. Lima [4] and Luciana V. Amorim [1]

1 Academic Unit of Petroleum Engineering, Federal University of Campina Grande, Campina Grande 58429-900, Brazil; anap.olvs@gmail.com (A.P.O.S.); mariocesar.slima@gmail.com (M.C.S.L.); warodriguespc@gmail.com (W.R.P.C.); lvamorim@gmail.com (L.V.A.)
2 Academic Unit of Santo Agostinho, Rural Federal University of Pernambuco, Cabo de Santo Agostinho 54518-430, Brazil; nalenascimento@gmail.com
3 Institute of R&D in Sctructures and Construction (CONSTRUCT-GFC), Department of Civil and Georesources Engineering, Faculty of Engineering, University of Porto, 4200-465 Porto, Portugal
4 Department of Mechanical Engineering, Federal University of Campina Grande, Campina Grande 58429-900, Brazil; antonio.gilson@ufcg.edu.br
* Correspondence: jdelgado@fe.up.pt; Tel.: +351-225-081-404

Abstract: Glycerin-based fluids are proposed as a promising alternative to inhibited fluids in the drilling of highly-reactive formations. However, even with the use of these fluids, it is still possible to observe the occurrence of problems related to the balling of drill bits and drill pipes, such as the agglomeration and accretion of cuttings. This study aims to analyze how the interaction between expandable minerals from reactive formations and glycerin-based drilling fluids affects the stability of oil wells, focusing on the occurrence and extent of the accretion phenomenon. For this purpose, bentonite pellets were characterized regarding their mineralogical composition and plastic behavior. In addition, accretion tests were performed in order to evaluate the interaction between bentonite pellets and glycerin-based drilling fluids containing different types of inhibitors. The results revealed that the pellets were predominantly composed of interstratified illite–smectite (IS) clay minerals and presented highly plastic properties with a high degree of expansion. Furthermore, it was found that the accretion percentages were significant for all the fluids studied, at higher than 58%. Therefore, it was found that using glycerin in drilling fluids did not stabilize expandable minerals in reactive formations, even with different expansion inhibitors, which were ineffective in reducing the rock expansibility.

Keywords: well stability; drilling fluids; glycerin; reactive formations; clay minerals; inhibitors; bentonite pellets

1. Introduction

The instability of oil wells, resulting from the presence of geological formations that are difficult to drill, represents one of the main adversities faced during drilling operations. The behavior of these formations is influenced by a complex combination of mechanical, chemical, thermal, and electrical processes, which are interconnected and can-not be understood in isolation [1].

Clay minerals found in sedimentary rocks can be categorized into three main groups: smectite, kaolinite, and illite. Each of these groups can lead to various issues in wells, including fine migration, increase in pore tortuosity, formation swelling, and decreased penetration rate [2].

Bentonite is a rock that generally presents a high concentration of expandable clay minerals, normally belonging to the smectite group. These smectite clays are characterized by the easy hydration capacity of cations and interlayer surfaces, giving them a remarkable expansion property. Due to this property, they are often called expansive clays [3].

The understanding of the agglomeration and accretion of cuttings remains limited due to the lack of field data, diverse drilling practices, and variations in the lithology of the formations. According to Reid and co-authors [4], theories based on the plasticity of clays have been proposed to explain this phenomenon. These theories are based on the idea that the reduced penetration of water into the rock fragments generated by the action of the bit decreases the hydration rate, resulting in a prolonged plasticity of the fragments. It is believed that this extended plasticity state of the clays allows the cuttings to be molded into the bottom hole assembly (BHA) components and incorporated close to the well wall.

In drilling operations where the formation presents a high degree of reactivity, it is necessary to use inhibited fluids. Most of the studies related to the prevention of agglomeration and accretion focus on the performance of additives in the formulation in which the continuous phase is composed of water. However, due to the thermal and oxidative stability characteristics and low interaction with clays during swelling tests, glycerin-based fluids have emerged as a promising alternative to inhibited fluids, potentially improving wellbore stability. Pormerlau [5] highlights several advantages of glycerin-based drilling fluids. However, even with the use of these fluids, it is still possible to observe the occurrence of problems related to the balling of drill bits and drill pipes, that can result in increased non-productive time during the drilling process.

In this context, the present study aims to analyze how the interaction between expandable minerals existing in reactive formations and glycerin-based drilling fluids affects the stability of oil wells, focusing on the occurrence and extension of the accretion phenomenon.

2. Materials and Methods

2.1. Materials

To carry out this work, a sample of pellets was used, which corresponded to the compacted form of bentonite. This sample was donated by Bentonite União Nordeste—BUN (Boa Vista, Brazil).

The glycerin-based drilling fluids were formulated with additives in the amounts specified in Table 1. All additives were supplied by Leopoldo Américo Miguez de Mello Research and Development Center—CENPES/PDDP/FCE from PETROBRAS (Rio de Janeiro, Brazil).

Table 1. Composition of the glycerin-based drilling fluid.

Product	Function	Amount
Water	Dispersing medium	262.5 mL
Sodium bicarbonate	Removal of divalent cations	0.5 g
Glycerin	Dispersing medium	87.5 mL
Xanthan gum	Rheological agent	0.5 g
PAC LV	Filtrate loss reducer	6 g
I1, I2, I3 **	Clay swelling inhibitor	15.7 g, 10.5 g, 10.5 g
Magnesium oxide	Alkalinity buffer	1 g
Limestone	Bridging agent	20 g
Silicone-based liquid	Anti-foaming	0.3 g
NaCl	Weighting agent	QS * 9.5 lb/gal

* Quantum satis—Enough amount to reach a density of 9.5 lb/gal. ** Inhibitors have different commercial chemical compositions.

Three different types of clay swelling inhibitors, named I1, I2 and I3, were incorporated into the fluid composition. In total, seven glycerin-based fluid formulations were prepared, for which there was variation in the use of inhibitors, as described in Table 2.

Table 2. Identification used for the drilling fluids studied.

Nomenclature	Glycerin Fluid
F1	No inhibitor added
F2	Added with I1
F3	Added with I1, I2, and I3
F4	Added with I1 and I2
F5	Added with I2
F6	Added with I1 and I3
F7	Added with I3

2.2. Methods

2.2.1. Pellets' Characterization

X-ray diffraction (XRD) was used to identify the mineralogical constituents of the samples using a Rigaku (Miniflex 600) X-ray diffractometer. The quantitative determination of the mineral composition was carried out by the Rietveld method, which performs a complete simulation of the diffractogram, refining the geometric parameters of the phases present, and taking into account the crystallographic aspects. Quantification is determined by comparing a calculated diffraction pattern with the observed pattern, performing a point-by-point analysis, and adjusting the differences found using the least squares method.

In addition, measurements of the liquid limit and plasticity limits of the bentonite pellets were carried out to evaluate their plastic behavior in the different liquid media used in the formulations of the drilling fluids under study: water, glycerin, and water with glycerin. The liquid limit was obtained in the Casagrande apparatus, according to the procedures of NBR 6459, while the plastic limit was obtained according to NBR 7180 [6,7]. The plasticity index of the pellets was calculated using Equation (1).

$$PI = (LL - PL) \times 100\%, \quad (1)$$

where PI = plasticity index; LL = liquid limit, and PL = plastic limit.

2.2.2. Drilling Fluids' Preparation

The drilling fluids used in this study were prepared in aliquots of 350 mL, using a Silverson L5M-A equipped with an emulsifying screen rotor/stator assembly, at a rotational speed of 8720 rpm.

The additives were added in the order presented in Table 1, with 5 min of homogenization for each one. The final homogenization time was 10 min.

2.2.3. Interaction Between Expandable Minerals and Drilling Fluids: Accretion Test

The accretion test was performed following the methodology established by Cliffe and Young [8]. For that, solid steel bars were positioned in the center of the test cell, which was filled with drilling fluid until reaching 50% of its volume. Subsequently, samples containing 50 g of pellets with a narrow particle size distribution (3.35–4.75 mm) were added to the cell. Gentle agitation was then applied with a spatula to ensure an even distribution of the solids in the fluid uniformly. Finally, the cell was completely filled with the fluid, sealed, and inverted 3 to 4 times to minimize the chances of pellets sticking to the bottom of the cell.

The test was conducted using a Rollen Over model 705S from Fann. During the procedure, the cells were rotated at predetermined time intervals (10, 20, 30, 40, and 50 min), maintaining room temperature. Figure 1 illustrates the test procedure.

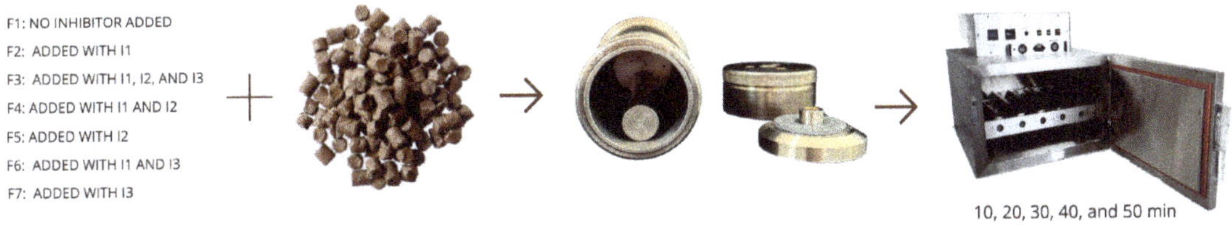

Figure 1. Equipment and methodology for accretion tests using bentonite pellets and inhibited drilling fluids.

After the rolling step, the bar was removed from the test cell, and the added solids were properly removed, finishing with a quick wash of the bar using a pipette containing water, in order to minimize any sample loss. Then, the solids were subjected to drying at 110 °C until they reached a constant mass. The accretion rate was then calculated according to Equation (2).

$$A\% = \frac{W_2}{\left[\frac{100-M_i}{100}\right] \times 50} \tag{2}$$

where A = accretion; W_2 = weight of solids, in grams, after drying at 110 °C, and M_1 = initial water content of the pellets sample, not exposed to the fluid.

3. Results and Discussion

3.1. Pellets' Characterization

3.1.1. Mineralogical Composition Section

Figure 2a displays the X-ray diffractogram of the bentonite pellet (total powder). To enhance the interpretation of the diffractogram, particularly for identifying clay minerals, the clay fraction underwent additional analyses, which included heating at 500 °C for 2 h and treatment with ethylene glycol, as shown in Figure 2b. These techniques were employed based on the principle that ethylene glycol treatment expands the basal spacing of smectites to 17Å, while heating at 500 °C dehydrates kaolinites and reduces the basal spacing of smectites. Therefore, the mineral identifications presented in Figure 2a incorporate the results from these additional tests.

In general, it is possible to observe, through the X-ray diffractogram, peaks indicative of the presence of interstratified clay minerals illite–smectite (IS), evidencing the reactive nature of the bentonite pellet sample with a high degree of expansion. Additionally, peaks associated with the presence of mica, feldspar, and quartz can be identified, as well as moderate peaks that denote the existence of clay minerals corresponding to kaolinite.

Figure 3 presents the qualitative and quantitative results of the X-ray diffraction analysis.

In the quantitative analysis, it was found that the sample was predominantly composed of an expandable clay mineral of illite interstratified with smectite, whose content reached 65%, evidencing the result presented in the diffractogram shown in Figure 1. The interstratified clay minerals are partial products resulting from a process of synthesis of the phyllosilicate layers, in which there is no predominance of a specific crystal, therefore presenting characteristics common to two or more clay minerals. According to Wilson and co-authors [9], the presence of smectite is directly related to the reactivity of the formation. Thus, due to the presence of smectite in the synthesized layer, the bentonite pellets have exhibited a notable sensitivity to the presence of water.

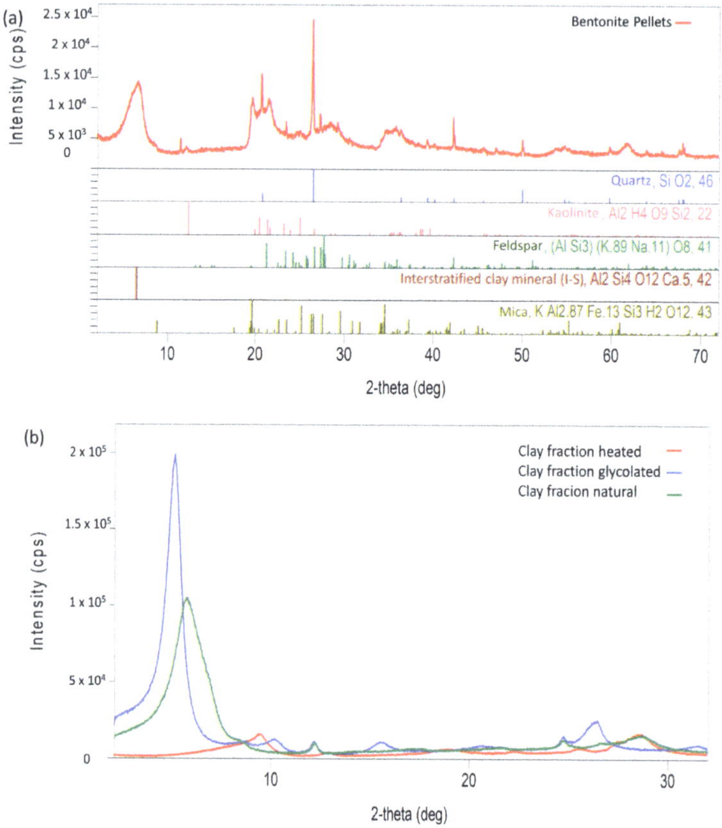

Figure 2. X-ray diffractogram of the pellet sample (total powder) (**a**) and XRD pattern of the heated, glycolated, and natural clay fraction of the pellet sample (**b**).

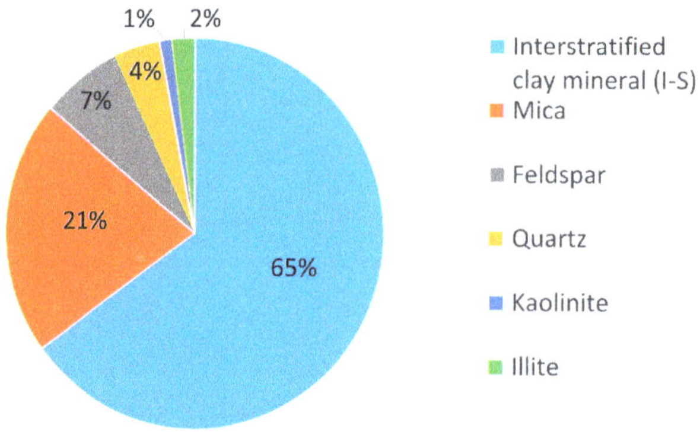

Figure 3. Qualitative and quantitative results of X-ray diffraction of the pellets (quantitative evaluation parameter: $0 \leq 1\%$ = possible presence; $1\% \leq 5\%$ = traces; >5% = presence).

Based on the data presented in Figure 2, it is also possible to verify that significant proportions of accessory minerals, such as quartz and feldspar, were observed. The mineral quartz, whose characteristics include a non-plastic nature and a brittle structure, presented a composition of 4%. For feldspar, a concentration of 7% was observed. According to Macedo et al. (2008), feldspathic accessory minerals have a direct influence on the plastic properties of the formations, since, together with quartz, they generally play a role in reducing plasticity [10]. However, Apolônio and co-authors [11] carried out a study on the reactivity of shales in the Northeast region of Brazil and found that the predominance of Na-smectite and Na-feldspar confers greater plasticity to the clay formations.

3.1.2. Atterberg Limits

The moisture contents of the pellet sample, in relation to each liquid medium that makes up the bases of the drilling fluids (water, glycerin, and water with glycerin), was investigated, as well as the number of drops (revolutions) used to determine the liquid limit. Those parameters were correlated and visually represented in the graphs shown in Figure 4. The highlighted points were obtained through a linearization of the other points, thus corresponding to the liquid limit for each medium.

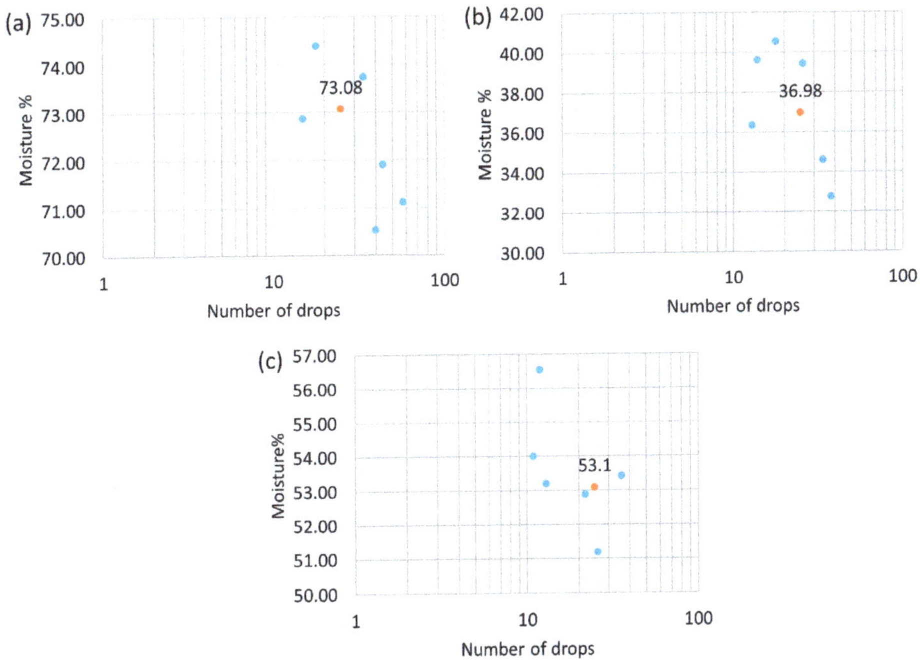

Figure 4. Liquid limit for the pellet sample with (**a**) water, (**b**) glycerin, and (**c**) water with glycerin.

The results in Figure 3 reveal that, in the presence of water, the pellets obtained a liquid limit of 73.08%, which indicates very high plasticity, according to the classification proposed by Bell [12] (Table 3). With glycerin, the pellets presented a liquid limit of 36.98%, and its plasticity was classified as intermediate. In the presence of water with glycerin, the pellet sample reached a liquid limit of 53.1%, which corresponded to the classification of high plasticity. Therefore, the pellets demonstrated liquid limit values that classify them as plastic materials in the three bases used in the drilling fluids.

The liquid limit is associated with the saturation point of the hydration state of the formations, at which the rock begins to behave like a liquid and loses the properties that would allow it to agglomerate in the BHA. Although the rock loses its fluid characteristics when it reaches the liquid limit, it still can be easily molded, being in a state known as plastic [13]. Therefore, a high liquid limit favors the permanence of the formation in the plastic state, increasing its plasticity.

Table 3. Plasticity classification according to the liquid limit [12].

Classification	Liquid Limit
Low plasticity	<35
Intermediate plasticity	35–50
High plasticity	50–70
Very high plasticity	70–90
Extra high plasticity	>90

The plasticity index, determined by the difference between the liquid limit and the plasticity limit values, was used as a reference for classifying the sample in relation to its degree of plasticity, following the classification described in Table 4 [14].

Table 4. Classification according to the degree of plasticity.

Classification	Plasticity Index (%)
Weakly plastic	$1 < IP \leq 7$
Moderately plastic	$7 < IP \leq 15$
Highly plastic	$IP > 15$

Table 5 presents the results of the plasticity index, together with the classification of the pellets according to their degree of plasticity.

Table 5. Plasticity index of pellets and their classifications according to the degree of plasticity.

Liquid Medium	Plasticity Index (%)	Degree of Plasticity (%)
Water	36	Highly plastic
Glycerin	5	Weakly plastic
Glycerin + water	22	Highly plastic

The results shown in Table 5 reveal that the pellets presented a plasticity index of 36% in the presence of water, which means that they are highly plastic in this medium. In glycerin, the plasticity index was 5%, resulting in a classification of weakly plastic. In water with glycerin, the plasticity index reached 22%, characterizing them as highly plastic in this medium. According to van Oort et al. (2000), the reactivity of the formations is more pronounced in rocks that present plastic characteristics [15].

Thus, the data obtained from X-ray diffraction (Figures 1 and 2) and the Atterberg limits reveal that the pellets analyzed were predominantly composed of the clay mineral smectite, being classified as highly plastic when exposed to water and water with glycerin. In this sense, the pellets can be considered as representative of reactive formations for drilling fluids composed of these bases. These results corroborate the studies of Hajjaji et al. (2010), who observed that formations in which smectite is the dominant clay mineral have a greater predisposition to reactivity in terms of their plastic characteristics [16]. Additionally, Fontoura (2002) observed that the rock–fluid system is directly influenced by the physicochemical properties of the clay–water system, since clays have a series of characteristics that make them reactive when in contact with water-based drilling fluids, such as high specific area and the expansibility of these minerals [17].

3.2. Interaction Between Expendable Minerals and Drilling Fluids

Figures 5–11 show the appearance of the pellets accreted to the steel bar, for each fluid studied, for times of 10, 20, 30, 40, and 50 min.

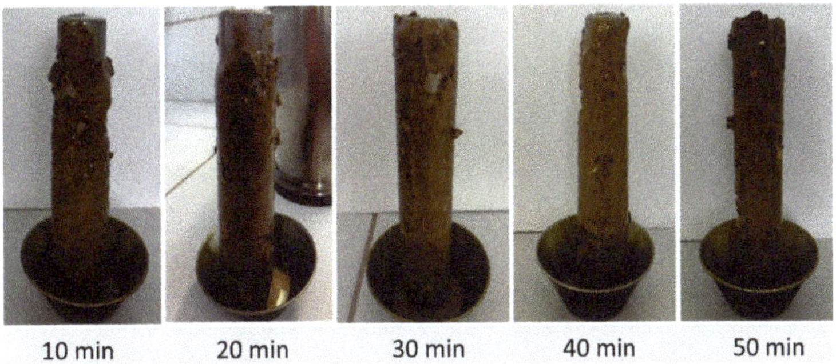

Figure 5. Pellet accretion using fluid F1 (without inhibitor).

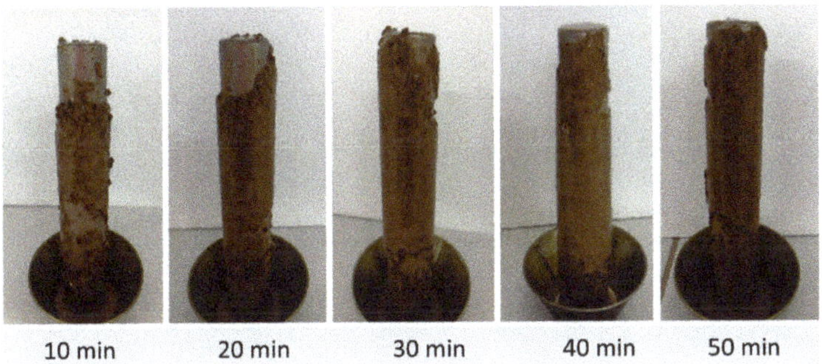

Figure 6. Accretion of pellets using fluid F2 (added with I1).

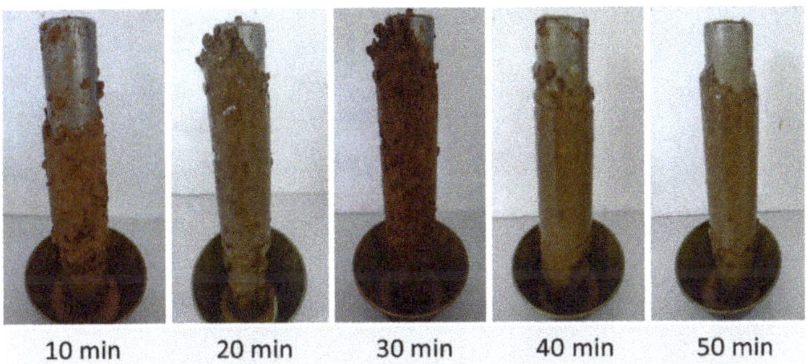

Figure 7. Accretion of pellets using fluid F3 (added with I1, I2, and I3).

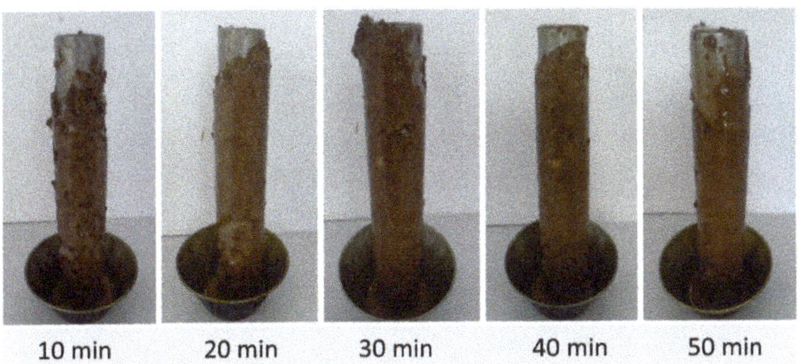

Figure 8. Accretion of pellets using fluid F4 (added with I1 and I2).

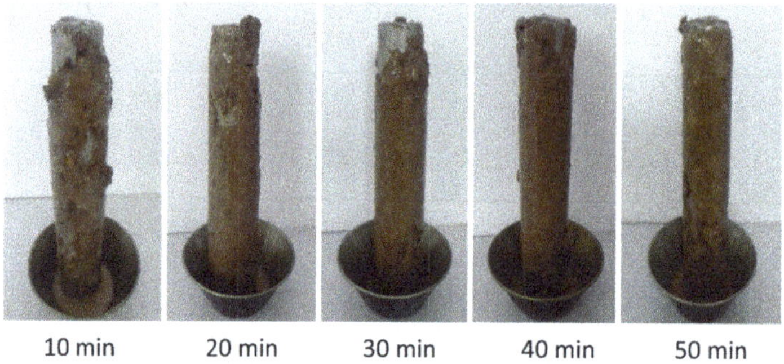

Figure 9. Accretion of pellets using fluid F5 (with I2).

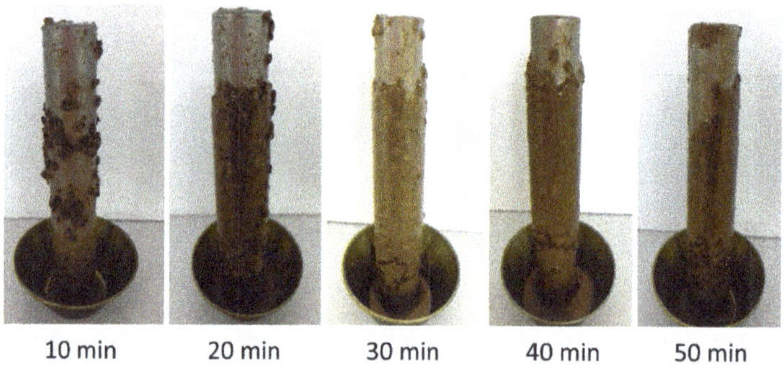

Figure 10. Accretion of pellets using fluid F6 (with I1 and I3).

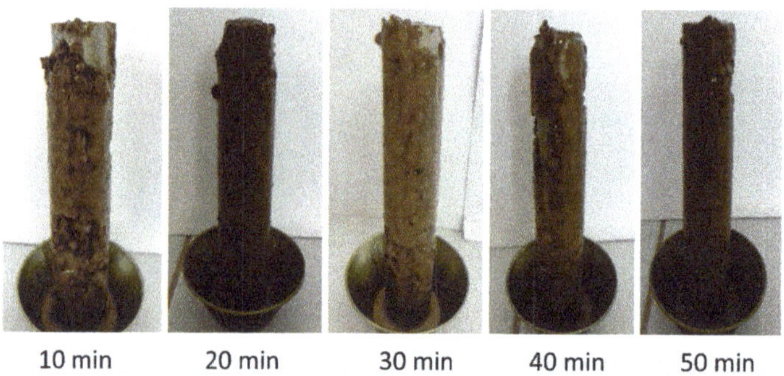

Figure 11. Accretion of pellets using fluid F7 (with I3).

It is possible to see in Figures 4–10 that the pellet accretion process was more evident after a period of 10 min, with the highest accretion percentages occurring between 20 and 50 min. The pellets used underwent a deformation process before aggregating and formed a more uniform layer, thus resulting in the accretion phenomenon.

Table 6 presents the pellets accretion values as a function of time, after performing the test with the analyzed drilling fluids.

According to the results, the accretion of the pellets using fluid F5 was the most significant, reaching 83.39% after 20 min of exposure. Fluid F7 showed a maximum accretion of 81.70% after 10 min of exposure. The maximum accretion values for fluids F1 and F3 were 79.92% and 71.27%, respectively. Fluid F4 showed a maximum accretion of 66.89% after 30 min. Samples F2 and F6 showed the lowest accretions, with maximum values of 61.93% and 58.91%, respectively, after 40 min of exposure.

Table 6. Pellet accretion percentages.

Glycerin Fluids	Accretion (%)				
	10 min	20 min	30 min	40 min	50 min
F1	75.67	79.92	78.61	71.77	64.86
F2	42.55	59.91	61.89	61.93	50.70
F3	44.35	71.27	71.25	69.58	65.19
F4	47.36	62.38	66.89	62.94	55.47
F5	80.20	83.39	77.26	72.88	66.35
F6	15.88	51.76	54.41	58.91	44.17
F7	81.70	78.63	79.26	70.18	67.94

After the pellets reached the maximum accretion value, they formed a consistent layer around the bars, which reduced the amount of water moving into the cuttings. This slowed down the rate at which the formation absorbed water, keeping it in a plastic state. Over time, the accretion values decreased for all fluids after reaching their peak, indicating a prolonged period of hydration for the pellets. This caused the pellets to exceed their liquid limits, resulting in reduced accretion percentages for all drilling fluids.

Studies conducted by Pivovarski and co-authors [18] revealed that as the time of exposure of bentonite to water increased, a reduction in the measured swelling rate was observed. This means that, in the initial stages in which the pellets absorbed a small amount of water, a significant volumetric increase occurred per unit of time. However, over time, this volumetric increase decreased exponentially.

The relationship between pellet behavior in the accretion test and time can be understood through the clay plasticity and Atterberg limits. Clay plasticity arises from the interaction of electrical charges of expandable mineral particle and water, which also lubricates these particles. The plastic state occurs when sufficient water allows mineral surfaces to slide and align under tangential stress. Higher clay content in a formation enhances its plastic properties, increasing the likelihood of the accretion phenomenon.

The interaction between glycerin and pellets is largely due to significant electrostatic interactions between the hydroxyl groups of glycerin and the ionic and polar sites in clay materials [19,20]. Glycerol molecules preferentially adsorb to the active sites of clay, displacing water molecules [21,22]. This adsorption may influence the aggregation and dispersion states of the system. In addition to well-known hydrogen bond interactions with water, relevant electrostatic interactions occur between the hydroxyl groups of glycerin and clay's ionic and polar sites.

The addition of glycerin—a tri-alcohol—into water-based mud increases the interlayer spacing of aluminosilicate, similar to ethylene glycol. The hydroxyl groups in glycerin interact strongly with clay lamellae and water molecules, with each molecule capable of forming hydrogen bonds with up to eight water molecules. This "glycerin–water" complex penetrates the interlayer space, resulting in expansion and enhanced hydration due to its volume.

The accretion test conducted with fluid F1, without inhibitors, demonstrated the pronounced reactivity of the pellets in relation to their plastic characteristics, as expected due to the high levels of expandable minerals present. In addition, it was observed that the glycerin-based drilling fluid was not able to stabilize the reactive formation, resulting in the accretion phenomenon.

From the comparative analysis between fluids F1 and F2, it was found that the inclusion of inhibitor I1 showed a partial effectiveness, promoting a decrease in the accretion levels. However, despite the action of this inhibitor, accretion phenomena still occurred in considerable proportions.

Considering the analysis of fluids F6 and F7, it can be concluded that the use of inhibitor I3 in the fluid composition was not effective, resulting in the second highest accretion rate (81.70%) among all tests performed. However, when adding inhibitor I1 to the composition, a 22.79% reduction in the accretion percentage was observed. This same behavior was noted in fluids F4 and F5. Using inhibitor I2, the highest accretion value for the study was obtained (83.39%), evidencing the inefficiency of this inhibitor in preventing expansion. However, when adding inhibitor I1 to the fluid, there was a 16.5% reduction in the accretion rate.

For fluid F3, which contained the three inhibitors, an accretion percentage of approximately 71% was observed at 20 and 30 min. These results reinforce the findings obtained when using other fluids, indicating that inhibitors I1, I2, and I3, either individually or together, were not able to effectively inhibit the swelling of the expandable minerals present in the pellets, thus allowing the occurrence of the accretion phenomenon.

4. Conclusions

With the aim of analyzing how the interaction between expandable minerals existing in reactive formations and glycerin-based drilling fluids affected the stability of oil wells, focusing on the occurrence and extension of the accretion phenomenon, the following was concluded:

- The characterization of bentonite pellets revealed a predominance of clay minerals of the smectite group in their composition, which provides a strong indication of their reactivity with a high degree of expansion;
- According to the liquid limit, the pellets were categorized as presenting very high plasticity in the presence of water, high plasticity in the presence of water with glycerin, and intermediate plasticity in the presence of glycerin;
- Based on the degree of plasticity, it was found that the pellets were classified as highly plastic both in contact with water and water with glycerin;
- The accretion percentages were considerable for all drilling fluids at different times of the accretion test, indicating that the addition of glycerin to the fluids was not efficient in stabilizing the reactive formations;
- The use of inhibitors I1, I2, and I3, alone or together, did not demonstrate effectiveness in reducing the accretion phenomenon;
- For future research, it is recommended to investigate the compositions of fluids with varying glycerin concentrations, as well as to evaluate additional commercial inhibitors beyond those already analyzed.

Based on the above, it can be seen that the drilling fluid with added glycerin was not effective in preventing the accretion phenomenon. When used to drill highly plastic formations, where smectite clay minerals are predominant, such as reactive shales, there will be a greater probability of instability problems, such as the agglomeration and accretion of cuttings. Furthermore, the addition of clay swelling inhibitors to the fluids has not been shown to have a positive effect in preventing or reducing the accretion process.

Author Contributions: Conceptualization, R.C.A.M.N. and L.V.A.; methodology, A.P.O.S. and M.C.S.L.; software, A.P.O.S.; validation, L.V.A. and W.R.P.C.; formal analysis, R.C.A.M.N. and M.C.S.L.; investigation, A.P.O.S.; resources, L.V.A. and R.C.A.M.N.; data curation, M.C.S.L. and W.R.P.C.; writing—original draft preparation, A.P.O.S.; writing—review and editing, L.V.A., W.R.P.C., J.M.P.Q.D. and A.G.B.L.; visualization, A.P.O.S., J.M.P.Q.D. and A.G.B.L.; supervision, R.C.A.M.N.;

project administration, L.V.A.; funding acquisition, R.C.A.M.N., L.V.A., J.M.P.Q.D. and A.G.B.L. All authors have read and agreed to the published version of the manuscript.

Funding: This research received no external funding.

Data Availability Statement: Data are contained within the article.

Acknowledgments: The authors A. P. O. Sousa, M. C. S. Lima, W. R. P. Costa, R. C. A. M. Nascimento and L. V. Amorim would like to acknowledge Bentonit União Nordeste (Boa Vista, Brazil) and Petrobras (Brazil), grant numbers 0050.0120134.21.9 and 0050.0126178.23.9, for the support for this research. J.M.P.Q. Delgado is grateful to the Research Unit CONSTRUCT funded by national funds through the FCT/MCTES (PIDDAC) and FCT through the individual Scientific Employment Stimulus 2020.00828.CEECIND/CP1590/CT0004, with DOI: 10.54499/2020.00828.CEECIND/CP1590/CT0004.

Conflicts of Interest: The authors declare no conflicts of interest.

References

1. Frydman, M.; Fontoura, S.A.B. Modeling Aspects of Wellbore Stability in Shales. In Proceedings of the SPE Latin American and Caribbean Petroleum Engineering Conference, Buenos Aires, Argentina, 25–28 March 2001.
2. Grim, R.E. *Clay Mineralogy*, 2nd ed.; McGraw-Hill: New York, NY, USA, 1968.
3. Fuenkajorn, K.; Daemen, J.J.K. Drilling-Induced Fractures in Borehole Walls. *J. Pet. Technol.* **1992**, *44*, 210–216. [CrossRef]
4. Reid, P.I.; Minton, R.C.; Twynam, A. Field Evaluation of a Novel Inhibitive Water-Based Drilling Fluid for Tertiary Shales. In Proceedings of the European Petroleum Conference, Cannes, France, 16–18 November 1992.
5. Pormerlau, D.G. Glycerol Based Drilling Fluids. U.S. Patent No. 2009/0143254 A1, 4 June 2009.
6. Associação Brasileira de Normas Técnicas (ABNT). *NBR 6459: Solo-Determinação do Limite de Liquidez*; ABNT: Rio de Janeiro, Brasil, 2016.
7. Associação Brasileira de Normas Técnicas (ABNT). *NBR 7180: Solo-Determinação do Limite de Plasticidade*; ABNT: Rio de Janeiro, Brasil, 2016.
8. Cliffe, S.; Young, S. Agglomeration and Accretion of Drill Cuttings in Water-based Fluids. In Proceedings of the AADE Fluids Conference and Exhibition, Houston, TX, USA, 8–9 April 2008.
9. Wilson, M.J.; Wilson, L.; Patey, I. The influence of individual clay minerals on formation damage of reservoir sandstones: A critical review with some new insights. *Clay Min.* **2014**, *49*, 147–164. [CrossRef]
10. Macedo, R.S.; Menezes, R.R.; Neves, G.A.; Ferreira, H.C. Study of clays used in red ceramic. *Cerâmica* **2008**, *54*, 411–417. [CrossRef]
11. Apolônio, T.G.; Amorim, L.V.; Leal, C.A. Correlação entre a Composição Química e Mineralógica e as Características Plásticas de Folhelhos do Nordeste do Brasil. *Rev. Eletrônica Mater. Process.* **2020**, *15*, 102–109.
12. Bell, F.G. *Engineering Geology*, 2nd ed.; Elsevier: London, UK, 2007.
13. Couto, B.O.C.; Pereira, E.L.; Gomes, R.C.; Ferreira, L.D. Correlação entre os Valores do Limite de Liquidez Obtidos pelos Métodos de Casagrande e Cone de Queda Livre para Diferentes Materiais. In Proceedings of the XVIII Brazilian Congress of Soil Mechanics and Geotechnical Engineering–COBRAMSEG, Belo Horizonte, Brazil, 19–22 October 2016.
14. Caputo, H.M. *Mecânica dos Solos e Suas Aplicações*, 6th ed.; LTC: Rio de Janeiro, Brazil, 1988.
15. Van Oort, E.; Bland, R.; Pessier, R. Drilling more Stable Wells Faster and Cheaper with PDC Bits and Water Based Muds. In Proceedings of the IADC/SPE Drilling Conference, New Orleans, LA, USA, 23–25 February 2000.
16. Hajjaji, W.; Moussi, B.; Hachani, M.; Mounir, M.; Galindo, A.; Rocha, F.; Labrincha, J.; Jamoussi, F. The Potential Use of Tithonian–Barremian Detrital Deposits from Central Tunisia as Raw Materials for Ceramic Tiles and Pigments. *Appl. Clay Sci.* **2010**, *48*, 552–560. [CrossRef]
17. Fontoura, S.A.B. Lade and Modified Lade 3D Rock Strength Criteria. *Rock Mech. Rock Eng.* **2012**, *45*, 1001–1006. [CrossRef]
18. Pivovarski, R.G.; Daroz, V.; Lugarini, A.; Franco, A.T.; Loureiro, S.; Waldmann, A.T.A.; Martins, A.L. I Testes de Hidratação Estáticos e Dinâmicos de Pellets de Bentonita para P&A. In Proceedings of the Encontro Nacional de Construção de Poços de Petróleo e Gás, Serra Negra, Brazil, 9–22 August 2019.
19. Campos, L.F.A.; Macedo, R.S.; Kiyohara, P.K.; Ferreira, H.C. Plasticity characteristics of clays for use in structural clay products. *Cerâmica* **1999**, *45*, 140–145. [CrossRef]
20. Steiger, R.P. Nontoxic, Nonchloride, Water-Base, Inhibitive Fluid to Stabilize Water Sensitive Shale 1993. US Patent No. 5198415, 30 March 1993.

21. Bradley, W.F. Molecular Associations between Montmorillonite and Some Polyfunctional Organic Liquids. *J. Am. Chem. Soc.* **1945**, *67*, 975–981. [CrossRef]
22. Bloys, B.; Davis, N.; Smolen, B.; Bailey, L.; Houwen, O.; Reid, P.; Sherwood, J.; Fraser, L.; Hodder, M. Designing and managing drilling fluid. *Oilfield Rev.* **1994**, *6*, 33–43.

minerals

MDPI

Article

Physicochemical and Mineralogical Characterizations of Two Natural Laterites from Burkina Faso: Assessing Their Potential Usage as Adsorbent Materials

Corneille Bakouan [1,2,3], Louise Chenoy [3], Boubié Guel [2,*] and Anne-Lise Hantson [3]

1 Laboratoire de Recherche et de Développement (LRD), Université Lédéa Bernard Ouédraogo, Ouahigouya 01 BP 346, Burkina Faso; bakouancorneille@gmail.com
2 Laboratoire de Chimie Moléculaire et des Matériaux (LCMM), Université Joseph Ki-Zerbo, U.F.R–SEA/, Ouagadougou 03 BP 7021, Burkina Faso
3 Service de Génie des Procédés Chimiques et Biochimiques, Faculté Polytechnique, Université de Mons, Place du Parc 20, 7000 Mons, Belgium; louise.chenoy@umons.ac.be (L.C.); anne-lise.hantson@umons.ac.be (A.-L.H.)
* Correspondence: boubieguel@yahoo.fr or boubie.guel@ujkz.bf; Tel.: +226-76645044

Abstract: In the framework of lateritic material valorization, we demonstrated how the geological environment determines the mineralogical characterizations of two laterite samples, KN and LA. KN and LA originate from the Birimian and Precambrian environments, respectively. We showed that the geological criterion alone does not determine the applicability of these laterites as potential adsorbents but must be associated with their physicochemical properties. The characterizations were carried out using Fourier transform infrared spectroscopy (FTIR), X-ray diffraction (XRD), Thermal analysis, and Atomic Emission Spectrometry Coupled with an Inductive Plasma Source. The major mineral phases obtained by X-ray diffraction analysis coupled with infrared analysis showed that the KN and LA laterite samples were composed of quartz (33.58% to 45.77%), kaolinite (35.64% to 17.05%), hematite (13.36% to 11.43%), and goethite (7.44% to 6.31%). The anionic exchange capacity of the KN and LA laterites ranged from 86.50 ± 3.40 to 73.91 ± 9.94 cmol(-)·kg^{-1} and from 73.59 ± 3.02 to 64.56 ± 4.08 cmol(-)·kg^{-1}, respectively, and the cation exchange capacity values are in the order of 52.3 ± 2.3 and 58.7 ± 3.4 cmol(+)/Kg for the KN and LA samples, respectively. The specific surface values determined by the BET method were 58.65 m^2/g and 41.15 m^2/g for the KN and LA samples, respectively. The effects of adsorbent doses on As(III,V), Pb(II), and Cu(II) adsorption were studied. At 5 mg/L As and 15 g/L adsorbent (pH 6.5–7), arsenate removal was 99.72 ± 0.35% and 99.58 ± 0.45% for KN and LA, respectively, whereas arsenite removal reached 83.52 ± 2.21% and 98.59 ± 0.64% for LA and KN, respectively. The Pb(II) and Cu(II) removal rates were 74.20 ± 0.95% for 2.4 g/L KN and 54.18 ± 0.01% for 8 g/L KN, respectively. Based on their physicochemical and mineralogical characteristics, the KN and LA laterite samples were shown to possess a high potential as adsorbent material candidates for removing heavy metals and/or anionic species from groundwater.

Keywords: natural laterites; sorption properties; mineralogy; anionic exchange capacity

1. Introduction

Laterites constitute a large family of soils typical of humid tropical regions. They originate from the alteration process of a bedrock, which is depleted in silica and enriched in iron oxide and alumina. They are products of intense meteoric weathering and consist of a mineral assemblage of goethite, hematite, aluminum hydroxide, kaolinite, and quartz [1–3].

Chemically, the structure of lateritic materials contains a high percentage of iron oxide (Fe_2O_3), alumina (Al_2O_3), and silica (SiO_2) mineral phases, which are present as a combination consisting of $Fe_2O_3 - Al_2O_3 - SiO_2 - H_2O$ matrices [3]. The $SiO_2/(Al_2O_3 + Fe_2O_3)$ ratio compared to that of the parent rock must be in such a way that the laterite formation does not contain more silica than the one which is retained in the remaining quartz and which is only necessary for the formation of kaolinite [4]. Moreover, laterites occur in nature with various yellow, brown, and red residual solids of nodular gravels and fine-grained and/or cemented solids. Laterites can vary from loose red sand to massive hard rock; sometimes, both forms coexist. The characteristic red color appears due to the presence of iron compounds in laterite. The physicochemical and mineralogical properties of laterite vary considerably depending on the extent of lateralization, the parent rock, and the geological environment [5,6].

The mineralogical composition, including minerals such as kaolinite and hematite, influences the chemical reactivity of laterites, determining their ability to interact with metal ions. Moreover, the geological environment of certain laterites can make them more effective at removing pollutants. Understanding these properties is essential for assessing their potential in the remediation of contaminated waters.

Laterites have been the subject of diverse applications reported in the literature [7–12]. They have been widely used in road construction in tropical and equatorial African countries, as well as in South America, whether they are lateritic gravels, lateritic clays, lateritic shells, or crusts [1,13,14]. They were also shown to be efficient adsorbents in treating water contaminated by inorganic pollutants. In several countries such as India, Vietnam, and Bangladesh, they have been used with great satisfaction in the adsorption of arsenic [8,9,15–17]. We recently reported their application for arsenic removal in Burkina Faso [2]. According to the literature, the efficiency of these materials as the best adsorbent candidates for the removal of inorganic and organic pollutants is only highlighted by their physicochemical and structural characteristics: the specific surface area, the anionic exchange capacity, the cation exchange capacity, and the composition, which is made up mostly of iron and aluminum oxides. These characteristics appeared to be the most critical and were highlighted in several studies on clay and/or lateritic minerals used to remove inorganic or organic pollutants [11,12,18–20]. Their applications depend closely on their structure, composition, and physicochemical characteristics. Being aware of these characteristics is decisive for better exploitation and probably opens up new application areas [21]. On the one hand, upon reviewing papers published in the last ten years, it was found that most of the studies did not cover the gap between the environmental perspective of the laterites, as adsorbent materials, and their physicochemical properties. On the other hand, the development of lateritic material-based environmental applications constitutes a relevant research area that contributes to their economical valorization. From this environmental perspective, any research that demonstrates the great potential of laterites to address practical environmental problems, such as the removal of arsenic and/or heavy metal ions from aqueous solutions, should highlight the fact that those applications depend closely on their structure, their mineralogical composition as determined by the geological environment, and their physicochemical properties. The present work focuses on this correlation, unlike most publications in which the emphasis is not placed on physicochemical properties, such as the cation exchange capacity (CEC) and the anionic exchange capacity (AEC).

In the literature, various techniques, such as X-ray powder diffraction, Fourier transform infrared (FT-IR) spectroscopy, scanning electron microscopy (SEM), chemical analysis, cation exchange capacity (CEC), and specific surface area (BET) have been carried out to investigate the laterite characterization processes because of the removal of inorganic pollutants [7–11,22–24]. Although numerous studies have been conducted on the characterization of natural laterites for the sorption of heavy metal ions and/or metalloids [8,9,24–26], few of them have explored the relationship among their physicochemical and mineralogical properties, their adsorptive properties, and geological environment. Thus, given their interesting properties, laterite-based adsorbents can be particularly effective for the adsorption of heavy metal ions in aqueous solution.

Ghani et al. used a geopolymer based on lateritic clay as a potential adsorbent for the removal of heavy metals from aqueous solutions [23]. Zhu et al. effectively removed several heavy metals (As, Cr, Pb, Cu, Cd, and Hg) from wastewater using a new lateritic ceramic [27]. Mitra et al. also determined the characteristics of lateritic soil for its use in the removal of Pb(II) and Cr(VI) from synthetic wastewater [25]. He et al. found excellent behavior of a cost-effective limonitic laterite for the adsorption of Pb(II) and Cd(II) hydrates [24]. Mohapatra et al. studied the adsorption of lead (II) from contaminated waters using Indian nickel laterite ores [26]. Although numerous papers have been published in this field, we noted that the previous studies are not exhaustive. To the best of our knowledge, this study is the first to provide a complete description of laterite properties, in terms of specific surface area, pore volume, DSC/TGA, PZC, chemical composition, mineralogical characterization, cation exchange capacity (CEC), and anionic exchange capacity (AEC), in relation to the adsorption ability of the material. Indeed, these properties are helpful criteria that provide strong evidence of the adsorption capacity of the laterites regarding the removal of cationic and/or anionic pollutants from aqueous solutions.

The geological map of the country is divided into square degrees depending on the nature of the rocks and their structure, which facilitates the understanding of the distribution of resources, notably the lateritic sites. We initially selected four laterites (KN, LA, BN, and DA) based on the specific geological characteristics of each square degree. Experiments were conducted on these four laterites rich in iron and aluminum oxides, which are crucial for environmental applications, such as the removal of arsenic and/or heavy metal ions from aqueous solutions. The results obtained for the DA laterite were satisfactory since in the batch mode, we achieved an elimination rate of 97.30% for As(III) at a dose of 0.75 g of laterite [2]. The adsorption capacity was 0.30 mg/g [2]. These investigations paved the way for the first uses of local natural laterites from Burkina Faso for As(III) remediation from drinking water using fixed-bed columns [28]. We also achieved an elimination rate of 99.69% for As(V) at a dose of 0.75 g of laterite [2]. The adsorption capacity for DA laterite was 0,33 mg/g [2]. The results obtained with BN laterite were not satisfactory due to the low content of iron and aluminum oxides. The present paper provides the results obtained with the KN and LA laterites that were chosen on the basis of their geological environment, so as to be rich in iron and aluminum oxides.

This study aims to determine (i) the physicochemical and structural characteristics of the two KN and LA natural laterites, by using various techniques, such as XRD, FTIR, SEM + EDX, DSC, CEC, and AEC; and (ii) the correlation between their physicochemical and structural characteristics and their potential usage as adsorbent materials. This work makes it possible to determine how the physicochemical properties of these natural laterites predispose them to be potential adsorbent materials, thereby contributing to the sustainable management of the natural lateritic resources.

2. Materials and Methods

2.1. Origin of Samples

Laterite samples were collected from two distinct sites of Burkina Faso (Figure 1). The first one, named KN, was collected in the northern part of Kaya, at the following coordinates: 13°07′13.47″ N and 1°06′52.28″ W. The second one, named LA, was collected in Laye village, at the following coordinates: 12°31′27.05″ N and 1°47′07.22″ W. The laterite from KN is light red, and the one from LA is red-brown.

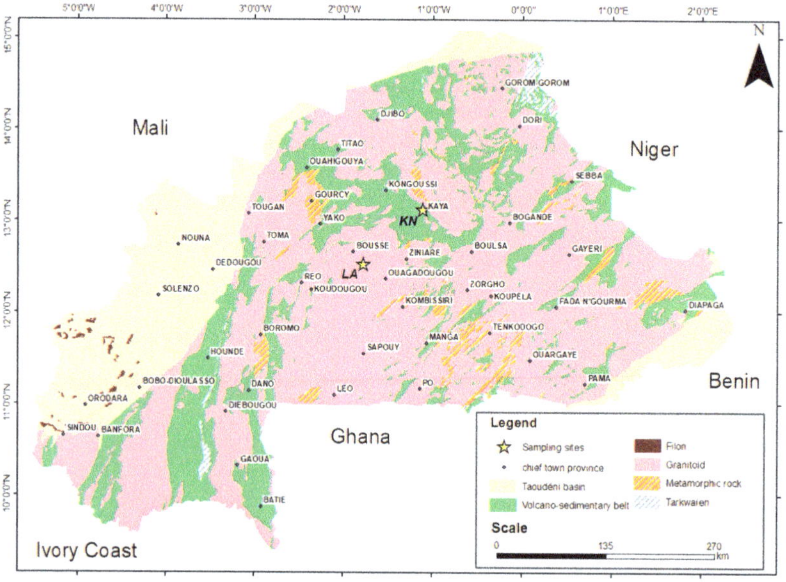

Figure 1. Location of collected laterite samples.

The term "Tarkwaien" refers to specific geological formations, typically associated with gold deposits, primarily located in West Africa. Tarkwaien deposits are usually metamorphic rock formations, often composed of quartzite and schist. These rocks are formed under high pressure and temperature conditions [29].

2.2. *The Specific Geological Contexts of the Sites*

The territory of Burkina Faso is divided into square degrees, and each site belongs to a square degree. The geological details of these square degrees are shown on the various positioning maps produced for the different sites.

2.2.1. Geological Context of the Northern Kaya Site

This site belongs to the Kaya square degree, located between latitudes 13° and 14° north and longitudes 1° and 2° west. The geological formations are primarily composed of volcanic and volcano-sedimentary rocks (Figure 2). These rocks form relief areas, while the flat vast regions, at altitudes ranging from 300 to 350 m, comprise granitic materials such as protoliths. From a geological point of view, the KN samples originate from the environment of Birimian rocks, stemming from andesine (with a calcic-alkaline affinity), basalt, and dacit alteration.

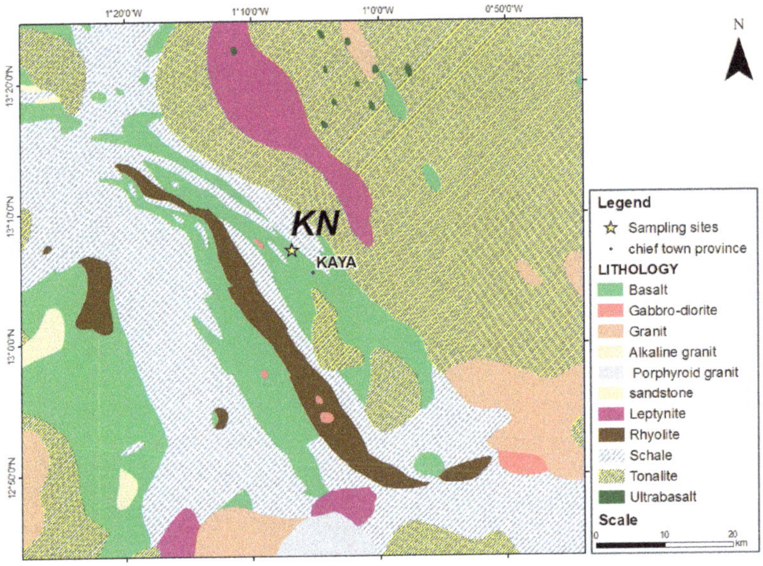

Figure 2. Geological map of northern Kaya laterite site.

2.2.2. Geological Context of the Laye Site

The Laye (LA) site belongs to the square degree of Ouagadougou and is located 35 km from Ouagadougou. The square degree of Ouagadougou is located between parallels 12 and 13° north latitude and meridians 1 and 2° west longitude. Unlike Kaya, this square degree is mainly composed of granite formations, corroborating its generally flat relief. The Laye site is located on highly indurated lateritic formations derived from a granitic protolith (alkaline granite). From a geological point of view, this area is located in an environment of Precambrian rocks (Figure 3).

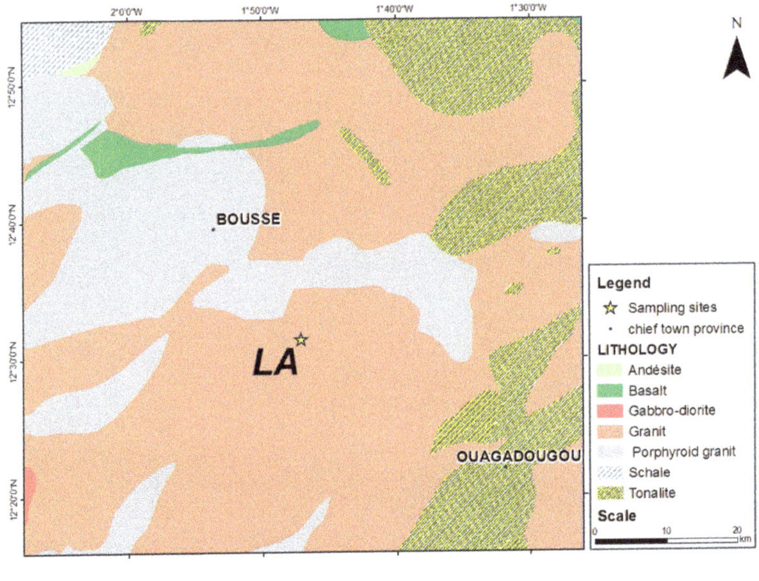

Figure 3. Geological map of Laye laterite site.

2.3. Raw Material Characterization

2.3.1. Chemical Composition

Elementary chemical analysis was performed by ICP (ICP-AES-IRIS Intrepid II XSP model, detection limit = 9 µg/L). A total of 0.25 g of laterite samples were digested in a microwave oven in 4 mL of HF (30% *w/w*), 3 mL of H_2SO_4 (96% *w/w*), and 3 mL of HNO_3 (65% *w/w*) [30].

2.3.2. Infrared Spectroscopy

Fourier transform infrared (FTIR) spectroscopy was performed using a Shimadzu FTIR-8400S spectrometer (FTIR-8400S spectrometer, Duisburg, Germany) to identify functional groups on the laterite samples. The spectra were acquired by accumulating 20 scans with a resolution of 4 cm^{-1} over a wavelength range of 400 to 4000 cm^{-1}. Infrared analysis of the laterite samples was performed using the potassium bromide pelletizing technique. The pellets were made by mixing 5 mg of each sample with 500 mg of potassium bromide (KBr, Merck, Darmstadt, Germany). The mixture was finely ground and subjected to a pressure of ten (10) tons. The pellet thus formed was analyzed with a spectrophotometer.

2.3.3. X-Ray Powder Diffraction (XRD)

The structure, characteristics, and phase composition of the laterite samples were analyzed by X-ray diffraction analysis (XRD) on a SIEMENS D5000 equipment (KS Analytical Systems, Plano, TX, USA) running with the DIFFRAC AT software version 2.0 (Co-Kα radiation (λ = 1.54060 Å); graphic monochromator; measurement in the 20 to 80 θ game; 40 kV voltage and 30 mA current).

2.3.4. Semi-Quantification

The semi-quantitative analysis of the different mineral phases was carried out by coupling the X-ray diffraction results with the chemical analysis results. This coupling allowed for the evaluation of the relative quantities of the minerals contained in the laterites using Equation (1) [2,31].

$$T(a) = \sum M(i) \times P_i(a) \quad (1)$$

T(a): content of the oxide "a" in the sample; Mi: content (%) of mineral "i" in the sample; P_i(a): proportion of oxide "a" in mineral "i" (this proportion is deduced from the ideal formula assigned to mineral "i"). The quantitative approach was performed on the following basis:

- Alumina is distributed in kaolinite;
- Iron oxide is distributed between goethite and hematite;
- Silicon oxide is distributed between quartz and kaolinite.

The mass percentages of mineral elements were obtained from Equations (2)–(5).

$$\%\text{Kaolinite} = \%\text{Al}_2\text{O}_3 \times \frac{M_{\text{Kaolinite}}}{M_{\text{Al}_2\text{O}_3}} \quad (2)$$

$$\%\text{Quartz} = \left(\%\text{SiO}_2 - \%\text{Kaolinite} \times \frac{M_{\text{SiO}_2} \times 2}{M_{\text{Kaolinite}}}\right) \times \frac{M_{\text{Quartz}}}{M_{\text{SiO}_2}} \quad (3)$$

$$\begin{gathered} M_{(\text{Goethite+Hematite})} \rightarrow \%\text{Fe}_2\text{O}_3 \\ M_{\text{Goethite}} \rightarrow \%\text{Goethite} \end{gathered} \quad (4)$$

$$\%\text{Hematie} = (\%\text{Al}_2\text{O}_3 - \%\text{Goethite}) \quad (5)$$

2.3.5. Differential Scanning Calorimetry (DSC) and Thermogravimetric Analysis (TGA)

Thermogravimetric and differential scanning calorimetry analysis was carried out on a TGA/DSC thermal analyzer (TGA/DSC1-STARe METTLER TOLEDO System, Zaventem, Belgium) under a nitrogen atmosphere at a heating rate of 10 °C/min. A mass of 5 g of raw laterite powder with a particle size of 106 µm was heated from 25 °C to 1500 °C. The TGA/DSC data were analyzed using the METTLER TOLEDO STARe software version 9.00.

2.3.6. Zeta Potential Measurements

Zeta potential measurements were performed to determine the electrostatic magnitude between the particles of the laterite samples. Zeta meter equipment (Zetasizer nano ZS, Malvern, Belgium) was used to measure the isoelectric point (IP) of the natural laterite samples. The pH solution was adjusted with 0.025 mol.L^{-1} HCl and 0.025 mol.L^{-1} NaOH solutions. The data were calculated using the software ZS (Malvern Panalytical, version 4.0).

2.3.7. Specific Surface Area and Porosity by Nitrogen Sorption Analysis

Nitrogen (N_2) sorption isotherms at 77 K were measured using the BelSorp-max instrument (Osaka, Japan) running with the Bel Japan Inc Surface Area and Porosity Analyzer. Before N_2 sorption measurements, the samples were vacuum-dried at room temperature for at least 24 h, followed by degassing under heating and vacuum, using a Sample Degas System. The value of each material's specific surface was determined with a device from Bel Sorp-max/Bel Japan.Inc (Osaka, Japan) brand controlled by Bel Japan software. Inc., version 1.3

2.4. *Cation Exchange Capacity (CEC)*

The cation exchange capacity was determined by shaking 0.5 g of the laterite sample with 30 mL of 0.05 mol/L hexaamminecobalt (III) chloride ($[Co(NH_3)_6]^{3+}$, $3Cl^-$) solution for 2 h. Before each determination of the CEC, a calibration curve was established using standard solutions ranging from 0.005 to 0.05 mol/L. This procedure creates a reference that links instrumental responses to known concentrations, allowing for reliable and reproducible measurements of CEC. The supernatant was collected and analyzed by SHIMADZU UV-visible spectrometry at 475 nm. The cation exchange capacity was calculated according to Equation (6).

$$\text{C.E.C} = \frac{(C_i - C_f) \times V}{m} \times 100 \tag{6}$$

C_i and C_f are the initial and equilibrium concentrations of the hexaamminecobalt (III) chloride solution (mol/L), respectively; m is the mass of the laterite sample; and V is the volume of the solution.

2.5. *Anionic Exchange Capacity (AEC)*

The anionic exchange capacity (AEC) was determined by adapting the method proposed by Zelazny et al. to our study context [32]. It was determined at different pH values (pH framing the isoelectric point of the material) using chloride ions as an anionic index. The AEC measurement methods are based on the same principle as the cation exchange capacity (CEC) and are subject to the same constraints. All tests were carried out in triplicate.

Thus, 2 g of laterite was placed in a series of centrifugation tubes with a capacity of 50 mL. Then, 20 mL of $CaCl_2$ (0.1 mol/L) was added and left under gentle stirring for 1 h to saturate the laterite with chloride. After 1 h of stirring, the mixtures were centrifuged at 3000 rpm for 5 min, and the residues were recovered. The residues were rewashed five times with 20 mL of $CaCl_2$ (0.002 mol/L), left in contact for 1 h, and centrifuged for 5 min at 3000 rpm. Then, 5 mL of $CaCl_2$ (0.002 mol/L), 1.5 mL of HCl (0.1 mol/L), and 3.5 to 5 mL of milli-Q water, respectively, were added so that the total volume was equal to 10 mL.

The suspensions were placed in a bath at 25° C for six days, shaking them manually, thrice daily. On the seventh day, the pH of each suspension was measured, and the mixture was centrifuged at 3000 rpm for 10 min. The chloride ions (C_1) were then evaluated. The mass of the residues and of the tube were weighed and recorded in order to determine the volume (V_1) of the solution that remained trapped.

The residues were resuspended in 30 mL of an ammonium nitrate solution (1 mol.L^{-1}), left in contact for 1 h, and then centrifuged at 3000 rpm for 10 min. The clear solution was recovered in a 100 mL volumetric flask (V_2). This operation was repeated twice, and the flasks' contents were adjusted with an ammonium nitrate solution (1 mol.L^{-1}). The chloride ions (C_2) were assayed in this final solution, and the anionic exchange capacity was determined by Equation (7).

$$\mathrm{AEC}\left(\mathrm{cmol.kg^{-1}}\right) = (C_2V_2 - C_1V_1) \times \frac{0.1 \times z}{M \times W} \tag{7}$$

were

C_1 is the concentration (mg.L^{-1}) of Cl^- in the final washing solution of 0.1 M $CaCl_2$.

C_2 is the concentration (mg.L^{-1}) of Cl^- in the displacing solution of 1 M NH_4NO_3.

V_1 is the volume (mL) of the solution contained in laterites after the final washing of 0.1 M $CaCl_2$.

V_2 is the total volume (mL) of the displacing solution 1 M NH_4NO_3.

M and z are the atomic weight and charge of Cl^-, respectively.

W is the laterite sample's weight (g).

2.6. Experimental Studies on Arsenic and Heavy Metal Ion Removal

All arsenic and heavy metal ion (Cu and Pb) solutions were prepared with ultra-pure water (milliQ-water, resistivity 18.2 MΩ.cm). The glass materials were preliminary immerged in diluted nitric acid (5% w/w) for at least 12 h and then rinsed with ultra-pure water. Arsenic (V) and arsenic (III) stock solutions of 1000 mg/L were prepared from $Na_2HAsO_4\ 7H_2O$ (98%, Alfa Aesar, Geel, Belgium) and $NaAsO_2$ (0.05 mol/L Alfa Aesar, Belgium), respectively. Pb(II) and Cu(II) stock solutions of 1000 mg/L were prepared from $Pb(NO_3)_2$ (≥99%, Merck, Darmstadt, Germany) and $Cu(NO_3)_2\ 3H_2O$) (≥99%, Merck), respectively. All solutions were stored at 4°C. The working solutions for the experiments were freshly prepared by diluting the stock solutions.

All experiences were carried out in triplicate. The influence of the laterite sample dose on the arsenic adsorption was carried out under equilibrium conditions. In 250 mL polyethylene bottles, 50 mL As(III), As(V), Pb(II), and Cu(II) solutions of known concentrations were used with laterite samples at various quantities. Working solutions of 5 mg/L arsenic (III, V) were mixed with laterite samples varying from 2.5 g/L to 45 g/L. The solutions were agitated for 24 h at temperature of 18 ± 2 °C (overhead shaker Heidolph REAX-15 rpm, Schwabach, Germany). The solutions of Pb(II) (10 mg/L) or Cu(II) (10 mg/L) were mixed with variable mass samples of KN laterite. The masses ranged from 0.4 g/L to 3.2 g/L for Pb(II) and from 0.8 g/L to 16 g/L for Cu(II). The mixtures were stirred for 1 h and 30 min using a Heidolph REAX 20 mechanical stirrer at room temperature (18 ± 2 °C). Then, 14 mL of each filtrate was placed in 15 mL bottles, then acidified with 1 mL of nitric acid and stored at 4 °C before being analyzed by ICP-AES.

Before each analysis, a ten-point calibration curve was established by successive dilution (5 ppm, 2.5 ppm, 1 ppm, 0.5 ppm, 0.25 ppm, 0.1 ppm, 0.05 ppm, 0.01 ppm, and 0.005 ppm, blank) using a commercial standard solution of 1000 ± 0.05 mg/L in 5% nitric acid.

The equilibrium solutions were filtered through a 0.45 μm nylon membrane (diameter 25 mm, VWR) after centrifugation (BECKMAN J2-MI model, Cincinnati, OH, USA) for 15 min at 3000 rpm and analyzed by ICP-AES (detection limit = 9 μg/L). The percentage of removed As(III, V), Pb(II), and Cu(II) was determined according to Equation (8).

$$\%Ads\ (As(III,V);Pb(II);Cu(II)) = \frac{C_0 - C_e}{C_0} \times 100 \tag{8}$$

C_0 and C_e are the initial and final concentrations (mg/L), respectively, of arsenic and heavy metal ions in the solutions.

3. Results and Discussion

3.1. Chemical Composition

The chemical composition in Table 1 indicates the major oxides (SiO_2, Al_2O_3, and Fe_2O_3) and the minor oxides (Na_2O, K_2O, TiO_2, MgO, MnO_2, and CaO) in the KN and LA samples. Other properties, such as the apparent density, total organic carbon (TOC), and organic matter (OM), are also shown in Table 1.

Table 1. Characteristics and composition (wt, %) of KN and LA laterite samples.

Properties	KN	LA
Total organic carbon (TOC) (%)	0.16	0.09
Organic matter (OM) (%)	0.73	1.32
Apparent density	2.9	2.6
Inorganic composition (wt.%)		
Fe_2O_3	20.8	17.65
Al_2O_3	14.09	6.74
SiO_2	50.16	53.70
K_2O	1.70	1.82
Na_2O	1.43	1.40
TiO_2	2.10	2.10
MgO	0.65	0.95
MnO_2	0.35	0.68
CaO	0.70	0.75
L.O.I	8.01 ± 0.02	12.55 ± 0.23
Total	99.99	98.34

L.O.I: loss on ignition.

According to the literature [13,33], iron is found as oxyhydroxides. Oxides of potassium, sodium, and titanium are in small amounts in both samples. These results suggest that quartz, aluminosilicates, and iron minerals are predominant in the samples studied. The low levels of potassium, sodium, and titanium oxides indicate that compounds containing titanium, potassium, and sodium are non-existent or present in tiny proportions. The elements in KN and LA are grouped according to their family in Table 2.

Table 2. Chemical composition of samples by type of elements in % by weight.

Family	Alkalis	Alkaline Earth	Metals	Silica
KN	3.1	0.1	37.3	50.2
LA	3.2	0.2	26.6	53.7

This study is the first to carry out this classification by type of elements, providing useful criteria of the elements present in lateritic materials that can influence the adsorption of inorganic pollutants in solution.

3.2. Mineralogical Characterization

3.2.1. X-Ray Diffraction (XRD)

Figure 4 shows the XRD pattern. The major mineral phases of the two laterite samples KN and LA include quartz (SiO_2), kaolinite ($Al_2Si_2O_5(OH)_4$), hematite (Fe_2O_3), and goethite ($FeO(OH)$). These mineral phases are commonly found in laterites [13,34]. The minerals present in KN and LA laterites were compared with other laterites presented in the literature and used for the adsorption of heavy metals and/or metalloids (Table 3).

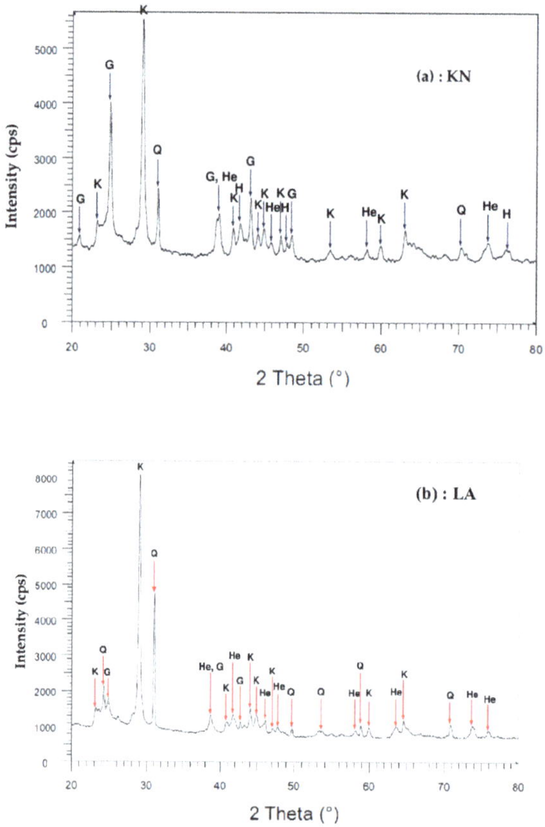

Figure 4. X-ray diffraction pattern of laterite samples: (**a**) KN sample; (**b**) LA sample. K = kaolinite; Q = quartz; He = hematite; G = goethite.

Table 3. Comparison of main minerals present in KN and LA laterites with other natural laterites reported in literature.

Samples	Main Minerals	References
red soil	quartz, hematite, goethite, aluminum oxides	[35]
raw laterite	quartz, hematite, goethite, aluminum oxides, iron oxides, titanium oxides	[22,36,37]
iron-rich laterite	quartz, hematite, goethite, aluminum oxides	[38]
laterite (Australia)	quartz, hematite, goethite, aluminum oxides	[39]
DA	quartz, hematite, goethite, aluminum oxides	[2]
laterite KN	quartz, hematite, goethite, aluminum oxides	This study
laterite LA	quartz, hematite, goethite, aluminum oxides	

Several authors [8,24–26], who have used laterites to adsorb inorganic pollutants, suggested that these laterites should be rich mainly in hematite, goethite, or aluminum oxides.

3.2.2. Semi-Quantification

Table 4 shows the results of the semi-quantitative analysis of the different mineral phases present in the KN and LA laterites. These results show that the laterites are composed of hematite (13.36% to 11.43%), goethite (7.44% to 6.31%), kaolinite (35.64% to 17.05%), and quartz (33.58% to 45.77%).

Table 4. Mineralogical composition of laterites in % by mass.

Mineral Phases		Hematite	Goethite	Kaolinite	Quartz
wt (%)	KN	13.36	7.44	35.64	33.58
	LA	11.43	6.31	17.05	45.77
	DA *	13.11	7.29	48.32	22.53

* DA: previous work on natural laterite (See Reference [2]).

The results from the chemical, mineralogical, cation exchange capacity, and anionic exchange capacity analyses led to the conclusion that the natural laterites investigated possess good adsorbent characteristics for removing inorganic and organic pollutants from aqueous matrices.

3.2.3. Infrared Spectrometry (IR)

Figure 5 shows the FT-IR spectra of the laterite samples. The FT-IR spectra results distinguish three spectral domains: 3700–3400 cm^{-1}, 1650–900 cm^{-1}, and 800–400 cm^{-1}, respectively. Table 5 gives the different attributions of the observed bands.

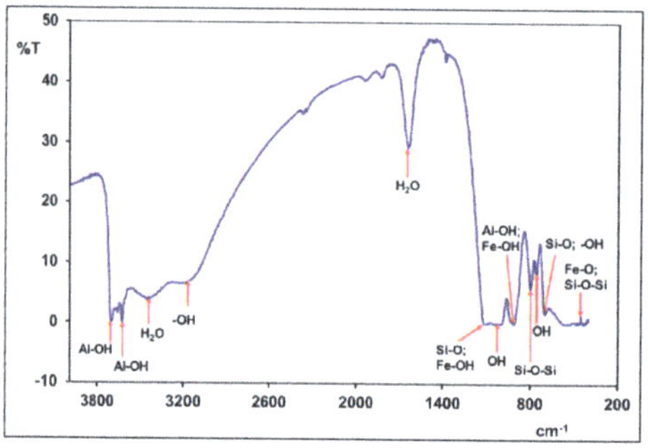

(a)

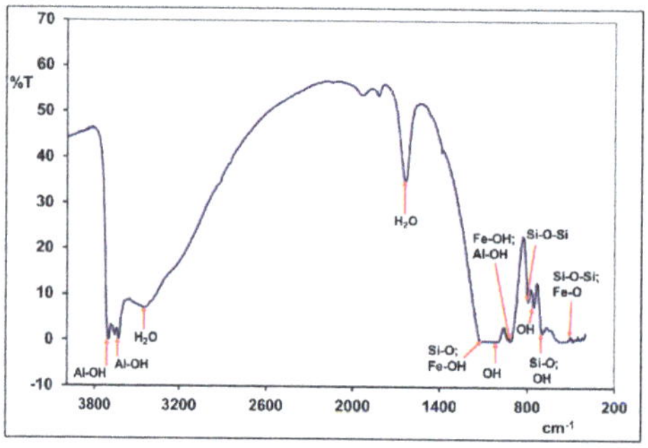

(b)

Figure 5. FT-IR spectrum of natural laterites: (**a**) KN sample; (**b**) LA sample.

Table 5. Assignments of FT-IR bands of KN and LA laterite samples.

(ν en cm^{-1})	Probable Bands Assignments	References
3695	Vibration bands linked to external hydroxyls (Al-OH) in kaolinite	[40]
3618	Vibration bands related to internal hydroxyls (Al-OH) in kaolinite, located between a tetrahedron sheet and an octahedron $Al_2(OH)_6$	[40]
3170	Band related to –OH bound vibrations in goethite	[41]
3430	Band related to water contained in the intersheet	[40]
1638	Band related to hygroscopic water	[41,42]
1112	Vibration band corresponding to Si-O bound of kaolinite	[40,43,44]
1034	Vibrations bands corresponding to Si-O bound of kaolinite and Fe-OH bound of goethite	[40,44,45]
1004	Vibrations bands related to OH bounds of kaolinite and Fe-OH bound of goethite	[44,45]
914	Band related to distortion vibrations of Al-OH bound of kaolinite and Fe-OH bound of goethite	[40,44,45]
791	Band corresponding to bending vibration of Si-O and Fe-OH bounds of kaolinite	[41,43]
752	Vibrations bands related to OH bounds of kaolinite and Fe-OH bound of goethite	[40,45]
694	Vibrations bands related to OH bound of kaolinite and Si-O bounds of quartz	[41]
539	Bands corresponding to distortions vibrations of Si-O-Al bound of kaolinite and Fe-O bound of hematite	[40,41,44]
470	Vibrations bands related to flexion of Si-O-Si and Fe-O bounds of hematite	[41,43]
421	Vibrations bands of Si-O-Si bounds of kaolinite	[44]

The results of the infrared spectrum analysis confirm the presence of mineral phases, such as goethite, hematite, quartz, and kaolinite, which we had already identified during the previous X-ray diffractogram analysis.

3.3. Thermogravimetric Analysis and Differential Scanning Calorimetry (TGA /DSC)

DSC/TGA techniques provide information on the thermal stability and phase transformations of lateritic materials. Figure 6 shows experimental TGA/DSC curves of the KN and LA samples. Several thermal processes are observed. The endothermic peak located between 94 °C and 110 °C in the DSC curves is related to hygroscopic water loss or hydration in the samples. This incident is associated with 1% and 1.5% weight loss in the TGA curves for the KN and LA samples, respectively. Endothermic peaks between 296 °C and 356 °C, followed by 4.1% and 1.3% weight losses for KN and LA, respectively, are due to goethite transformation in hematite (Equation (9)).

$$2\alpha - \mathrm{FeOOH} \longrightarrow \alpha - \mathrm{Fe_2O_3} + \mathrm{H_2O} \tag{9}$$

Endothermic peaks between 500 °C and 600 °C in the DSC curves are assigned to kaolinite deshydroxylation (Equation (10)), indicating 7.3% and 8.7% weight losses in the TGA curves for KN and LA, respectively. It is suggested that structural hydroxyls are removed during the chemical reaction, leading to the destruction of the mineral crystalline network. As for kaolinite, deshydroxylation forms an amorphous phase which is named metakaolinite [46,47].

$$\mathrm{Si_2O_5Al_2(OH)_4} \rightarrow \mathrm{2SiO_2 - Al_2O_3 + 2H_2O} \tag{10}$$

The only exothermal peaks observed in the DSC curves are located between 950 °C and 1000 °C. These peaks may be attributed to the structural reorganization of meta kaolinite in the spinel phase and amorphous silica (Equation (11)) [40,46].

$$2[\mathrm{2SiO_2 - Al_2O_3}] \rightarrow \mathrm{Si_3Al_4O_{12} + SiO_2} \tag{11}$$

The DSC/TGA results can indicate changes in the pore structure, influencing the available adsorption sites. DSC analyses reveal the stability of the mineral phases present in laterites. A stable phase is generally more effective at adsorbing anionic and/or cationic pollutants. Temperature variations can affect interactions between laterites and adsorbed molecules. DSC and TGA are essential for understanding the physical and chemical properties of materials, which may include their ability to adsorb contaminants such as heavy metals.

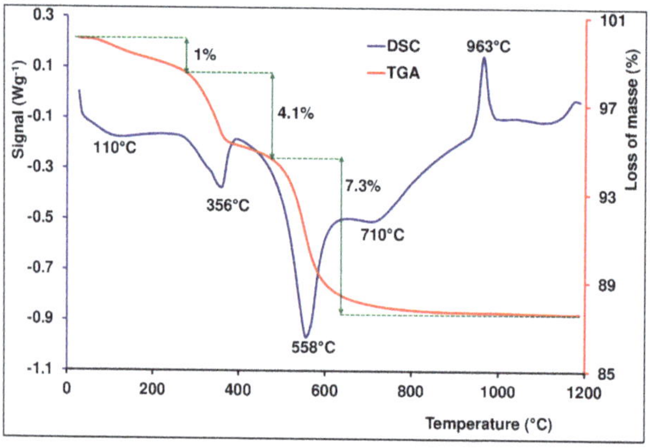

(a)

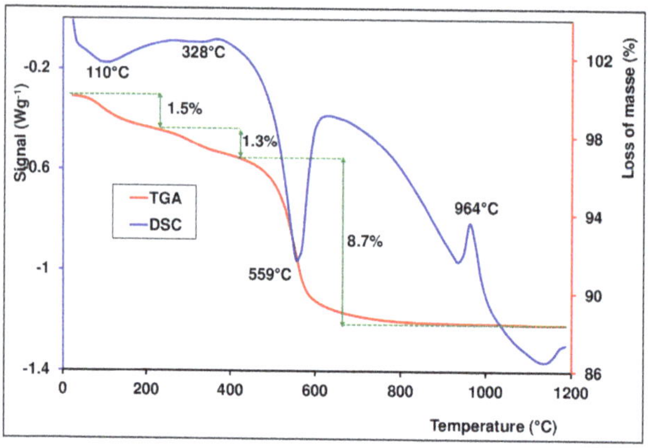

(b)

Figure 6. TGA/DSC curves of lateritic samples: (**a**) KN sample; (**b**) LA sample.

3.4. Specific Surface Area and Porosity Using Nitrogen Sorption Analysis

Figure 7 shows the N_2 adsorption–desorption isotherms of the raw laterite samples. The initial part of the isotherms for all laterites of a type I shape (IUPAC classification) indicates the presence of micro pores [48]. As determined by the B.E.T method from Figure 7, the specific surface values are 58.65 m^2/g and 41.15 m^2/g for the KN and LA samples, respectively. These specific surface area values are significantly high when compared to values reported in the literature for natural laterites in the range of 16–32 m^2/g [15,17,22,36–38] (Table 6).

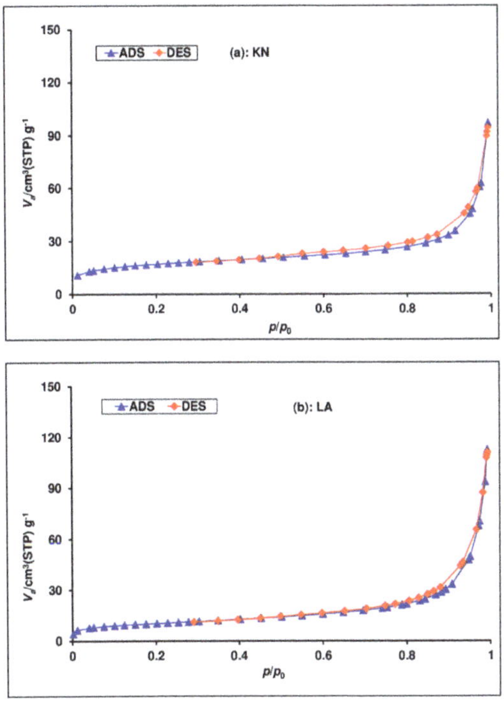

Figure 7. N_2 adsorption/desorption isotherms of (**a**) KN and (**b**) LA.

Table 6. Specific surface area B.E.T values, pore volume of two samples, and literature-reported specific areas of other laterites.

Laterites	Specific Surface Area by B.E.T (m^2/g)	Pore Volume (cm^3/g)	References
Laterite raw (India)	15.3	0.013	[15]
Red soil	16.1	-	[35]
Laterite raw (India)	17.5–18.5	0.011	[36]
Modified laterite	178–184	0.22	[36]
Laterite raw (Vietnam)	10.9	0.01	[49]
Laterite raw	24.7	0.08	[37]
Iron rich laterite	32	-	[38]
Calcined laterite	187.5	0.04	[50]
Laterite soil (DA)	35.08	0.10	[28]
Laterite soil (KN)	58.6	0.14	This study
Laterite soil (LA)	41.1	0.10	

We noted that our investigated lateritic materials showed high specific surface area values compared with most natural materials reported in the literature used to remove inorganic pollutants. These particular surface areas suggest that the natural materials we investigated are potential candidates for the adsorption of inorganic pollutants in solution. Several authors have highlighted the importance of having high specific surface area values in their studies using clay and/or lateritic materials to remove inorganic pollutants [50,51].

3.5. Determination of Anionic Exchange Capacity (AEC)

The anionic exchange capacity values (Table 7) of the measured laterites ranged from 86.50 ± 3.40 to 73.91 ± 9.94 cmol(-)$\cdot$kg^{-1} and 73.59 ± 3.02 to 64.56 ± 4.08 cmol(-)$\cdot$kg^{-1}, respectively. The anionic exchange capacity of laterites increased with the decreasing pH of the solution. This significant increase in AEC with decreasing pH is due to positive charges that are located on iron hydroxides associated with aluminum. The high values of the anionic exchange capacities of the investigated laterites would be an asset for using them as adsorbents to remove anionic pollutants.

Table 7. Anionic exchange capacity of laterites.

Laterite KN		Laterite LA	
pH	**AEC (cmol(-)·kg^{-1})**	**pH**	**AEC (cmol(-)·kg^{-1})**
3.47 ± 0.02	6.50 ± 3.40	3.67 ± 0.04	73.59 ± 3.02
3.67 ± 0.01	86.02 ± 8.29	3.84 ± 0.04	73.33 ± 3.03
5.51 ± 0.05	73.91 ± 9.94	5.22 ± 0.04	64.56 ± 4.08

The results obtained are similar to other materials studied in the literature. Cheng et al. determined the anionic exchange capacity of black carbon at pH 7.1 and 3.4 and obtained values of 84 and 18 cmol·kg^{-1}, respectively [52]. Lawrinenko et al. also determined the anionic exchange capacity of biochar produced from three raw materials at 500 and 700 °C at pH 4, 6, and 8 [53]. The measured AEC values ranged from 0.60 to 27.8 cmol(-)·kg^{-1}. Moreover, the study by Schell and Jordan [54], on kaolinite, pyrophyllite, halloysite, and bentonite materials showed that there are close relationships between the anionic exchange capacity of these materials and their physicochemical properties. Indeed, materials rich in iron oxide, aluminum oxide, and amorphous silicates showed a high anionic exchange capacity. Consequently, the investigated natural laterites containing important amounts of iron oxide and alumina are likely to exhibit high anionic exchange capacity values, which gives them interesting adsorbent properties.

3.6. Determination of Cation Exchange Capacity (CEC)

The CEC value, determined using the hexaamminecobalt (III) chloride method, is an important parameter in assessing the adsorption capacity of adsorbent materials for cations removal [18]. We found high CEC values in the order of 52.3 ± 2.3 and 58.7 ± 3.4 cmol(+)/kg (dried matter) for the KN and LA samples, respectively. We cannot compare these results with previous literature data because we did not find extensive documentation of this parameter for natural laterites that remove inorganic pollutants from solution. To our knowledge, this study appears to be the first to provide scientific documentation of this parameter for laterites used in the field of adsorption of inorganic pollutants. However, we compare the CEC values obtained in this study with other adsorbents in Table 8 below. These specific surface area values are significantly high when compared to values reported in the literature for other natural laterites. We noted that the CEC values of the KN and LA laterite samples are comparable with those reported for other natural adsorbents. Considering the CEC and the specific surface area values, compared with the literature values on natural adsorbents, these materials could be considered potential candidates for heavy metal cation removals through adsorption processes [55,56].

The laterites of Burkina are rich in iron oxide and alumina, like the laterites of other countries used to eliminate heavy metals and/or metalloids in ground waters [36,37,57]. Considering all of these physical and chemical characteristics, we can conclude that the natural laterites of Burkina Faso prove to be potential candidates for the adsorption of heavy metals and/or metalloids in groundwater.

Table 8. Comparison of CEC values for KN and LA laterites with other adsorbents.

Adsorbents	C.E.C (cmol(+)/kg)	References
Clay mineral	42.38	[58]
Peat soil	33–48	[59]
Bauxite	24–33	[59]
Iron concretion	59–65	[59]
Natural clay	18.66	[60]
Kaolinite	13.00	[61]
Bentonitic clay	67.00	[62]
Ivory Coast clay	35.47	[63]
Laterite soil (KN)	52.33	This study
Laterite soil (LA)	58.70	

3.7. Isoelectric Point (IP) of Laterite Samples

Figure 8 shows variations in the zeta potential versus pH of the laterite sample solutions. These variations allowed the determination of isoelectric points (IP). For zeta potentials whose values equal zero, the IP values are 3.82 and 3.78 for the KN and LA samples, respectively. According to the literature [64], the theoretical value of IP, calculated based on the silica and alumina percentage, is approximatively 4.6. The experimental values of IP are not far from the theoretical value. The observed deviations are the result of the iron percentage being higher than the one of alumina and also by the presence of other oxides in the samples.

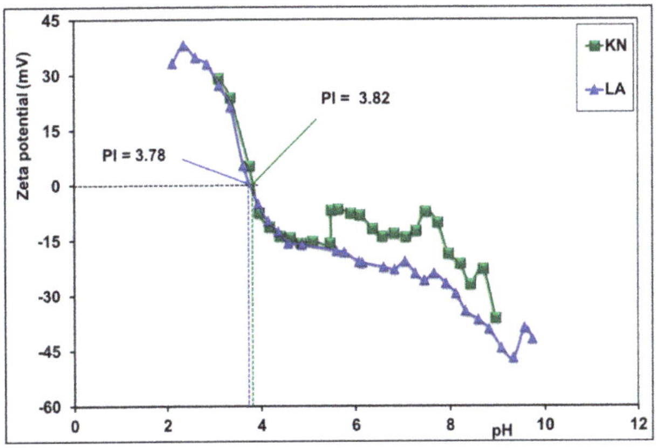

Figure 8. Isoelectric point of laterite soils determined from variations in zeta potential versus pH.

The values of the isoelectric point (IP) of laterites, KN (3.82) and LA (3.78), mean that the surfaces of these laterites are positively charged at pH levels below these values, which favors the adsorption of anionic pollutants. The increase in adsorption below IP is due to electrostatic attraction between the positive surface and the negative charge of the anions. Similarly, the decrease in adsorption with increasing pH is due to electrostatic repulsion between the opposing surface and the negative charge of anions. The electric surface charge plays an important role in the adsorption of inorganic pollutants on laterites.

3.8. Batch Adsorption Experiments: Application to Arsenic (III, V) and Pb(II) and Cu(II) Removal

Figure 9a shows the removal percentage of As(III, V) versus adsorbent doses, at a concentration of 5 mg/L and at a pH varying between 6.57 and 7.15 and 6.5–7 for As(V) and As(III), respectively. An increase in adsorbent dose from 2.5 to 15 g/L leads to arsenic (III) removal efficiency from 80% to 98% for LA and KN samples, respectively. As for arsenic (V), the removal efficiency is 99% for both adsorbents KN and LA samples.

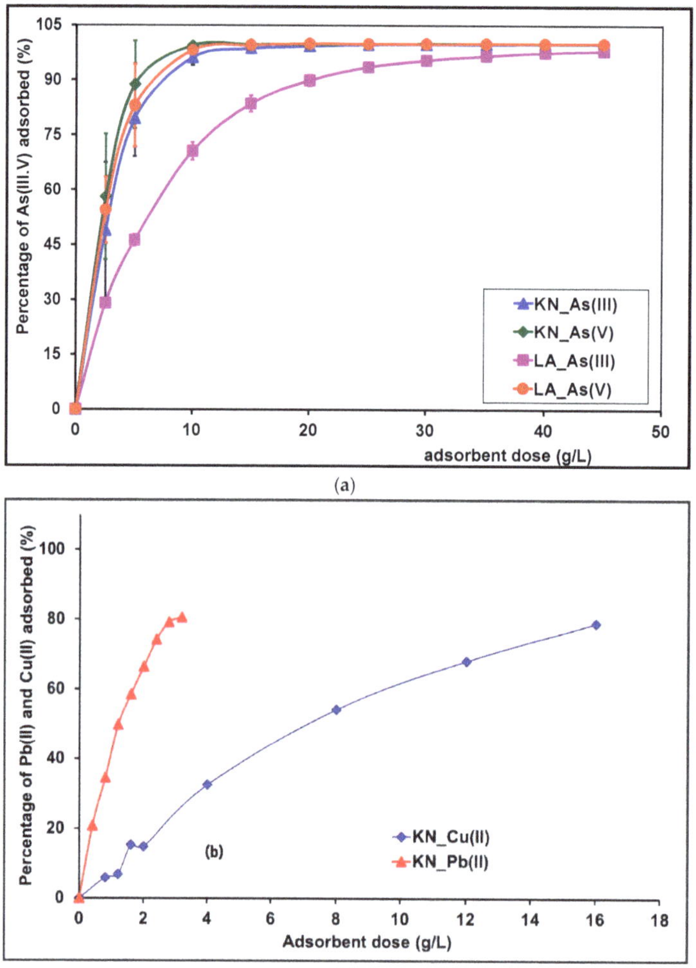

Figure 9. (**a**) Effect of adsorbent dose on adsorption of As(III, V). (**b**) Effect of KN adsorbent dose on adsorption of Pb(II) and Cu(II).

With an amount of 15 g/L of the laterites from Burkina Faso, As(III) removal percentages, reaching 98% and 80%, are achieved for the laterites KN and LA, respectively, with an initial arsenite concentration of 5 mg/L. With an initial arsenite concentration of 1 mg/L, Pravin D. Nemade et al. obtained a rate of elimination of 90% for an amount of 25 g laterite/L [36]. With an amount of 15 g/L of the KN and LA samples and an initial arsenate concentration of 5 mg/L, a removal rate of As(V) of more than 95% was achieved. The results in the present work are better than those obtained by Abhijit Maiti et al. on a natural laterite coming from the zone of Midnapore of Western Bengal (India) for which a removal rate of approximately 80% was achieved when using 15 g laterite/L [22]. For an amount of 5 g/L, a removal rate of approximately 70% is achieved for both KN and LA laterites of Burkina in the case of As(V) adsorption (initial arsenate concentration of 5 mg/L), and a removal rate of approximately 80% for laterite KN in the case of As(III) adsorption (initial arsenite concentration of 5 mg/L). Trivalent arsenic (As(III)) is less adsorbed on LA laterite than on KN laterite for the following reasons. KN laterite has a higher proportion of iron oxides (20.8%) and aluminum (14.09%), providing more adsorption sites for As(III) and promoting stronger interactions. Additionally, KN laterite's specific surface area (58.6 m^2/g) and porosity (0.14 cm^3/g) exceed those of LA laterite, which are 41.1 m^2/g and 0.10 cm^3/g, respectively, allowing better accessibility for As(III) adsorption. Finally, KN laterite's anionic exchange capacity is also higher than that of the LA laterite in the studied pH range, further enhancing its effectiveness as an adsorbent for arsenic.

The efficiency of the removal may be due to an increase in adsorption sites which are made available when the dosage of adsorbent dose increases [65]. For both the KN and LA samples, a high removal percentage is achieved with an amount of natural laterites, which is lesser than those described in the literature. This result could be attributed to the high specific area and anionic exchange capacity of the laterites KN and LA of Burkina Faso. These results are in agreement with the literature data showing that better results are achieved when using adsorbents with large specific surface area [36]. However, contrary to the previous reported literature, the present work shows that the specific surface area must be combined to the anionic exchange capacity for a better interpretation.

The results for the removal of Pb(II) and Cu(II) metal ions at different doses of adsorbent are presented in Figure 9b. A general trend observed was that the adsorption of heavy metal ions increased with the increase in adsorbent dose. The results show that the removal rates of Pb(II) reached 74.20 $\pm$ 0.95% (3.09 $\pm$ 0.04 mg/g) when the amount of the KN adsorbent was 2.4 g/L. For Cu(II), the adsorption rates were 54.18 $\pm$ 0.01% (0.68 $\pm$ 0.01 mg/g) when the dose of the KN adsorbent was 8 g/L. As the dose of laterite increased, the removal of Pb(II) and Cu(II) from the solutions also increased. This can be explained by the increase in functional groups on the surface of the laterite, providing a greater number of available sites to form complexes through surface complexation, or to exchange cations due to the ion exchange process [66–68].

Arsenic is better adsorbed on laterites than copper and lead due to its stronger chemical affinity for the adsorption sites present on laterites, such as iron and aluminum oxides, which promote specific interactions. Additionally, the size and charge of arsenic allow it to be attached more effectively to these surfaces compared to copper and lead ions.

3.9. A Comparison of the Main Physicochemical Properties Related to the Adsorption of Laterites from Burkina Faso with Those Reported in the Literature

This study investigated the main physicochemical properties of two laterites (LA and KN) from Burkina Faso, showing how to link these properties to an adsorption process. The results showed that these laterites possess interesting properties, allowing them to be used to treat water contaminated by anionic and/or cationic pollutants. The physicochemical properties of the two laterites with regard to their adsorption properties were compared with other results reported in the literature concerning laterites used for the adsorption of anionic and/or cationic pollutants. Table 9 shows the main observations of this comparison. From the results of Table 9, we noted that the CEC and AEC determinations of lateritic materials are omitted in most studies dealing with the removal of cationic and/or anionic pollutants from aqueous solutions. However, these parameters are crucial to elucidate the adsorption mechanisms, in terms, for example, of surface complexation (inner-sphere and outer-sphere surface complexation). While CEC value is an important parameter to determine the adsorption of heavy metal cations, AEC values are more indicated for adsorbents like As(III) and As(V), which appear in solution as neutral or anionic-charged species.

Table 9. Comparison of physicochemical properties of several laterites in relation to pollutant adsorption.

Adsorbent	Adsorbats	AEC cmol(-)/kg)	C.E.C cmol(+)/kg)	Specific Surface Area (B.E.T) (m^2/g)	Pore Volume (cm^3/g)	DSC /TGA	IEP or PZC	References
Laterite soil	cationic dye	-	-	66.97	-	-	6.6	[69]
Raw laterite	arsenic and fluoride	-	-	31.6037	0.0097	-	-	[70]
Raw laterite	Phosphate	-	-	29.54	0.0676	-	-	[7]
Laterite	Arsenic	-	-	155	0.5489	-	7.1	[8]
Laterite clay	Ni(II) and Co(II)	-	-	17.441	0.005	-	-	[23]
Laterite soil	Arsenic	-	-	15.365	0.013	-	6.96	[15]
Natural laterite	Arsenic	-	-	18.05	-	-	7.49	[16]
Treated laterite	Led	-	-	75.5	0.02	-	6.0	[71]
Plateau laterite ceramic	Pb, Cd, Hg, As, Cu and Cr	-	-	26.73	0.15	-	-	[27]
Limonitic laterite	Pb(II) and Cd(II)	-	-	62.73	0.62	-	-	[24]
Lateritic nickel	Pb(II)	-	-	68.39	-	-	6.70	[26]
Laterite soil	Pb(II) and Cr(VI)	-	-	23.015	0.011	-	-	[25]
Laterite DA **	As(III,V)	40.61–230.80	-	35.08	0.10	-	4.75	[2,28]
Laterite LA *	As(III,V)	64.56–73.59	58.7 ± 3.4	58.80	0.14	Det ***	3.78	This study
Laterite KN *	As(III,V)	73.90–86.50	52.3 ± 2.3	41.10	0.10	Det **	3.82	

* Composition mineralogical characterization of laterites are common characteristics provided in most studies (see Table 3). ** References [2,28] describe our previous study on a natural laterite; Det ***: determined.

Table 9 shows that our study is the first to provide a complete description of laterite properties, in terms of specific surface area, pore volume, DSC/TGA, PZC, chemical composition*, mineralogical characterization*, cation exchange capacity (CEC), and anionic exchange capacity (AEC), in relation to the adsorption ability of the material. Indeed, these properties are helpful criteria that provide strong evidence of the adsorption capacity of the laterites regarding the removal of cationic and/or anionic pollutants from aqueous solutions [18–20,54].

It is worth noting that the adsorptive properties of laterites are closely related to the geology of the deposit ore they originate from. Indeed, for a naturally occurring laterite ore, the type of parent rock it originates from determines its chemical composition, particularly the percentages of iron, aluminum, silica, and titanium oxides, as well as its mineralogical composition. The mineralogical composition is generally characterized by mineral phases such as goethite, hematite, and kaolinite. However, laterite ores possess a large variability in their composition. Consequently, even if different geological environments are qualitatively characterized by the same chemical and mineralogical composition, the percentage of oxides can be different from one site to another. The physicochemical properties may also vary. In these circumstances, it is appropriate to focus on quantitative criteria that we need to forecast the ability of the material to remove pollutants in the adsorption process. Moreover, the extent of these criteria may lead to different adsorption capacities. As we have already stated, these critical criteria are the following: specific surface area, pore volume, PZC or IEP, cation exchange capacity (CEC), and anionic exchange capacity (AEC).

The laterites LA and KN in this study are from two sites of different geological environments (Table 10). The first site is an environment of Birimian rocks, resulting from the weathering of andesite (with a calcic-alkaline affinity), basalt, and dacit. On the other hand, the second site is a weathering site of alkaline granites located in a Precambrian rock environment. Preliminary investigations showed (Table 10) that the removal percentage for arsenic (III) removal at a concentration of 5 mg/L and an adsorbent dose of 15 g/L leads to arsenic (III) removal of 80 ± 0.15% and 98 ± 0.05% for the LA and KN samples, respectively [72]. As for arsenic (V), the removal efficiency was 99 ± 0.02% for the two samples, KN and LA, at an adsorbent dose of 15 g/L [72]. Although the two laterites (LA and KN) do not have the same geological environments, they showed a high efficiency for arsenic removal due to the combination of their AEC, specific surface area, and pore volume values. In addition, their low IEP or PZC values also favored the adsorption of neutral or anionic-charged arsenic species. Adsorption of anionic arsenic species will only be important when the charges on the laterites become positive, which happens at low soil solution pH where anionic-charged arsenic species occur [2,28]. Noting that the surfaces of the samples are positively charged under their PZC values, it is clear these materials better adsorb arsenate (V) anions at pH values under their PZC values. It is, therefore, important to measure the anionic exchange capacity (AEC) of the laterites rather than the CEC when we are dealing with the adsorption of neutral or anionic-charged arsenic species.

Table 10. Adsorptive properties of laterites used for As(III, V) removal in relation to their physico-chemical characteristics.

Laterite	Geological Environment	Mineralogical Characterizations XRD, FT-IR	AEC cmol(-)/kg)	Specific Surface Area (B.E.T) (m^2/g)	Pore Volume (cm^3/g)	IEP or PZC	Efficiency (%)
KN	Environment of Birimian rocks, resulting from the weathering of andesite (with a calcic-alkaline affinity), basalt, and dacit	Det *	73.90–86.50	58.80	0.14	3.82	98 ± 0.05% for As(III) 99 ± 0.02% for As(V)
LA	Environment of precambrian rocks and alteration of alkaline granites	Det *	64.56–73.59	41.10	0.10	3.78	80 ± 0.15% for As(III) 99 ± 0.02% for As(V)
DA	Lateritic plateau, well indurated and resulting from the alteration of a neutral basic rock	Det *	40.61–230.80	35.08	0.10	4.75	99.69% for As(V)) 97.30% for As(III)

Det *: determined (Tables 3 and 5).

At this step, the geological environment cannot constitute the sole criterion that would justify the adsorption capacity of negatively charged arsenic species. This criterion may not be decisive in predicting the adsorption capacity of laterite. It must be associated with the physicochemical properties of the laterites. Table 10 outlines the criteria that affect the adsorptive properties of laterites (LA, KN) in connection with their physicochemical and geological characteristics.

To date, it has been demonstrated that when particular geological environments, such as Birimian/Precambrian and lateritic plateaus, are combined with appropriate physicochemical properties, a potential lateritic material for adsorption is expected (Table 10). Unfortunately, in the literature, no direct link has yet been established on this subject. This study constitutes a preliminary finding, which we will validate by extending our investigations to other lateritic sites in Burkina Faso.

4. Conclusions

Physicochemical and mineralogical analyses were carried out on two natural laterites from the country. The properties of these two natural laterites were determined with regard to their ability to adsorb heavy metals and/or metalloids from aqueous solutions. These laterite samples were characterized using several physical and chemical techniques, including XRD, FTIR, elementary chemical analyses, and SEM. The results obtained by

X-ray diffraction analysis coupled with infrared showed that the laterites are composed of hematite (13.36% to 11.43%), goethite (7.44% to 6.31%), kaolinite (35.64% to 17.05%), and quartz (33.58% to 45.77%). Chemical analysis showed that these natural laterites are rich in iron and aluminum oxide. The specific surface areas and cation exchange capacity values, as determined by the BET and cobalt hexamine chloride methods, were shown to have suitable values compared to previously determined values in the literature. The anionic exchange capacity of laterites KN and LA ranged from 86.50 ± 3.40 to 73.91 ± 9.94 cmol(-)·kg^{-1} and from 73.59 ± 3.02 to 64.56 ± 4.08 cmol(-)·kg^{-1}, respectively. Furthermore, the investigations on these laterite samples showed that they could remove heavy metals and/or metalloids from contaminated ground waters. The main minerals identified in these two Burkina Faso laterites were consistent with those described in the literature. This study led to some valuable criteria we can use as a basis for classifying natural laterites as potential adsorbent materials for the removal of inorganic and/or organic pollutants from aqueous matrices. At a dose of 15 g/L, the elimination of arsenic (III) is 80% for sample LA and 98% for sample KN, while the elimination efficiency of arsenic (V) reaches 99% for both samples. Regarding Pb(II), the elimination rate with 2.4 g/L of KN adsorbent is 74.20 ± 0.95% (3.09 ± 0.04 mg/g), and for Cu(II), it is 54.18 ± 0.01% (0.68 ± 0.01 mg/g) with 8 g/L of KN adsorbent. The adsorption of As(III,V), Pb(II), and Cu(II) species on the surfaces of laterites would primarily occur through the formation of surface complexes, which can be either inner-sphere or outer-sphere complexes. This study showed how the geological environment determined the mineralogical characteristics of the laterites and their chemical composition. Combining the geological environment with appropriate criteria related to the physicochemical properties of the materials opens up interesting perspectives regarding the rapid valorization of the laterites in Burkina Faso as potential adsorbents. To deepen the understanding of the adsorption mechanisms, detailed studies will be conducted using various techniques to reduce the uncertainty associated with the hypotheses regarding the nature of the formed complexes. Additionally, the analysis of the FTIR spectra of the residues will help to identify the chemical modifications that occur during adsorption, thereby providing crucial information on the interactions between the metals and the lateritic surfaces.

Author Contributions: Conceptualization, C.B., A.-L.H., and B.G.; methodology, C.B., L.C. and A.-L.H.; software, C.B.; validation, C.B. and L.C.; formal analysis, C.B.; investigation, C.B.; resources, L.C., A.-L.H., and B.G.; data curation, C.B., L.C.; A.-L.H., and B.G.; writing—original draft preparation, C.B.; writing—review and editing, C.B., L.C., A.-L.H., and B.G.; visualization, C.B. and A.-L.H.; supervision, B.G. and A.-L.H.; project administration; A.-L.H. and B.G.; funding acquisition, B.G. and A.-L.H. All authors have read and agreed to the published version of the manuscript.

Funding: The authors thank the ARES-CCD/PIC 2012 (Belgium) and the International Science Program (ISP, Uppsala, Sweden, BUF 02) for providing financial support.

Data Availability Statement: All data generated or analyzed during this study are included in this published article.

Acknowledgments: The authors thank the following services for their assistance: the departments of materials science, of metallurgy, and of thermodynamic and physical mathematics of UMONS and Materia Nova Research Center (Belgium).

Conflicts of Interest: The authors declare no conflicts of interest regarding the publication of this paper.

References

1. Ndiaye, M.; Magnan, J.P.; Cissé, I.K.; Cissé, L. Étude de l'amélioration de latérites du Sénégal par ajout de sable. *Bull. Lab. Ponts Chaussees* **2013**, *280*, 123–137.

2. Ouedraogo, R.D.; Bakouan, C.; Sorgho, B.; Guel, B.; Bonou, L.D. Characterization of a natural laterite of Burkina Faso for the elimination of arsenic (III) and arsenic (V) in groundwater. *Int. J. Biol. Chem. Sci.* **2019**, *13*, 2959–2977. [CrossRef]
3. Najar, M.; Sakhare, V.; Karn, A.; Chaddha, M.; Agnihotri, A. A study on the impact of material synergy in geopolymer adobe: Emphasis on utilizing overburden laterite of aluminium industry. *Open Ceram.* **2021**, *7*, 100163. [CrossRef]
4. Lawane, A.; Pantet, A.; Vinai, R.; Hugues, J. Etude géologique et géomécanique des latérites de Dano (Burkina Faso) pour une utilisation dans l'habitat, XXIXe Recontres Univ. *Genie Civ.* **2011**, *86*, 206–215.
5. Bourman, R.P.; Ollier, C.D. A critique of the Schellmann definition and classification of laterite. *Catena* **2002**, *47*, 117–131. [CrossRef]
6. Maiti, A.; Thakur, B.K.; Basu, J.K.; De, S. Comparison of treated laterite as arsenic adsorbent from different locations and performance of best filter under field conditions. *J. Hazard. Mater.* **2013**, *262*, 1176–1186. [CrossRef] [PubMed]
7. Huang, W.Y.; Zhu, R.H.; He, F.; Li, D.; Zhu, Y.; Zhang, Y.M. Enhanced phosphate removal from aqueous solution by ferric-modified laterites: Equilibrium, kinetics and thermodynamic studies. *Chem. Eng. J.* **2013**, *228*, 679–687. [CrossRef]
8. Nguyen, T.H.; Tran, H.N.; Vu, H.A.; Trinh, M.V.; Nguyen, T.V.; Loganathan, P.; Vigneswaran, S.; Nguyen, T.M.; Trinh, V.T.; Vu, D.L.; et al. Laterite as a low-cost adsorbent in a sustainable decentralized filtration system to remove arsenic from groundwater in Vietnam. *Sci. Total Environ.* **2020**, *699*, 134267. [CrossRef]
9. Nguyena, T.H.; Nguyen, A.T.; Loganathan, P.; Nguyen, T.V.; Vigneswaran, S.; Nguyen, T.H.H.; Trand, H.N. Low-cost laterite-laden household filters for removing arsenic from groundwater in Vietnam and waste management. *Process Saf. Environ. Prot.* **2021**, *152*, 154–163. [CrossRef]
10. Thanakunpaisit, N.; Jantarachat, N.; Onthong, U. Removal of Hydrogen Sulfide from Biogas using Laterite Materials as an Adsorbent. *Energy Procedia* **2017**, *138*, 1134–1139. [CrossRef]
11. Kamagate, M.; Assadi, A.A.; Kone, T.; Giraudet, S.; Coulibaly, L.; Hanna, K. Use of laterite as a sustainable catalyst for removal of fluoroquinolone antibiotics from contaminated water. *Chemosphere* **2018**, *195*, 847–853. [CrossRef] [PubMed]
12. Karki, S.; Timalsina, H.; Budhathoki, S.; Budhathoki, S. Arsenic removal from groundwater using acid-activated laterite. *Groundw. Sustain. Dev.* **2022**, *18*, 100769. [CrossRef]
13. Millogo, Y.; Traoré, K.; Ouedraogo, R.; Kaboré, K.; Blanchart, P.; Thomassin, J.H. Geotechnical, mechanical, chemical and mineralogical characterization of a lateritic gravels of Sapouy (Burkina Faso) used in road construction. *Constr. Build. Mater.* **2008**, *22*, 70–76. [CrossRef]
14. Lawane, A.; Vinai, R.; Pantet, A.; Thomassin, J.-H.; Messan, A. Hygrothermal Features of Laterite Dimension Stones for Sub-Saharan Residential Building Construction. *J. Mater. Civ. Eng.* **2014**, *26*, 05014002. [CrossRef]
15. Maji, S.K.; Pal, A.; Pal, T.; Adak, A. Adsorption Thermodynamics of Arsenic on Laterite Soil. *J. Surf. Sci. Technol.* **2007**, *22*, 161–176.
16. Maiti, A.; Dasgupta, S.; Basu, J.; De, S. Adsorption of arsenite using natural laterite as adsorbent. *Sep. Purif. Technol.* **2007**, *55*, 350–359. [CrossRef]
17. Kadam, A.M.; Nemade, P.D.; Oza, G.H.; Shankar, H.S. Treatment of municipal wastewater using laterite-based constructed soil filter. *Ecol. Eng.* **2009**, *35*, 1051–1061. [CrossRef]
18. Xu, D.; Tan, X.L.; Chen, C.L.; Wang, X.K. Adsorption of Pb(II) from aqueous solution to MX-80 bentonite: Effect of pH, ionic strength, foreign ions and temperature. *Appl. Clay Sci.* **2008**, *41*, 37–46. [CrossRef]
19. Eren, E.; Afsin, B. An investigation of Cu(II) adsorption by raw and acid-activated bentonite: A combined potentiometric, thermodynamic, XRD, IR, DTA study. *J. Hazard. Mater.* **2008**, *151*, 682–691. [CrossRef]
20. Melichová, Z.; Hromada, L. Adsorption of Pb^{2+} and Cu^{2+} Ions from Aqueous Solutions on Natural Bentonite, Polish. *J. Environ. Stud.* **2012**, *22*, 457–464.
21. Moutou, J.M.; Foutou, P.M.; Matini, L.; Samba, V.B.; Mpissi, Z.F.D.; Loubaki, R. Characterization and Evaluation of the Potential Uses of Mouyondzi Clay. *J. Miner. Mater. Charact. Eng.* **2018**, *06*, 119–138. [CrossRef]
22. Maiti, A.; DasGupta, S.; Basu, J.K.; De, S. Batch and Column Study: Adsorption of Arsenate Using Untreated Laterite as Adsorbent. *Ind. Eng. Chem. Res.* **2008**, *47*, 1620–1629. [CrossRef]
23. Ghani, U.; Hussain, S.; Amin, N.U.; Imtiaz, M.; Khan, S.A. Laterite clay-based geopolymer as a potential adsorbent for the heavy metals removal from aqueous solutions. *J. Saudi Chem. Soc.* **2020**, *24*, 874–884. [CrossRef]
24. He, F.; Ma, B.; Wang, C.; Chen, Y.; Hu, X. Adsorption of Pb(II) and Cd(II) hydrates via inexpensive limonitic laterite: Adsorption characteristics and mechanisms. Sep. *Purif. Technol.* **2023**, *310*, 123234. [CrossRef]
25. Mitra, S.; Thakur, L.S.; Rathore, V.K.; Mondal, P. Removal of Pb(II) and Cr(VI) by laterite soil from synthetic waste water: Single and bi-component adsorption approach. *Desalin. Water Treat.* **2016**, *57*, 18406–18416. [CrossRef]
26. Mohapatra, M.; Khatun, S.; Anand, S. Kinetics and thermodynamics of lead (II) adsorption on lateritic nickel ores of Indian origin. *Chem. Eng. J.* **2009**, *155*, 184–190. [CrossRef]
27. Zhu, D.; He, Y.; Zhang, B.; Zhang, N.; Lei, Z.; Zhang, Z.; Chen, G.; Shimizu, K. Simultaneous removal of multiple heavy metals from wastewater by novel plateau laterite ceramic in batch and fixed-bed studies. *J. Environ. Chem. Eng.* **2021**, *9*, 105792. [CrossRef]

28. Ouedraogo, R.D.; Bakouan, C.; Sakira, A.K.; Sorgho, B.; Guel, B.; Somé, T.I.; Hantson, A.L.; Ziemons, E.; Mertens, E.; Hubert, P.; et al. The Removal of As(III) Using a Natural Laterite Fixed-Bed Column Intercalated with Activated Carbon: Solving the Clogging Problem to Achieve Better Performance. *Separations* **2024**, *11*, 129. [CrossRef]
29. Koffi, Y.H.; Wenmenga, U.; Djro, S.C. Tarkwaian Deposits of the Birimian Belt of Houndé: Petrological, Structural and Geochemical Study (Burkina-Faso, West Africa). *Int. J. Geosci.* **2016**, *7*, 685–700. [CrossRef]
30. Njopwouo, D.; Orliac, M. Note sur le comportement de certains minéraux à l'attaque triacide. *Cah. ORSTOM Sér. Pédol.* **1979**, *17*, 329–337.
31. Laibi, A.B.; Gomina, M.; Sorgho, B.; Sagbo, E.; Blanchart, P.; Boutouil, M.; Sohounhloule, D.K.C. Caractérisation physico-chimique et géotechnique de deux sites argileux du Bénin en vue de leur valorisation dans l'éco-construction. *Int. J. Biol. Chem. Sci.* **2017**, *11*, 499–514. [CrossRef]
32. Zelazny, L.W.; He, L. Chapter 41 Charge Analysis of Soils and Anion Exchange. In *Methods of Soil Analysis: Part 3 Chemical Methods—SSSA Book Series*; Sparks, D.L., Page, A.L., Helmke, P.A., Loeppert, R.H., Soltanpour, P.N., Tabatabai, M.A., Johnston, C.T., Sumner, M.E., Eds.; Wiley: Hoboken, NJ, USA, 1996; Volume 5, pp. 1231–1253. [CrossRef]
33. Konan, K.L. Interactions Entre des Matériaux Argileux et un Milieu Basique Riche en Calcium. Ph.D. Thesis, Université de Limoge, Limoges, France, 2006; pp. 1–144.
34. Mbumbia, L.; De Wilmars, A.M.; Tirlocq, J. Performance characteristics of lateritic soil bricks fired at low temperatures: A case study of Cameroon. *Constr. Build. Mater.* **2000**, *14*, 121–131. [CrossRef]
35. Nemade, P.D.; Kadam, A.M.; Shankar, H.S.; Bengal, W. Adsorption of arsenic from aqueous solution on naturally available red soil. *J. Environ. Biol.* **2009**, *30*, 499–504.
36. Maiti, A.; Basu, J.K.; De, S. Experimental and kinetic modeling of As(V) and As(III) adsorption on treated laterite using synthetic and contaminated groundwater: Effects of phosphate, silicate and carbonate ions. *Chem. Eng. J.* **2012**, *191*, 1–12. [CrossRef]
37. Glocheux, Y.; Pasarín, M.M.; Albadarin, A.B.; Allen, S.J.; Walker, G.M. Removal of arsenic from groundwater by adsorption onto an acidified laterite by-product. *Chem. Eng. J.* **2013**, *228*, 565–574. [CrossRef]
38. Partey, F.; Norman, D.; Ndur, S.; Nartey, R. Arsenic sorption onto laterite iron concretions: Temperature effect. *J. Colloid Interface Sci.* **2008**, *321*, 493–500. [CrossRef]
39. Jahan, N.; Guan, H.; Bestland, E.A. Arsenic remediation by Australian laterites. *Environ. Earth Sci.* **2011**, *64*, 247–253. [CrossRef]
40. Joussein, E.; Petit, S.; Decarreau, A. Une nouvelle méthode de dosage des minéraux argileux en mélange par spectroscopie IR. *Comptes Rendus L'académie Sci.-Ser. IIA-Earth Planet.* **2001**, *332*, 83–89. [CrossRef]
41. Lakshmipathiraj, P.; Narasimhan, B.R.V.; Prabhakar, S.; Raju, G.B. Adsorption of arsenate on synthetic goethite from aqueous solutions. *J. Hazard. Mater.* **2006**, *136*, 281–287. [CrossRef]
42. Madejová, J. FTIR techniques in clay mineral studies. *Vib. Spectrosc.* **2003**, *31*, 1–10. [CrossRef]
43. ALzaydien, A.S. Adsorption of methylene blue from aqueous solution onto a low-cost natural Jordanian Tripoli. *Am. J. Appl. Sci.* **2009**, *6*, 1047–1058. [CrossRef]
44. Kloprogge, J.T.; Frost, R.L.; Hickey, L. Infrared emission spectroscopic study of the dehydroxylation of some hectorites. *Thermochim. Acta* **2000**, *345*, 145–156. [CrossRef]
45. Ristić, M.; Musić, S.; Godec, M. Properties of γ-FeOOH, α-FeOOH and α-Fe_2O_3 particles precipitated by hydrolysis of Fe^{3+} ions in perchlorate containing aqueous solutions. *J. Alloys Compd.* **2006**, *417*, 292–299. [CrossRef]
46. Chen, Y.F.; Wang, M.C.; Hon, M.H. Phase transformation and growth of mullite in kaolin ceramics. *J. Eur. Ceram. Soc.* **2004**, *24*, 2389–2397. [CrossRef]
47. Sarkar, M.; Banerjee, A.; Pramanick, P.P.; Sarkar, A.R. Design and operation of fixed bed laterite column for the removal of fluoride from water. *Chem. Eng. J.* **2007**, *1*, 329–335. [CrossRef]
48. Gallios, G.P.; Tolkou, A.K.; Katsoyiannis, I.A.; Stefusova, K.; Vaclavikova, M.; Deliyanni, E.A. Adsorption of arsenate by nano scaled activated carbon modified by iron and manganese oxides. *Sustainability* **2017**, *9*, 1684. [CrossRef]
49. Sanou, Y.; Pare, S.; Nguyen, T.T.; Phuoc, N.V. Experimental and Kinetic modeling of As (V) adsorption on Granular Ferric Hydroxide and Laterite. *J. Environ. Treat. Tech.* **2016**, *4*, 62–70.
50. Nguyen, P.T.N.; Abella, L.C.; Gaspillo, P.D.; Hinode, H. Removal of arsenic from simulated groundwater using calcined laterite as the adsorbent. *J. Chem. Eng. Jpn.* **2011**, *44*, 411–419. [CrossRef]
51. Uddin, M.K. A review on the adsorption of heavy metals by clay minerals, with special focus on the past decade. *Chem. Eng. J.* **2017**, *308*, 438–462. [CrossRef]
52. Cheng, C.H.; Lehmann, J.; Engelhard, M.H. Natural oxidation of black carbon in soils: Changes in molecular form and surface charge along a climosequence. *Geochim. Cosmochim. Acta* **2008**, *72*, 1598–1610. [CrossRef]
53. Lawrinenko, M.; Laird, D.A. Anion exchange capacity of biochar. *Green Chem.* **2015**, *17*, 4628–4636. [CrossRef]
54. Schell, W.R.; Jordan, J.V. Anion-exchange studies of pure clays. *Plant Soil* **1959**, *10*, 303–318. [CrossRef]
55. Achour, S.; Youcef, L. Elimination du Cadmium Par Adsorption Sur Bentonites Sodique et Calcique. *Larhyss J.* **2003**, *2*, 68–81.

56. Youcef, L.; Achour, S. Elimination du cuivre par des procédés de précipitation chimique et d'adsorption. *Courr. Savoir* **2006**, *7*, 59–65.
57. Maji, S.K.; Pal, A.; Pal, T. Arsenic removal from real-life groundwater by adsorption on laterite soil. *J. Hazard. Mater.* **2008**, *151*, 811–820. [CrossRef]
58. Sorgho, B.; Paré, S.; Guel, B.; Zerbo, L.; Traoré, K.; Persson, I. Etude d'une argile locale du Burkina Faso à des fins de décontamination en Cu^{2+}, Pb^{2+} et Cr^{3+}. *J. Société Ouest-Africaine Chim.* **2011**, *31*, 49–59.
59. Alshaebi, F.Y.; Yaacob, W.Z.W.; Samsudin, A.R. Removal of arsenic from contaminated water by selected geological natural materials, Aust. *J. Basic Appl. Sci.* **2010**, *4*, 4413–4422.
60. Ghorbel-Abid, I.; Galai, K.; Trabelsi-Ayadi, M. Retention of chromium (III) and cadmium (II) from aqueous solution by illitic clay as a low-cost adsorbent. *Desalination* **2010**, *256*, 190–195. [CrossRef]
61. Tekin, N.; Kadinci, E.; Demirbaş, Ö.; Alkan, M.; Kara, A. Adsorption of polyvinylimidazole onto kaolinite. *J. Colloid Interface Sci.* **2006**, *296*, 472–479. [CrossRef]
62. Ayari, F.; Srasra, E.; Trabelsi-Ayadi, M. Characterization of bentonitic clays and their use as adsorbent. *Desalination* **2005**, *185*, 391–397. [CrossRef]
63. Kouadio, L.M.; Lebouachera, S.I.; Blanc, S.; Sei, J.; Miqueu, C.; Pannier, F.; Martinez, H. Characterization of Clay Materials from Ivory Coast for Their Use as Adsorbents for Wastewater Treatment. *J. Miner. Mater. Charact. Eng.* **2022**, *10*, 319–337. [CrossRef]
64. Zhang, X.; Hong, H.; Li, Z.; Guan, J.; Schulz, L. Removal of azobenzene from water by kaolinite. *J. Hazard. Mater.* **2009**, *170*, 1064–1069. [CrossRef] [PubMed]
65. Mostafapour, F.K.; Bazrafshan, E.; Farzadkia, M.; Amini, S. Arsenic Removal from Aqueous Solutions by Salvadora persica Stem Ash. *J. Chem.* **2013**, *2013*, 740847.
66. Ayach, J.; Duma, L.; Badran, A.; Hijazi, A.; Martinez, A.; Bechelany, M.; Baydoun, E.; Hamad, H. Enhancing Wastewater Depollution: Sustainable Biosorption Using Chemically Modified Chitosan Derivatives for Efficient Removal of Heavy Metals and Dyes. *Materials* **2024**, *17*, 2724. [CrossRef] [PubMed]
67. Khater, D.; Alkhabbas, M.; Al-Ma'abreh, A.M. Adsorption of Pb, Cu, and Ni Ions on Activated Carbon Prepared from Oak Cupules: Kinetics and Thermodynamics Studies. *Molecules* **2024**, *29*, 2489. [CrossRef]
68. Chuang, Y.; Chen, J.; Lu, J.; Su, L.; Jiang, S.Y.; Zhao, Y.; Lee, C.H.; Wu, Z.; Ruan, H.D. Sorption studies of Pb(II) onto montmorillonite clay. *IOP Conf. Ser. Earth Environ. Sci.* **2022**, *1087*, 012007. [CrossRef]
69. Pham, T.D.; Pham, T.T.; Phan, M.N.; Ngo, T.M.V.; Dang, V.D.; Vu, C.M. Adsorption characteristics of anionic surfactant onto laterite soil with differently charged surfaces and application for cationicdye removal. *J. Mol. Liq.* **2020**, *301*, 112456. [CrossRef]
70. Rathore, V.K.; Dohare, D.K.; Mondal, P. Competitive adsorption between arsenic and fluoride from binary mixture on chemically treated laterite. *J. Environ. Chem. Eng.* **2016**, *4*, 2417–2430. [CrossRef]
71. Chatterjee, S.; De, S. Application of novel, low-cost, laterite-based adsorbent for removal of lead from water: Equilibrium, kinetic and thermodynamic studies. *J. Environ. Sci. Health—Part A Toxic/Hazard. Subst. Environ. Eng.* **2016**, *51*, 193–203. [CrossRef]
72. Bakouan, C. Caractérisation de Quelques Sites Latéritiques du Burkina Faso: Application à L'élimination de L'arsenic (III) et (V) Dans Les Eaux Souterraines. Ph.D. Thesis, Université de Ouaga I Pr Joseph KI-ZERBO, Ouagadougou, Burkina Faso, Université de Mons, Mons, Belgium, 2018; pp. 1–241. Available online: https://orbi.umons.ac.be/bitstream/20.500.12907/31806/1/Th%C3%A8se (accessed on 1 February 2018).

minerals

MDPI

Article

Thermal Behaviour of a Carbonatic Clay: A Multi-Analytical Approach

Corina Ionescu [1], Viorica Simon [2,3], Volker Hoeck [4,†] and Ágnes Gál [5,*]

1 Electron Microscopy Center, Babeş-Bolyai University, 5-7 Clinicilor Str., 400006 Cluj-Napoca, Romania; corina.ionescu@ubbcluj.ro
2 Department of Physics, Babeş-Bolyai University, 1 Kogălniceanu Str., 400084 Cluj-Napoca, Romania; viosimon@phys.ubbcluj.ro
3 Institute of Interdisciplinary Research on Bio-Nano-Sciences, Babeş-Bolyai University, 1 Kogălniceanu Str., 400084 Cluj-Napoca, Romania
4 Division Geography and Geology, Paris Lodron University, 34 Hellbrunner Str., A-5020 Salzburg, Austria
5 Department of Geology, Babeş-Bolyai University, 1 Kogălniceanu Str., 400084 Cluj-Napoca, Romania
* Correspondence: agnes.gal@ubbcluj.ro; Tel.: +40-744644894
† 20 June 1943 Innsbruck–16 March 2025 Salzburg.

Abstract: A Miocene carbonatic clay quarried in Transylvania (Romania) has been used for more than 100 years for the production of traditional ceramic ware, bricks, and tiles. To investigate the mineralogical and microstructural changes of this clay when heated between 700 °C and 1200 °C, a combination of polarized light optical microscopy, X-ray powder diffraction, scanning electron microscopy coupled with energy dispersive X-ray spectrometry, and Fourier transform infrared spectroscopy was applied. Primary mineral phases such as illite, muscovite, feldspar, carbonate, Fe oxyhydroxides, and quartz undergo a gradual thermal alteration and form, besides a glassy phase, a wide range of minerals such as gehlenite, clinopyroxene, feldspar, maghemite, hematite, mullite, and α-cristobalite. These firing phases can be regarded as 'ceramic markers'. A comparison between the data obtained by several methods is discussed. The combination of the optical appearance and the microstructure on one side, and the specific associations of primary phases and ceramic markers on the other side, can be used as a 'ceramic thermometer' in estimating the firing temperature for ancient ceramics.

Keywords: carbonatic clay; thermal treatment; ceramic markers; Romania

1. Introduction

Argillaceous rocks or mudstones, commonly called 'clays', have been deeply involved in the history of humankind since the Late Paleolithic–Early Neolithic [1,2], in particular in the production of ceramics, "the first technological revolution in the human history" [3]. Objects made of ceramic material, such as pottery, statuettes, bricks, and tiles, bear information which can be used in archaeometry as well as in mineralogical and petrological studies. A wealth of literature deals with the mineralogical, chemical, and physical processes which are involved in the transformation of a mixture of clay and temper (e.g., ash, sand, shells, straw, chamotte etc.) into a coherent, strong, and insoluble ceramic body. The ceramic technology includes several stages, from extraction and preparation of the raw materials, modelling and shaping, surface treatment, decoration, painting and glazing, to drying and finally firing. Among these, of utmost importance for deciphering the technological level of an ancient society are the firing conditions.

The firing issues related to clay-based ceramics can be solved in several ways. The simplest is to study the ceramic sherds based on the assumption that mineralogical, physical, and chemical characteristics are the result of the technology applied to clayey raw materials. Another way is the experimental one, when ceramic clays are heated [4–7]. Special attention is given to the clays that can be used for ceramics without tempering because the latter significantly influence the final ceramic microstructure and composition [5,8–12]. Since prehistory, the illitic clays, either carbonatic or non-carbonatic, have been the most used material in producing pottery. Among many types of clays, the carbonatic ones have been most used for experiments because they are highly reactive and may produce upon heating a large range of Si-Ca, Si-Al-Ca, and Si-Al-Ca-Fe firing phases, such as wollastonite, gehlenite, clinopyroxenes, larnite, monticellite, garnet, and feldspars, among others. These firing minerals, as well as the newly-formed glassy phase, play a crucial role in connecting the clay components into a coherent ceramic material [13].

A large range of analytical methods can be applied to study the compositional and structural changes undergone by clays and to solve technological issues such as firing conditions for both modern and ancient ceramics [14]. Each of these analytical methods, taken alone, has advantages as well as limitations. For example, by means of optical microscopy (OM), essential information on the optical characteristics, mineralogical composition, granulometry, and texture are obtained [15–17]. Nevertheless, the results are limited by microscope resolution/magnification, the small size of the particles, and the covering effect of the finely distributed Fe compounds. The X-ray powder diffraction (XRPD) helps in identifying the mineral species, but it is not equally useful for low crystallinity and amorphous phases (for nomenclature and details, see [18–21]). Scanning electron microscopy coupled with energy dispersive X-ray spectrometry (SEM-EDX) is used to obtain data on the physical state and chemistry of a thin surface layer [22,23]. Being able to identify the vibrational frequencies of atomic bonds, the Fourier transform infrared spectroscopy (FTIR) is useful in identifying the crystalline and amorphous phases occurring in low quantity, below 1 vol.% [24]. FTIR is widely used to study both clay minerals [25,26] and mineral phases occurring in the ceramic body [19,24,27–32]).

The aim of this study was to find, experimentally, which are the structural and compositional changes which take place in a common clay upon heating, and allow obtaining high quality fine wares. The carbonatic illitic clay was extracted from a single quarry and was used without any temper or additives which might alter the normal path of thermal changes. However, "there is still much to be done to clarify what happens to a ceramic paste when it is fired" [10]. A combination of OM, XRPD, SEM-EDX, and FTIR has been applied in order: (i) to ascertain the characteristic mineral association for each temperature interval, (ii) to investigate the evolution of mineralogy and microstructure with increasing temperature, (iii) to define a 'ceramic thermometer' which can be used for ancient ceramics, and (iv) to demonstrate the advantages and limitations of analytical methods used in the study.

2. Materials and Analytical Methods

The studied Miocene clay deposits outcrop in the eastern part of the Transylvanian Basin (Romania) as a thick layer of several tens of meters [33]. The clay is quarried in an open pit at Bodogaia (coordinates: 46°16′15.9″ N and 24°59′3.24″ E), a village located 17 km east of the city of Sighişoara, and transported a couple of km to a small village where a brick factory has been functioning since 1905. Minimal enhancement of the clay starts with a light grinding aimed to disintegrate the larger clay lumps. After mixing with water, the clay is passed through a 1 mm-sieve to separate out the coarse clasts. Following filtering at 15 bars pressure to remove the added water, the material is finally packed under vacuum in

2, 5 and 10 kg blocks. The processed clay has a greenish-grey colour, an excellent plasticity, and a low shrinkage rate. It fires red when there is a constant supply of fresh air at the top of the kiln (oxidising atmosphere). When the kiln is kept closed until the end of firing, the atmosphere is reducing and the pottery is black. This clay is used for hand-made rustic style bricks and traditional pottery, as well as for art modelling. Chemically, the enhanced raw clay contains ~52 wt.% SiO_2, 15 wt.% Al_2O_3, 6 wt.% FeO_{TOT}, 2 wt.% MgO, 3 wt.% K_2O, 1 wt.% Na_2O, and 1 wt.% TiO_2. The loss of ignition is 13 wt.% [7]. As the limit between carbonatic and non-carbonatic clays is conventionally established at 6 wt% CaO [23], our Miocene clay (having 7 wt.% CaO) may be classified as (low)-carbonatic.

For the experiments, the raw clay was hand-shaped into nine briquettes, 5 × 5 × 1 cm in size, which were left for a couple of days to dry at room temperature. Subsequently the briquettes were heated in an electric Nabertherm furnace (Institute of Interdisciplinary Research on Bio-Nano-Sciences, Babeş-Bolyai University Cluj-Napoca), from 700 °C to 1200 °C, with a 10 °C/min rate and 2 h soaking time. The material of the heated clay briquettes is regarded here as 'ceramics'. The original greenish-grey colour of the raw clay changed upon heating into light reddish brown at 700 and 800 °C, red at 900 and 1000 °C, and again to reddish brown at 1100 °C [7]. Heated for 10 min at 1200 °C, the whole clay briquette melted, then solidified into a black glassy mass of 'ceramic slag' [34].

From a lump of raw clay impregnated with epoxy resin to avoid disintegration, as well as from each of the nine heated briquettes, a few mm thick slice was cut with a diamond saw in order to prepare standard petrographic thin sections for OM. The thin sections were investigated with an Axio Imager.A2m Zeiss transmitted light polarizing microscope (Electron Microscopy Center at Babeş-Bolyai University Cluj-Napoca). The images were captured with a Zen 2011 Axio high resolution digital video camera.

A few grams of each sample, hand-milled in an agate mortar, were analysed with a Bruker D8 Advance X-ray diffractometer with Bragg-Brentano geometry (Department of Geology, Babeş-Bolyai University in Cluj-Napoca). The equipment worked at 40 kV and 40 mA, with $CuK_{\alpha 1}$ radiation (λ = 1.5418 Å), a Ni filter, and a LynxEye one-dimensional detector. The data were collected between 5° and 65° 2θ, with 0.02° 2θ step interval and 2 s/step counting time. The mineral identification was based on the Bruker's Diffrac.Eva 2.1 software and the International Centre for Diffraction Data 2016 database [35]. Corundum NIST SRM1976a (US National Institute of Standards and Technology reference material [36]) served as lab internal standard. Oriented and ethylene glycol-treated raw clay sample were also subject to XRPD in order to distinguish between chlorite and smectite.

The rough surface of small chips taken from each sample was coated with a 20 nm thick layer of Au using a BALTEC-SCD-005 sputter. The JSM-6610 LV scanning electron microscope (University of Belgrade—Faculty of Mining and Geology) worked at an acceleration potential of 20 kV, a 5–10 µm diameter spot, and 50 s live time. The spectra were obtained by an Oxford X-Max energy dispersive spectrometer. The detection limit for most elements was ~0.1 wt.%.

The FTIR spectroscopy was performed at room temperature, with a Bruker Equinox 55 spectrophotometer (Institute of Interdisciplinary Research on Bio-Nano-Sciences at Babeş-Bolyai University Cluj-Napoca). Three milligrams of each powdered sample were mixed with 250 mg of spectroscopically-pure KBr, and analysed immediately after preparation. No hydration risk was encountered and consequently no supplementary thermal treatment was necessary. The spectra were recorded with a 2 cm^{-1} resolution, in the 400 cm^{-1} to 4000 cm^{-1} range. The experimental spectra were checked with fitted spectra using five and six component bands. The interpretation of FTIR data and the identification of mineral phases are based on [37–42], as well as the relevant literature.

3. Results

Firstly, the results obtained for the raw clay will be presented, in order to obtain a general image of the composition and structure of the starting material. The data obtained from the heated clay briquettes by using the same analytical methods as those involved in the raw material investigation will follow.

3.1. Raw Clay

3.1.1. Optical Microscopy

A groundmass of undifferentiated fine-grained phyllosilicates, as well as larger lamellae of muscovite, chlorite, and biotite, are the main mineral components of the raw clay (Figure 1a). The clay mineral species (the term 'clay mineral' refers to phyllosilicate minerals and to minerals which impart plasticity to clay and which harden upon drying or firing" [43]) could not be identified solely by OM, thus XRPD (Figure 1b) was also involved. Microscopically, quartz occurs as subangular grains, less than 20 µm in size. Micritic carbonate (<5 µm) is unevenly dispersed in the clay mass. Sparite carbonate (up to ~100 µm in size) is rarely found. Primary feldspar (feldspar$_p$ = Fsp$_p$) occurs as plagioclase and K-feldspar, about 20–25 µm in size. Brown pellets consisting of Fe oxyhydroxides, as well as rare grains of heavy minerals such as zircon, garnet, apatite, monazite and amphibole are also present. Sandstone, mudstone, biotite schist, quartzite, gneiss, and andesite clasts are extremely rare. The raw clay components are below 1 mm in size, as a result of grinding and sieving.

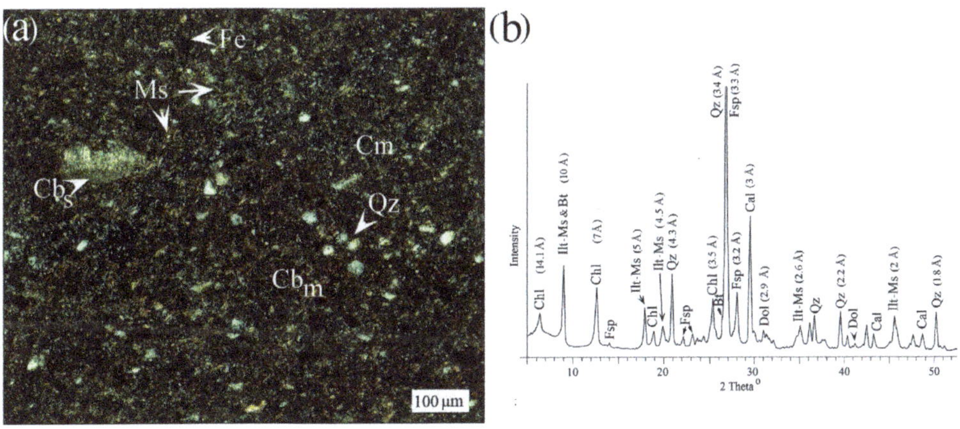

Figure 1. Raw clay composition: (**a**) Polarized-light microphoto showing a mass of fined-grained clay minerals (Cm) embedding fine grains of quartz (Qz), micrite carbonate (Cb$_m$), sparite carbonate (Cb$_s$), muscovite lamellae (Ms) and Fe-rich pellets (Fe) crossed polarizers. (**b**) X-ray diffractogram displaying peaks of quartz, feldspar (Fsp), calcite (Cal), illite-muscovite (Ilt-Ms), chlorite (Chl), biotite (Bt), and dolomite (Dol). Mineral abbreviations according to [44] are used throughout this paper.

3.1.2. X-Ray Powder Diffraction

All diffraction peaks displayed in Figure 1b are sharp and narrow, reflecting the crystalline state of the material composing the raw clay. The strongest peaks belong to quartz (3.4 Å, 4.3 Å, and 2.2 Å). The peaks at 10 Å, 5 Å, and 4.5 Å are common for illite and muscovite; therefore, they are treated together as 'illite-muscovite'. Feldspar (3.3 Å and 6.3 Å), calcite (3.0 Å and 2.1 Å), dolomite (2.9 Å), and biotite (3.4 Å) occur as well. The small 14.1 Å peak can be assigned to chlorite.

3.1.3. FTIR Spectroscopy

The FTIR absorption bands shown in Figure 2a,b originate from Si–O, Si–O–Si, Si–O–Al, Si–OH, Al–OH, and C–O vibrations [37–42,45–51] from various mineral phases (Table S1). Roughly, the signals can be grouped in four domains of frequency: (i) from 3700 to 3400 cm^{-1}, (ii) from 1200 to 875 cm^{-1}, (iii) from 800 to 625 cm^{-1}, and (iv) from 550 to 425 cm^{-1}.

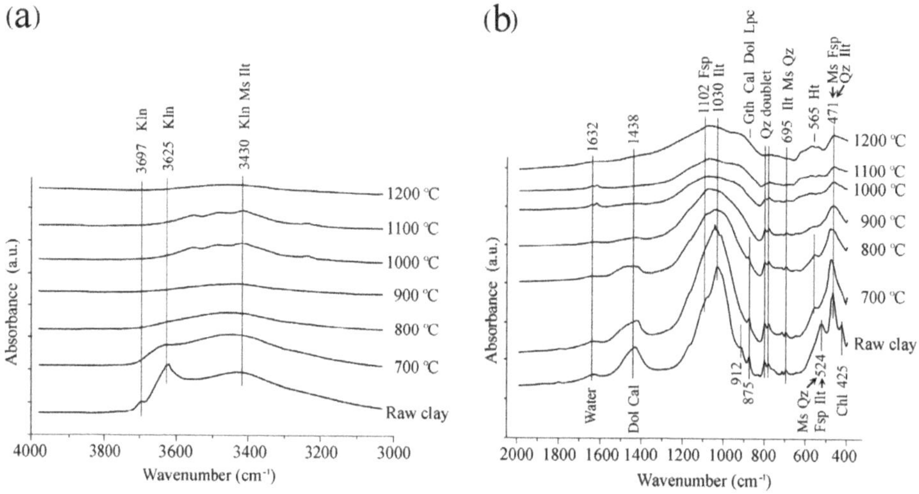

Figure 2. FTIR spectra of raw and heated clay in 4000 cm^{-1} to 3000 cm^{-1} (**a**), and 2000 cm^{-1} to 400 cm^{-1} (**b**) domains. Abbreviations as previously stated; additionally: Ilt for illite; Kln for kaolinite, Gth for goethite, Lpc for lepidocrocite; Cpx for clinopyroxene; Ht for hematite.

The main mineral phases are the same as those found by OM and XRD, i.e., quartz, feldspar, and carbonate, besides illite-muscovite. The FTIR allowed illite and kaolinite, but not muscovite and chlorite, to be identified.

The deconvolution of the most intense and sharp signal centred at 1030 cm^{-1} is due to Si-O-Al stretching vibrations and reveals the main contribution of illite, quartz, feldspar$_p$ and less kaolinite to the composition of the raw clay (Figure 3). The characteristic doublet at 798–779 cm^{-1}, as well as the signals at 524 cm^{-1} and 471 cm^{-1}, are assigned to the Si-O stretching bands of quartz [27,52]. The vibrations of other minerals are recorded at 1102 cm^{-1}, 524 cm^{-1} and 471 cm^{-1} (feldspar [53]); and at 1438 cm^{-1} and 875 cm^{-1} (carbonate [25,54]). The iron compounds are goethite and lepidocrocite, with vibrations at 912 cm^{-1} and 875 cm^{-1} [55]. The presence of kaolinite is shown also by signals at ~3697 and 3625 cm^{-1} [56]. The OH stretching bands are weak, except that at 3625 cm^{-1}, which might include also signals produced by absorbed water.

3.2. *Heated Clay*

3.2.1. Optical Microscopy

For the microscopic description of the heated clay (i.e., ceramics), we follow here [16,57,58], who defined 'matrix' as those parts of the ceramic body containing only particles smaller than ~15–20 μm. The 'clasts' or 'macroclasts' are non-plastic components larger than 15–20 μm. The matrix consists of clay minerals, muscovite, chlorite, and some biotite, as well as 'micro-clasts' i.e., tiny grains of quartz, feldspar$_p$, carbonate, and Fe compounds. The polarized light microphotos (Figure 4a–f) display a gradual change of the both the matrix and clasts, with increasing temperature.

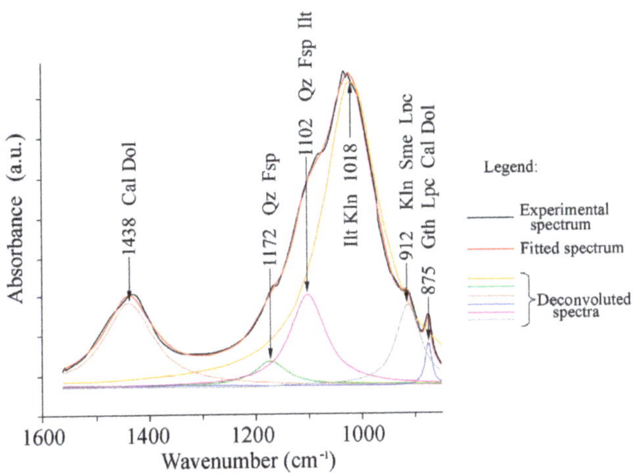

Figure 3. Deconvoluted FTIR spectrum of raw clay, in 1600–850 cm^{-1} range and according mineral assignement.

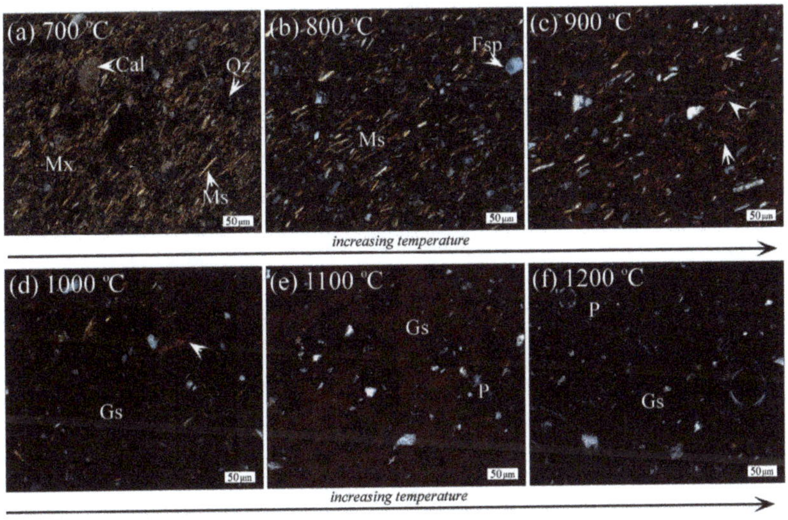

Figure 4. Polarized-light microphotos of clay briquettes heated: (**a**) at 700 °C: Birefringent matrix (Mx) containing clay minerals (yellow elongated grains), muscovite, quartz and calcite; (**b**) at 800 °C: Mixture of birefringent (lighter areas) and low birefringent matrix (darker areas), embedding micas and a plagioclase grain (dark grey) overgrown by K-feldspar; (**c**) at 900 °C: Formation of orange clusters; (**d**) at 1000 °C: Relics of quartz, muscovite and biotite (white arrows) in a mostly isotropic matrix; (**e**) at 1100 °C: Partly melted quartz grains in a glass (Gs) incorporating hematite and probably maghemite; (**f**) at 1200 °C: Glass and relics of quartz (white spots). Crossed polarizers for all images. Abbreviations as previously stated; additionally: Mx for matrix; Gs for glass; P for pores. Parallel orientation of lamellar minerals as seen in (**a**–**c**) is due to the pressure applied when shaping the clay briquettes.

At 700 °C, the matrix is still microcrystalline and consequently highly birefringent (Figure 4a). Quartz, illite, muscovite, feldspar$_p$, chlorite, biotite, and carbonates are not yet essentially affected by heating. Very thin, isotropic rims formed at the contact between phyllosilicates reflect a 'low degree' of sintering [59,60].

At 800 °C, the matrix consists of a mix of birefringent and low birefringent areas (Figure 4b). The latter mark the beginning of the collapse of illite and muscovite structure. Quartz, feldspar, illite, muscovite, rare chlorite, and biotite, as well as fine grains of carbonate are still easily distinguishable. Grains of plagioclase (feldspar$_p$) are overgrown by newly-formed K-feldspar (feldspar$_h$), a process characteristic for heating at temperatures over 800–850 °C [61,62]. The overall colour of the ceramic body is reddish, most likely due to a finely-distributed iron-rich phase.

At 900 °C, the isotropic areas are prevalent due to advanced collapse of the crystalline structure of most of the phyllosilicates and beginning of extended vitrification. Still, quartz, feldspar$_p$, and Fe-rich pellets, as well as scarce mica (muscovite and biotite), are present. A notable change is marked by the occurrence of bright orange small clusters with high relief and polycrystalline microstructure (Figure 4c), which can be related to newly formed phases such as gehlenite and pyroxene, coloured by Fe trapped in their structure [63].

Heated up to 1000 °C, the ceramic body shows a predominantly isotropic matrix (Figure 4d). Small grains of quartz and feldspar (most-likely primary and probably also newly-formed) are identifiable. The patchy appearance of some of the feldspar is due to their forming during heating [62]. "Ghost relics" of micas, showing only a slight contour, are characteristic.

At 1100 °C (Figure 4e), the matrix has a reddish colour, an overall vitreous appearance, and is optically completely isotropic. Dark brown spots replace the former Fe-pellets. The brownish-orange patches with diffuse rims observed in the glassy matrix are most likely relics of firing minerals such as gehlenite and clinopyroxene, coloured by Fe [63]. Quartz is relatively frequent. Feldspar$_p$ grains show melting at the rims, whereas feldspar$_h$ has a patchy appearance. The grains showing low refractive index and very low birefringence are probably α-cristobalite, a high temperature silica polymorph. The pores are irregular, larger, and more frequent than in ceramic material obtained at lower temperatures. The increased porosity is due to a gaseous phase released upon decomposition processes.

The vitreous material formed at 1200 °C (Figure 4f) is completely isotropic and contains isolated and very small fragments of quartz and feldspar (probably both primary and newly-formed). The pores are large and numerous.

3.2.2. X-Ray Powder Diffraction

The XRPD patterns of the heated clay (Figure 5) show both primary minerals (quartz, illite-muscovite, feldspar$_p$, calcite, chlorite, dolomite, and biotite), and numerous firing minerals (gehlenite, clinopyroxene, feldspar$_h$, hematite, maghemite, mullite, and α-cristobalite), as well as glass. The most important peaks belong to quartz, followed by illite-muscovite, feldspar, and calcite.

The 3.4, 4.3, 2.5, and 2.2 Å peaks of quartz dominate the diffractograms and remain almost unchanged up to 1000 °C, when they start to decrease in intensity, due to melting, chemical reactions as well as transformation into α-cristobalite (4.1 Å).

The 10, 5, 4.5, 2.5, 2.1, and 1.9 Å peaks which may be assigned to 'illite-muscovite' (Figure 1b) gradually decrease in intensity, up to complete disappearance between 900 and 1000 °C, when it records irreversible structural changes [64] and breaks down into alumina-rich and silica-rich domains [65]. Chlorite was not identified anymore in the briquette fired at 800 °C. The 3.4 Å peak of primary feldspar remains unchanged up to 900 °C. Starting with 1000 °C, this peak spectacularly increases, most likely due to the newly-formed feldspar.

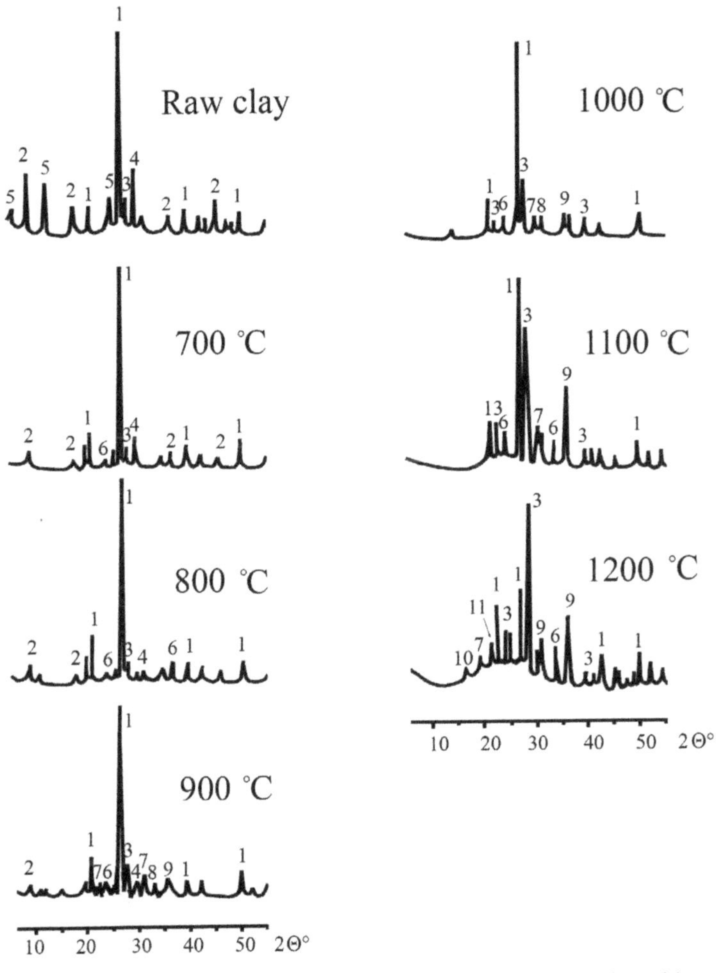

Figure 5. X-ray diffractograms of the raw clay and clay briquettes heated between 700 and 1200 °C. Legend: 1—quartz, 2—illite-muscovite, 3—feldspar, 4—carbonate, 5—chlorite, 6—hematite, 7—clinopyroxene, 8—gehlenite, 9—maghemite, 10—mullite, 11—α-cristobalite.

Calcite (3.0 Å, 2.1 Å, and 1.9 Å) and dolomite (2.9 Å and 2.2 Å) X-ray patterns show obvious alteration upon heating. At 800 °C, the calcite peak is low in intensity, whereas the dolomite peak does not appear anymore. At 900 °C, the carbonates are replaced by newly formed phases, such as gehlenite (1.9 Å) and clinopyroxene (2,9 and 3.0 Å). After 1000 °C the reaction between gehlenite and quartz led to the formation of feldspar [66]. The hematite (2.7 Å) and maghemite (2.5 Å) peaks were identified in the clay heated at 800 °C and 900 °C, respectively, and their intensity slowly increases with temperature. At 1100 °C, α-cristobalite peaks (4.1 Å) occur as well. Small amounts of mullite (3.4 Å–3.5 Å) were detected at 1100 °C and 1200 °C. The diffractograms of the material heated over 900 °C show a lump between 20 and 35 °2θ, due to amorphous material and glass.

3.2.3. SEM-EDX

The secondary electron images (Figure 6a–f) and energy dispersive spectra (Figure 7a,b) offer detailed information on the microstructure and mineralogy of the heated clay. The gradual thermal changes affecting the matrix range from: (i) 'low sintering'—when phyllosilicate and aplastic particles are glued together at intergranular boundary [67]—and (ii) 'incipient vitrification'—with thin filaments of glassy material occurring in isolated places—to (iii) 'partial vitrification'—when there are frequent and areas with a glassy material—and finally (iv) 'advanced vitrification'—where large parts of the ceramic body are transformed into a porous, net-like, filamentous vitreous material, with a smooth surface. The clay heated at 1200 °C (Figure 6f) records a so-called 'continuous vitrification' [23].

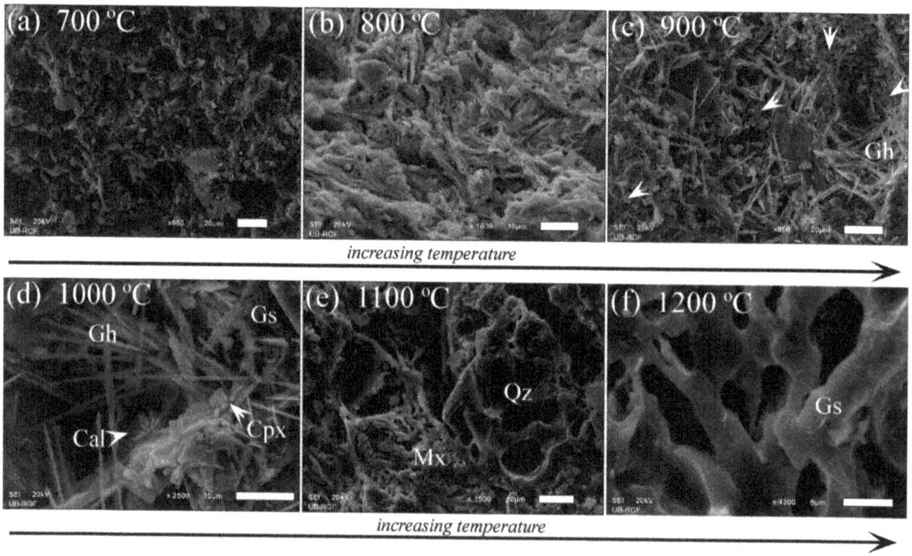

Figure 6. Secondary electron images (SEM) of heated clay samples, showing mineralogy and microstructure of heated clay: (**a**) Phyllosilicates displaying a low degree of sintering, at 700 °C; (**b**) sintering and incipient vitrification, at 800 °C; (**c**) partly glassy matrix (indicated by white arrows), with voids containing needle-like crystals of gehlenite, at 900 °C; (**d**) gehlenite, clinopyroxene, and calcite, at 1000 °C; (**e**) various vitrification stages of the matrix embedding partly melted quartz, at 1100 °C; (**f**) detailed image of glassy filaments formed at 1200 °C. Abbreviations as previously stated. Value of scale bars (white lines at bottom-right of each image) is: 20 μm for (**a**,**c**), 10 μm for (**b**,**d**,**e**), and 5 μm for (**f**).

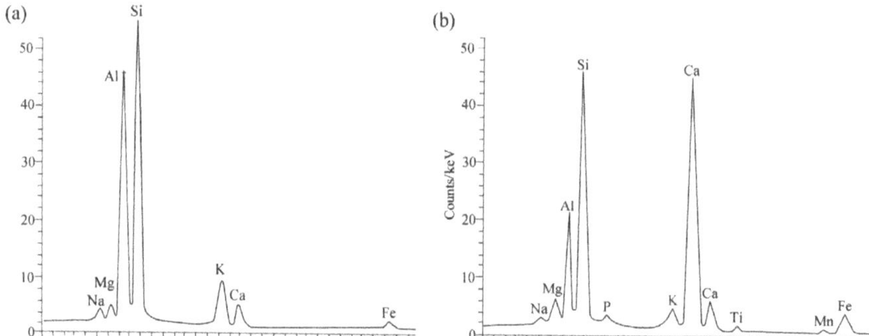

Figure 7. ED spectra of illitic matrix (**a**) and gehlenite (**b**) in clay briquettes heated at 1000 °C.

The samples heated at 700 and 800 °C show a microcrystalline structure, with phyllosilicates sintered together by narrow areas of unstructured material (Figure 6a,b). Between 900 and 1100 °C, the increased amount of glass and formation of new minerals are main processes. Gehlenite and clinopyroxene occur in the voids left by carbonate decomposition or are randomly distributed in the ceramic mass. Gehlenite forms tabular to needle-like crystals with a skeletal appearance, whereas clinopyroxene occurs as isometric grains (Figure 6c,d). The gehlenite and clinopyroxene crystal morphologies are similar to those described by [68].

At 1000 °C (Figure 6d), the increased amount of gehlenite is accompanied, besides clinopyroxene, by small prismatic crystals of feldspar$_h$ and needle-like crystals with Al-Si composition, most likely mullite. In between these phases there are bouquets of calcite crystals, resulting from recarbonation during cooling [69]. At 1100 °C, only 'ghosts' of former quartz (partly melted), muscovite, and feldspar$_p$ are preserved (Figure 6e). At 1200 °C, the ceramic consists of a mass of glassy filaments (Figure 6f).

Various amounts of Ca, Mg, and Fe, as displayed by the ED spectrum in Figure 7a, were measured in the K-rich alumosilicate matrix, as well in the glass. Both the matrix and the glass may contain a small amount of P. The variable amount of each element in the ED spectra of the newly-formed phases indicates complex compositions, and possible solid solutions. For example, the distribution of elements in the needle-like and isometric crystals shown in Figure 6d is compatible with gehlenite (Figure 7b) and clinopyroxene, respectively. The intermediate composition of Ca-rich plagioclase with some K content (1000 °C) is in agreement with their formation by firing [62].

3.2.4. FTIR Spectroscopy

In the following section, the thermal evolution of the FTIR signals and the assignment of mineral phases from the heated clay will be presented in detail. A gradual flattening coupled with widening, as well as the shifting of the main signals is observed in the spectra collected from the heated clay and displayed in Figure 2a,b and Figure 8a–f. The large band of Si-O stretching vibrations centred at 1030 cm^{-1}, is related to the main component of the clay i.e., illite, and appears at all heating temperatures.

The deconvolution for the 1600–850 cm^{-1} interval (Figure 8a–f) is in good agreement with the experimental and fitted spectra. Several newly-formed mineral phases were identified in the FTIR spectra of the heated clay: meta-clay (Si-O vibrations), hematite, gehlenite, feldspar$_h$, clinopyroxene, maghemite, mullite and glass (Figure 8a–f). The meta-clay is a 'pseudoamorphous phase' [52] which may have been formed already at 700 °C from a predominant illite-like clay mineral. Hematite appears before 700 °C, as a result of Fe-oxihydroxides and chlorite transformation. At 1200 °C, part of hematite transforms into spinel (1425 cm^{-1} signal).

The bands at 1126, 1029, and 1019 cm^{-1}, visible only between 1000 and 1200 °C (Figure 8d–f), can be linked to feldspar$_h$ formed by reaction between gehlenite and quartz. Figure 2b displays there a signal at 471 cm^{-1}, related to Si–O, Al–O deformation in several minerals such as illite-muscovite, feldspar, and quartz. This band gradually flattens and broadens with increasing temperature.

The quartz doublet at 798 and 779 cm^{-1} at 1000 °C coalescences into a single band (Figure 2b), which reflects the melting and formation of glass [70]. The quartz signals at 1172, 1101, and 524 cm^{-1} shift towards higher wave numbers and strongly diminish after 1100 °C. This is in agreement with the reducing of quartz grains size (OM), the diminishing of quartz peaks (XRPD), and their melting (SEM). Quartz melts and 'dissolves' into the matrix, it is consumed in reactions with calcite and clay minerals, and also transforms into α-cristobalite.

The 875 cm^{-1} band disappearing between 800 and 900 °C (Figure 2a,b) marks the destruction of calcite and dolomite [18]. The other calcite main signal, at 1438 cm^{-1}, flattens progressively and is no longer found at 900 °C (Figure 8b). Thus, an interval between

800 and 850 °C can be assigned to decomposition of primary calcite and forming of CaO—lime. The 1259, 1306, and 1384 cm^{-1} signals recorded above 900 °C (Figure 8c–e) can be assigned to recarbonation of lime [69].

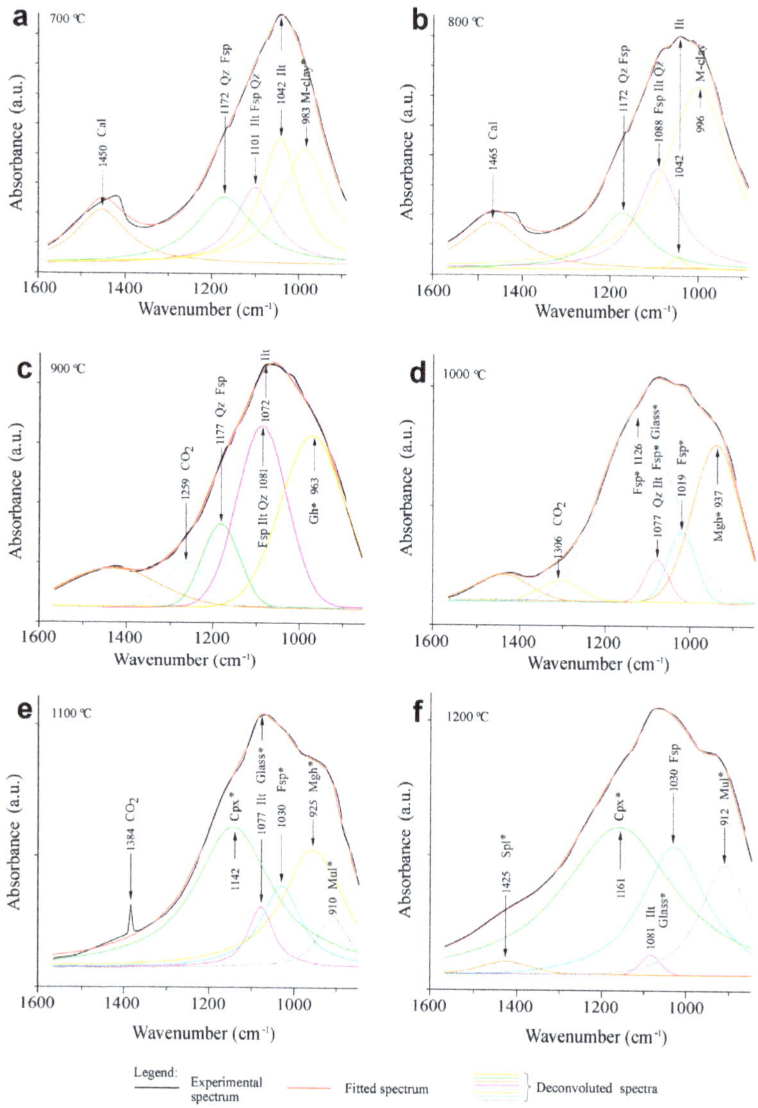

Figure 8. FTIR spectra of heated clay in the 1600–850 cm^{-1} range. Experimental, fitted, and deconvoluted spectra at 700 °C (**a**), 800 °C (**b**), 900 °C (**c**), 1000 °C (**d**), 1100 °C (**e**), and 1200 °C (**f**). Abbreviations as previously stated; additionally: M-clay for meta-clay; Spl for spinel. Newly-formed phases are marked by (*).

4. Discussion

Subject of a wealth of papers, e.g., ref. [10] and references therein, the firing constraints of the clay-based ceramics, in particular the ancient one, rely basically on thermal changes which include dehydration, dehydroxylation, structural breakdown, chemical reaction, recrystallisation, and forming new phases [10,65,71,72]. Each of these processes stretches over various temperature intervals, which may overlap.

Despite the low amount, the small grain size, the non-stoichiometric composition and often the deformed crystalline structure, the newly-formed phases may be regarded as 'ceramic markers' [73] and can be used in inferring technological constraints of ceramics [74]. Nevertheless, the temperature values given in the literature for the formation of the ceramic markers are stretched over large intervals. For example, gehlenite is thought to start forming at 800 °C [13,75], 850 °C [73], 900 °C [4], between 800 and 900 °C [76], or after 900 °C [18]. Onset of clinopyroxene formation is also recorded at various temperatures, i.e., 825 °C [77], 850 °C [78–80], and 900 °C [76]. The values given for the upper stability limit for these firing phases vary also largely. In a similar way, different formation/disappearance temperatures are recorded for hematite, feldspar, mullite, and α-cristobalite, among others. They may form successively or concomitantly, but may exist only within certain temperature limits.

Besides the technological choices such as kiln type, heating/cooling gradient, firing atmosphere, timing, loading or the location within the kiln, there are a high number of other variables influencing the final composition and the structure of a ceramic body. Among these variables, the most important are the mineralogy and chemistry of the raw materials, their homogeneity, possible mixing and refinement, the size of the components and the presence of temper. Thus, the resulting ceramic body may show an extremely wide range of possible compositions and structures.

Table S2 displays the correspondence between the optical characteristics of the matrix (Figure 4), the pattern of the illite-muscovite XRPD peaks (Figure 5), and the appearance of the ceramic body as seen in SEM images (Figure 6). Up to 800 °C, the ceramic material—still highly birefringent—shows sharp peaks corresponding to a crystalline structure. However, "the most significant textural and mineralogical changes are observed in samples with carbonates when fired at T > 800°C" [15].

The main physical and chemical reactions upon heating are linked to the phyllosilicates, e.g., illite and muscovite. As the temperature increases, after losing water (dehydration), the phyllosilicates start to dehydroxylate before 650–700 °C [72], and their structure collapses and grades into an amorphous state. Small amounts of glass occur around 800 °C [22]. The increase of the amount of glass at T > 900 °C is in good agreement with previous studies [75,81–83]. The illite-muscovite XRPD peaks diminish in intensity and disappear above 900 °C (see also [84]). At 1100 °C and 1200 °C, only FTIR shows small amounts of illite (Figure 8e,f), most likely present as relics trapped in the glassy mass.

Maghemite could not be detected by OM, but XRPD shows the presence of its peaks between 900 °C and 1200 °C. The FTIR recorded maghemite only at 1000 °C and 1100 °C. Although it can nucleate earlier [4,85], hematite was not detected by XRPD and FTIR before 800 °C, possibly due to its small grain size and low amount. The reddish hue of the matrix argues for the presence of hematite starting with 700 °C and its persistence up to 1200 °C.

The most significant reactions involve carbonates. The micrite carbonate (mostly calcite) decomposes already at 800–850 °C. To react and form Ca-Al-Si minerals (gehlenite, clinopyroxene and feldspar in our case), a temperature high enough to also decompose the clay minerals—the source for Si and Al [86]—is necessary. The minerals resulting from the reaction between calcite, clay minerals, and quartz have a high impact in assessing the firing temperature.

Illite and muscovite in contact with quartz leads to the formation of K-feldspar [58,62] as well as some mullite after 900 °C. This is in agreement with the obvious increase of the XRPD peaks at 1000 °C (Figure 5). Formation of K-feldspar after 800–850 °C [61,62] is proved also by primary feldspar grains overgrown by K-feldspar coronas (Figure 4b).

As gehlenite and clinopyroxene have been found at 900 °C, most likely they start to form before, e.g., between 800 °C [13] and 850 °C (see also [71–73,79,80,87]). According to [83], gehlenite results by the reaction between illite and calcite, whereas clinopyroxene (diopside, augite, hedenbergite) may form by reaction between calcite, dolomite, and quartz. Gehlenite and clinopyroxene can be found until a maximum temperature of 950–1000 °C [52]. Nevertheless, traces of clinopyroxene have been detected by FTIR in the clay heated at 1100 °C and 1200 °C. The disappearance of gehlenite over 1000 °C can be explained in two ways: it reacts with quartz, resulting in wollastonite and feldspar anorthite [83], or it reacts with the Si-Al melt resulting from decomposition of clay minerals, and thus leading also to feldspar anorthite [83]. However, no traces of wollastonite have been detected by any of the techniques used. The large-scale formation of feldspar is also marked by the strong increase of the 3.4 Å XRPD peak of feldspar with 1100 °C. Most feldspar detected above 1000 °C (Figure 5) is most likely have been formed by the reaction between calcite and illite, not from gehlenite.

Feldspar is an important primary component of the raw clay and its 1172 cm^{-1} band remains almost unchanged upon heating until after 900 °C (Figure 8a–c). The 1101 cm^{-1} band, which shifts towards lower wavenumber (1088 cm^{-1}) and disappears beyond 1100 °C, is related to newly-formed feldspar. The newly-formed Ca-rich feldspar (anorthite) on muscovite expenses, feldspar, and small amounts of mullite and feldspar may appear also before 1000 °C (Figure 5).

Quartz is partly consumed by melting and by reactions with illite and calcite. Only the XRPD (Figure 5) recorded the transformation of quartz into α-cristobalite 1100 °C (see also [83]).

The FTIR investigation show that it is complementary to other techniques, e.g., XRPD. The reducing of the intensity of FTIR bands and their shift with increasing temperature is to be expected (see also [41,45]). The structural changes of the illite-like clay mineral as observed by means of FTIR cover a wide range of temperatures. The large band centred at 1030 cm^{-1} which is intense and narrow at 700 °C, with increasing temperature flattens and shifts to higher wavelengths (1082 cm^{-1}). The destruction of muscovite, evidenced by the diminishing of the 3430 and 471 cm^{-1} signals, takes place at higher temperatures, specifically between 900 °C and 1100 °C. The band at 3430 cm^{-1} (Figure 2a,b), which was still recorded from the OH vibrations at higher temperatures (1200 °C), can be due to the short time of heating [80], which allowed relic illite and muscovite to be trapped in the glass. Chlorite survived only until 800 °C.

The above-described processes are well illustrated by the secondary electron images obtained by SEM. Mineral phases and glass have been identified and their association and distribution proved to be specifically related to the temperature. Due to the compositional inhomogeneity of the clay (Figure 1a), the 'ceramic markers' resulting upon firing are unevenly distributed in the heated clay mass, some of them being found mainly in the voids left by the decomposed carbonate. An unstructured material ranging from intergranular sintered material up to a glassy material connects primary compounds as well as the ceramic markers.

Despite some differences, which are due to the method and the amount and size of the components, the data obtained by several methods are consistent (Figure 9) and correlates well. This allows drawing an indicative frame in estimating the firing temperature undergone by a clay-based ceramic body (Table S2).

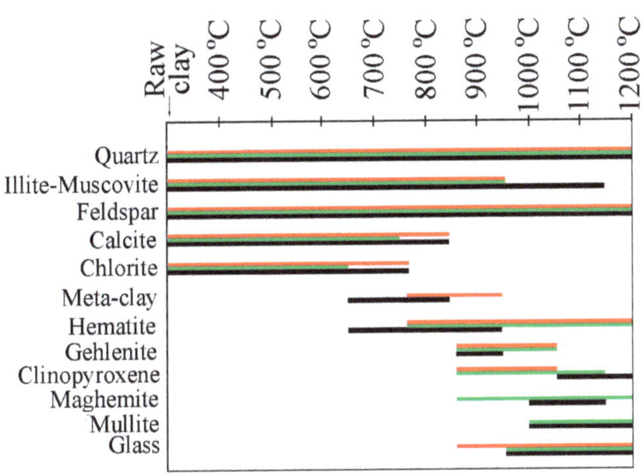

Figure 9. Comparative chart showing sequential presence of compounds as differently recorded by OM (red bar), XRPD (green bar), and FTIR (black bar), in raw and heated clay.

5. Conclusions

The combination of various analytical techniques contributes to a better understanding of small-scale processes occurring in a carbonatic clay when heated up to 1200 °C. The resulting ceramic body consists of a sintered-to-glassy network that glues together both relics of primary and newly formed minerals. The weight of various phases as detected by different techniques may slightly differ. Some of the methods (FTIR) revealed the survival of phyllosilicates at high temperature (1100 °C). Overall, the most significant structural and mineralogical changes and the formation/disappearance of the 'ceramic markers' are recorded within the 850 °C to 950 °C temperature interval.

The corroboration of the optical characteristics and microstructure of the matrix on one side, and certain associations of primary phases and 'ceramic markers' on the other side, may be used as a 'ceramic thermometer' in inferring the technological conditions for clay-based archaeological ceramics. Even more, this particular clay bears a high potential as a reference material for ethno-archaeological experiments.

Supplementary Materials: The following supporting information can be downloaded at: https://www.mdpi.com/article/10.3390/min15040390/s1, Table S1: Raw and heated clay. Wavenumber (in cm^{-1}), infrared bands (vibrations), and mineral assignment; based on [45–51]; Table S2: Summary of phase characteristics for heated clay briquettes obtained from carbonatic illitic clay, as determined by multi-analytical investigation. Shadowed rows mark intervals with the most significant thermal changes.

Author Contributions: Conceptualization: C.I., V.H. and V.S.; data curation: C.I.; formal analysis: C.I., V.H., V.S. and Á.G.; funding acquisition, C.I., Á.G.; methodology, C.I. and V.H. (OM, SEM), V.S. (FTIR), Á.G. (XRPD); C.I. and V.H. are responsible for OM and SEM data acquisition and interpretation of results, V.S. is responsible for FTIR data acquisition and interpretation of results, and Á.G. is responsible for XRPD sample preparation, data acquisition, and interpretation of results. C.I. and V.H. wrote the original and revised manuscript. All authors have read and agreed to the published version of the manuscript.

Funding: The study was financed by the UEFISCDI/CNCS projects PN-II-ID-PCE-2011-3-0881 (FTIR data) and PN-III-ID-PCE-2016-0229 (OM, XRPD, SEM-EDX investigations, as well as the manuscript editing) granted to C.I., in the framework of the Excellence Research Program of the Romanian

Ministry of Education and Research. C.I. and Á.G. also acknowledge support of the Babeş-Bolyai University (Grants ACG 31119/2024 and AGC30938/2024, respectively).

Data Availability Statement: The data presented in this study are partly available in Supplementary Material; all other data are available on request from the corresponding author.

Acknowledgments: Suzana Erić from University of Belgrade is thanked for the help with SEM-EDX investigations, and Monica Mereu from Babeş-Bolyai University Cluj-Napoca for the computer-assisted graphics. The anonymous reviewers are sincerely thanked for their comments and suggestions which significantly helped to improve the quality of the manuscript.

Conflicts of Interest: The authors declare no conflicts of interest. The funders had no role in: (i) the design of the study; (ii) the collection, analyses, or interpretation of data; (iii) the writing of the manuscript, or (iv) the decision to publish the results.

Abbreviations

The following abbreviations are used in this manuscript:

Cm	clay minerals
Qz	quartz
Cb_m	micrite carbonate
Cb_s	sparite carbonate
Ms	muscovite
Fsp	feldspar
Cal	calcite
Ilit-Ms	illite-muscovite
Chl	chlorite
Bt	biotite
Dol	dolomite
Kln	kaolinite
Gth	goethite
Lpc	lepidocrocite
Gs	glass
P	pore
Mx	matrix
Gh	gehlenite
Crs	α-cristobalite
Cpx	clinopyroxene
Mgh	maghemite
Ht	hematite
Ilt	illite
Mul	mullite
Spl	spinel
M-clay	meta-clay
Fsp_p	primary feldspar
Fsp_h	newly-formed feldspar

References

1. Vandiver, P.B.; Soffer, O.; Klima, B.; Svoboda, J. The origins of ceramic technology at Dolni Věstonice, Czechoslovakia. *Science* **1989**, *246*, 1002–1008. [CrossRef]
2. Shelach, G. On the invention of pottery. *Science* **2012**, *336*, 1644–1645. [CrossRef]
3. Maniatis, Y. The emergence of ceramic technology and its evolution as revealed with the use of scientific techniques. In *From Mine to Microscope: Advances in the Study of Ancient Technology*; Shortland, A.J., Freestone, I.C., Rehren, T., Eds.; Oxbow Books: Oxford, UK, 2009; pp. 11–27.
4. Maniatis, Y.; Simopoulos, A.; Kostikas, A. Moessbauer study of the effect of calcium content on iron oxide transformations in fired clays. *J. Am. Cer. Soc.* **1981**, *64*, 263–269. [CrossRef]

5. Müller, N.S.; Vekinis, G.; Kilikoglou, V. Impact resistance of archaeological ceramics: The influence of firing and temper. *J. Archaeol. Sci. Rep.* **2006**, *7*, 519–525. [CrossRef]
6. Tămăşan, M.; Mocuţa, H.; Ionescu, C.; Simon, V. Structural changes induced in mineral clays by high temperature heat treatments. *Stud. Univ. Babes Bolyai Phys.* **2009**, *54*, 3–8.
7. Ionescu, C.; Hoeck, V.; Gruian, C.; Simon, V. Insights into the EPR characteristics of heated carbonate-rich illitic clay. *Appl. Clay Sci.* **2014**, *97–98*, 138–145. [CrossRef]
8. Tite, M.S.; Kilikoglou, V.; Vekinis, G. Strength, toughness and thermal shock resistance of ancient ceramics, and their influence on technological choice. *Archaeometry* **2001**, *43*, 301–324. [CrossRef]
9. Hein, A.; Müller, N.; Day, P.M.; Kilikoglou, V. Thermal conductivity of archaeological ceramics: The effect of inclusions, porosity and firing temperature. *Thermochim. Acta* **2008**, *480*, 35–42. [CrossRef]
10. Gliozzo, E. Ceramic technology. How to reconstruct the firing process. *Archaeol. Antropol. Sci.* **2020**, *12*, 260. [CrossRef]
11. Neff, H.; Bishop, L.; Sayre, E.V. Simulation approach to the problem of tempering in compositional studies of archaeologial ceramics. *J. Archaeol. Sci.* **1988**, *15*, 159–172. [CrossRef]
12. Grifa, C.; Cultrone, G.; Langella, A.; Mercurio, M.; De Bonis, A.; Sebastián, E.; Morra, V. Ceramic replicas of archaeological artefacts in Benevento area (Italy): Petrophysical changes induced by different proportions of clays and temper. *Appl. Clay Sci.* **2009**, *46*, 231–240. [CrossRef]
13. Cultrone, G.; Rodriguez-Navarro, C.; Sebastian, E.; Cazalla, O.; De la Torre, M.J. Carbonate and silicate phase reactions during ceramic firing. *Eur. J. Miner.* **2001**, *13*, 621–634. [CrossRef]
14. Heimann, R. Firing technologies and their possible assessment by modern analytical method. In *Archaeological Ceramics*; Olin, J.S., Franklin, A.D., Eds.; Smithsonian Institution Press: Washington, DC, USA, 1982; pp. 89–96.
15. Shepard, A.O. Ceramics for the Archaeologist. Carnegie Institute: Washington, DC, USA, 1976; p. 414.
16. Maggetti, M. Phase analysis and its significance for technology and origin. In *Archaeological Ceramics*; Olin, J.S., Franklin., A.D., Eds.; Smithsonian Institution Press: Washington, DC, USA, 1982; pp. 121–133.
17. Degryse, P.; Braekmans, D. Petrography: Optical Microscopy. In *The Oxford Handbook of Archaeological Ceramic Analysis*; Hunt, A.M.W., Ed.; Oxford University Press: Oxford, UK, 2017; pp. 288–304.
18. Shoval, S. Mineralogical changes upon heating calcitic and dolomitic marl rocks. *Thermochim. Acta* **1988**, *135*, 243–252. [CrossRef]
19. Shoval, S. Using FT-IR spectroscopy for study of calcareous ancient ceramics. *Opt. Mater.* **2003**, *24*, 117–122. [CrossRef]
20. Heimann, R. X-ray powder diffraction (XRPD). In *The Oxford Handbook of Archaeological Ceramic Analysis*; Hunt, A.M.W., Ed.; Oxford University Press: Oxford, UK, 2017; pp. 288–304. [CrossRef]
21. Miras, A.; Galán, E.; González, I.; Romero-Baena, A.; Martín, D. Mineralogical evolution of ceramic clays during heating. An ex/in situ X-ray diffraction method comparison study. *Appl. Clay Sci.* **2018**, *161*, 176–183. [CrossRef]
22. Tite, M.S.; Maniatis, Y. Examination of ancient pottery using the scanning electron microscope. *Nature* **1975**, *257*, 122–123. [CrossRef]
23. Maniatis, Y.; Tite, M.S. Technological examination of Neolithic-Bronze Age pottery from central and southeast Europe and from the Near East. *J. Archaeol. Sci.* **1981**, *8*, 59–76. [CrossRef]
24. Shoval, S. Fourier transform infrared spectroscopy (FT-IR) in archaeological ceramic analysis. In *The Oxford Handbook of Archaeological Ceramic Analysis*; Hunt, A.M.W., Ed.; Oxford University Press: Oxford, UK, 2017; pp. 288–304. [CrossRef]
25. Madejová, J.; Komadel, P. Baseline studies of the clay minerals society source clays: Infrared methods. *Clays Clay Miner.* **2001**, *49*, 410–432. [CrossRef]
26. Madejová, J. FTIR techniques in clay mineral studies. *Vibr. Spectr.* **2003**, *31*, 1–10. [CrossRef]
27. De Benedetto, G.E.; Laviano, R.; Sabbatini, L.; Zambonin, P.G. Infrared spectroscopy in the mineralogical characterization of ancient pottery. *J. Cult. Herit.* **2002**, *3*, 177–186. [CrossRef]
28. Barone, G.; Crupi, V.; Galli, S.; Majolino, D.; Migliardo, P.; Venuti, V. Spectroscopic investigation of Greek ceramic artefacts. *J. Mol. Str.* **2003**, *651–653*, 449–458. [CrossRef]
29. Barone, G.; Crupi, V.; Longo, F.; Majolino, D.; Mazzoleni, P.; Tanasi, D.; Venuti, V. FT-IR spectroscopic analysis to study the firing processes of prehistoric ceramics. *J. Mol. Struct.* **2011**, *993*, 147–150. [CrossRef]
30. Shoval, S.; Beck, P. Thermo-FTIR spectroscopy analysis as a method of characterizing ancient ceramic technology. *J. Therm. Anal. Calorim.* **2005**, *82*, 609–616.
31. Kaufhold, S.; Hein, M.; Dohrmann, R.; Ufer, K. Quantification of the mineralogical composition of clays using FTIR spectroscopy. *Vibr. Spectr.* **2012**, *59*, 29–39. [CrossRef]
32. Germinario, C.; Cultrone, G.; De Bonis, A.; Izzo, F.; Langella, A.; Mercurio, M.; Morra, V.; Santoriello, A.; Siano, S.; Grifa, C. The combined use of spectroscopic techniques for the characterisation of Late Roman common wares from Benevento (Italy). *Measurement* **2018**, *114*, 515–525. [CrossRef]
33. Răileanu, G.; Marinescu, F.; Popescu, A. *Geological Map of Romania, 1:200000 Scale*; Geological Institute of Romania: Bucharest, Romania, 1968.

34. Hoeck, V.; Ionescu, C.; Metzner-Nebelsick, C.; Nebelsick, L.D. Mineralogy of the ceramic slags from the Bronze Age funerary site in Lăpuş, NW Romania. *Geol. Quat.* **2012**, *56*, 649–664. [CrossRef]
35. International Centre for Diffraction Data. Available online: https://www.icdd.com (accessed on 8 October 2024).
36. US National Institute of Standards and Technology Reference Material. Available online: www.nist.gov/programs-projects/powder-diffraction-srms (accessed on 8 October 2024).
37. Moenke, H. *Spektralanalyse von Mineralien und Gesteinen*; Akademische Verlagsgesellschaft Geest & Portig K.G.: Leipzig, Germany, 1962; p. 222. [CrossRef]
38. Moenke, H. Silica, the three-dimensional silicates, borosilicates, and beryllium silicates. In *The Infrared Spectra of Minerals*; Farmer, V.C., Ed.; Mineralogical Society of London: London, UK, 1974; pp. 365–382. [CrossRef]
39. Farmer, V.C. Vibrational spectroscopy in mineral chemistry. In *The Infrared Spectra of Minerals*; Farmer, V.C., Ed.; Mineralogical Society of London: London, UK, 1974; Monograph 4; pp. 1–10. [CrossRef]
40. Farmer, V.C. The layer silicates. In *The Infrared Spectra of Minerals*; Farmer, V.C., Ed.; Mineralogical Society of London: London, UK, 1974; Monograph 4; pp. 331–363. [CrossRef]
41. Freund, F. Ceramics and thermal transformations of minerals. In *The Infrared Spectra of Minerals*; Farmer, V.C., Ed.; Mineralogical Society of London: London, UK, 1974; Monograph 4; pp. 465–482. [CrossRef]
42. Kiefer, S.W. Thermodynamics and lattice vibrations of minerals: 2. Vibrational characteristics of silicates. *Rev. Geophys. Space Phys.* **1979**, *17*, 20–34. [CrossRef]
43. Guggenheim, S.; Martin, R.T. Definition of clay and clay mineral: Joint report of AIPEA and CMS nomenclature committees. *Clay Miner.* **1995**, *30*, 257–259. [CrossRef]
44. Warr, L.N. IMA–CNMNC approved mineral symbols. *Miner. Mag.* **2021**, *85*, 291–320. [CrossRef]
45. Chester, R.; Elderfiel, H. An infrared study of clay minerals, 2. The identification of kaolinite-group clays in deep-sea sediments. *Chem. Geol.* **1973**, *12*, 281–288. [CrossRef]
46. Rochester, C.H.; Topham, S.A. Infrared study of surface hydroxyl groups on goethite. *J. Chem. Soc. Faraday Trans. 1* **1979**, *75*, 591–602.
47. Skogby, H.; David, R.; Bell, D.R.; Rossman, G.R. Hydroxide in pyroxene: Variations in the natural environment. *Am. Miner.* **1990**, *75*, 764–774.
48. Weckler, B.; Lutz, H.D. Lattice vibration spectra. Part XCV. Infrared spectroscopic studies on the iron oxide hydroxides goethite (α), akaganéite (β), lepidocrocite (γ), and feroxyhite (δ). *Eur. J. Solid State Inorg. Chem.* **1998**, *35*, 531–544. [CrossRef]
49. Vargas, M.A.; Diosa, J.E.; Mosquera, E. Data on study of hematite nanoparticles obtained from Iron(III) oxide by the Pechini method. *Data Brief* **2019**, *25*, 104183. [CrossRef]
50. Deju, R.; Mazilu, C.; Stanculescu, I.; Tuca, C. Fourier transform infrared spectroscopic characterization of thermal treated kaolin. *Rom. Rep. Phys.* **2020**, *72*, 806.
51. Sočo, E.; Domoń, A.; Papciak, D.; Michel, M.M.; Pająk, D.; Cieniek, B.; Azizi, M. Characteristics of Adsorption/Desorption Process on Dolomite Adsorbent in the Copper(II) Removal from Aqueous Solutions. *Materials* **2023**, *16*, 4648. [CrossRef]
52. Shoval, S.; Yadin, E.; Panczer, G. Analysis of thermal phases in calcareous Iron Age pottery using FT-IR and Raman spectroscopy. *J. Therm. Analys. Calorim.* **2011**, *104*, 515–525. [CrossRef]
53. Hlavay, J.; Jonas, K.; Elek, S.; Inczedy, J. Characterization of the particle size and the crystallinity of certain minerals by IR spectrophotometry and other instrumental methods—II. Investigations on quartz and feldspar. *Clays Clay Miner.* **1978**, *26*, 139–143. [CrossRef]
54. Fabbri, B.; Gualtieri, S.; Shoval, S. The presence of calcite in archaeological ceramics. *J. Eur. Ceram. Soc.* **2014**, *34*, 1899–1911. [CrossRef]
55. Delineau, T.; Allard, T.; Muller, J.-P.; Barges, O.; Yvon, J.; Cases, J.-M. FTIR Reflectance vs. EPR studies of structural iron in kaolinites. *Clays Clay Miner.* **1994**, *42*, 308–320. [CrossRef]
56. Vaculícová, L.; Plevová, E. Identification of clay minerals and micas in sedimentary rocks. *Acta Geodynam. Geomater.* **2005**, *2*, 167–175.
57. Maggetti, M. Mineralogisch-petrographische Untersuchung des Scherbenmaterials der urnenfelderzeitlichen Siedlung Elchinger Kreuz, Ldkr. Neu-Ulm/Donau. *Kat. Prähistorischen Staatssamml.* **1979**, *19*, 141–172.
58. Ionescu, C.; Hoeck, V.; Ghergari, L. Electron microprobe analysis of ancient ceramics: A case study from Romania. *Appl. Clay Sci.* **2011**, *53*, 466–475. [CrossRef]
59. Gliozzo, E.; Fortina, C.X.; Memmi Turbanti, I.; Turchiano, M.; Volpe, G. Cooking and painted ware from San Giusto (Lucera, Foggia): The production cycle, from the supply of raw materials to the commercialization of products. *Archaeometry* **2005**, *47*, 13–29. [CrossRef]

60. Ionescu, C.; Hoeck, V.; Simon, V. Effect of the temperature and the heating time on the composition of an illite-rich clay: An XRPD study. *Studia Univ. Babeș-Bolyai Phys.* **2011**, *56*, 69–78.
61. Duminuco, P.; Messiga, B.; Riccardi, M.P. Firing process of natural clays: Some microtextures and related phase compositions. *Thermochim. Acta* **1998**, *321*, 185–190. [CrossRef]
62. Ionescu, C.; Hoeck, V. Firing-induced transformations in Copper Age ceramics from NE Romania. *Eur. J. Miner.* **2011**, *23*, 937–958. [CrossRef]
63. Molera, J.; Pradell, T.; Vendrell-Saz, M. The colours of Ca-rich ceramic pastes: Origin and characterization. *Appl. Clay Sci.* **1998**, *13*, 187–202. [CrossRef]
64. Carroll, D.L.; Kemp, T.F.; Bastow, T.J.; Smith, M.E. Solid-state NMR characterisation of the thermal transformation of a Hungarian white illite. *Solid State Nucl. Magn. Reson.* **2005**, *28*, 31–43. [CrossRef]
65. Drachman, S.R.; Roc, G.E.; Smith, M.E. Solid state NMR characterisation of the thermal transformation of Fuller's Earth. *Solid State Nucl. Magn. Reson.* **1997**, *9*, 257–267. [CrossRef]
66. Amadori, M.L.; Pallante, P.; Fermo, P.; Emami, M.A.; Chaverdi, A.A.; Callieri, P.; Matin, E. Advances in Achaemenid brick manufacturing technology: Evidence from the monumental gate at Tol-e Ajori (Fars, Iran). *Appl. Clay Sci.* **2018**, *152*, 131–142. [CrossRef]
67. Bordia, R.K.; Kang, S.-J.L.; Olevsky, E.A. Current understanding and future research directions at the onset of the next century of sintering science and technology. *J. Am. Cer. Soc.* **2017**, *100*, 2314–2352. [CrossRef]
68. Gál, Á.; Ionescu, C.; Bajusz, M.; Codrea, V.A.; Hoeck, V.; Barbu-Tudoran, L.; Simon, V.; Mureşan-Pop, M.; Csók, Z. Composition, technology and provenance of Roman pottery from Napoca (Cluj-Napoca, Romania). *Clay Miner.* **2018**, *53*, 621–641. [CrossRef]
69. Shoval, S.; Gaft, M.; Beck, P.; Kirsh, Y. Thermal behaviour of limestone and monocrystalline calcite tempers during firing and their use in ancient vessels. *J. Therm. Analys.* **1993**, *40*, 263–273. [CrossRef]
70. Berna, F.; Behar, A.; Shahack-Gross, R.; Berg, J.; Boaretto, E.; Gilboa, A.; Sharon, I.; Shalev, S.; Shilstein, S.; Yahalom-Mack, N.; et al. Sediments exposed to high temperatures: Reconstructing pyrotechnological processes in late Bronze and Iron Age Strata at Tel Dor (Israel). *J. Archaeol. Sci.* **2007**, *34*, 358–373. [CrossRef]
71. Guggenheim, S.; Chang, Y.-W.; van Groos, K. Muscovite dehydroxylation: High-temperature studies. *Am. Mineral.* **1987**, *72*, 537–550.
72. Abbott, R.N. Energy calculations on the dehydroxylation of muscovite. *Can. Miner.* **1994**, *32*, 87–92.
73. Padeletti, G.; Fermo, P. A scientific approach to the attribution problem of renaissance ceramic productions based on chemical and mineralogical markers. *Appl. Phys. A* **2010**, *100*, 771–784. [CrossRef]
74. Viani, A.; Cultrone, G.; Sotiriadis, K.; Ševčík, R.; Šašek, P. The use of mineralogical indicators for the assessment of firing temperature in fired-clay bodies. *Appl. Clay Sci.* **2018**, *163*, 108–118. [CrossRef]
75. Dondi, M.; Ercolani, G.; Guarini, G.; Marsigli, M.; Venturi, I. Evolluzione della microstruttura durante la cottura rapida di impasti per piastrelle porose. *Ceramurgia* **1995**, *25*, 301–314.
76. Trindade, M.J.; Dias, M.I.; Coroado, J.; Rocha, F. Mineralogical transformations of calcareous rich clays with firing: A comparative study between calcite and dolomite rich clays from Algarve, Portugal. *Appl. Clay Sci.* **2009**, *42*, 345–355. [CrossRef]
77. Maritan, L.; Nodari, L.; Mazzoli, C.; Milano, A.; Russo, U. Influence of firing conditions on ceramic products: Experimental study on clay rich in organic matter. *Appl. Clay Sci.* **2006**, *31*, 1–15. [CrossRef]
78. Dondi, M.; Ercolani, G.; Fabbri, B.; Marsigli, M. An approach to the chemistry of pyroxenes formed during firing of Ca-rich silicate ceramics. *Clay Miner.* **1999**, *33*, 443–445. [CrossRef]
79. Rathossi, C.; Pontikes, Y. Effect of firing temperature and atmosphere on ceramics made of NW Peloponnese clay sediments: Part II. Chemistry of pyrometamorphic minerals and comparison with ancient ceramics. *J. Eur. Ceram. Soc.* **2010**, *30*, 1853–1866. [CrossRef]
80. De Bonis, A.; Cultrone, G.; Grifa, C.; Langella, A.; Morra, V. Clays from the Bay of Naples (Italy): New insight on ancient and traditional ceramics. *J. Eur. Ceram. Soc.* **2014**, *34*, 3229–3244. [CrossRef]
81. Maritan, L.; Mazzoli, C.; Nodari, L.; Russo, U. Second Iron Age grey pottery from Este (northeastern Italy): Study of provenance and technology. *Appl. Clay Sci.* **2005**, *29*, 31–44. [CrossRef]
82. Khalfaoui, A.; Hajjaji, M. A chloritic-illitic clay from Morocco: Temperature–time–transformation and neoformation. *Appl. Clay Sci.* **2009**, *45*, 83–89. [CrossRef]
83. Grapes, R. *Pyrometamorphism*; Springer: Berlin/Heidelberg, Germany, 2011; p. 365.
84. Bauluz, B.; Mayayo, M.J.; Yuste, A.; Fernnadez-Nieto, C.; Gonzalez Lopez, J.M. TEM study of mineral transformation in fired carbonated clays: Relevance to brick making. *Clay Miner.* **2004**, *39*, 333–344. [CrossRef]
85. Nodari, L.; Marcuz, E.; Maritan, L.; Mazzoli, C.; Russo, U. Hematite nucleation and growth in the firing of carbonate-rich clay for pottery production. *J. Eur. Ceram. Soc.* **2007**, *27*, 4665–4673. [CrossRef]

86. Maggetti, M.; Heimann, R.B. Bildung und Stabilität von Gehlenit in römischer Feinkeramik. *Schweiz. Miner. Petrogr. Mitt.* **1979**, *59*, 413–417.
87. Hajjaji, M.; Kacim, S. Clay-calcite mixes: Sintering and phase formation. *British Ceram. Trans.* **2004**, *103*, 29–32. [CrossRef]

minerals

MDPI

Article

Controls on the Transformation of Clay Minerals in the Miocene Evaporite Deposits of the Ukrainian Carpathian Foredeep

Yaroslava Yaremchuk [1], Sofiya Hryniv [1] and Tadeusz Peryt [2,*]

[1] Institute of Geology and Geochemistry of Combustible Minerals, National Academy of Sciences of Ukraine, Naukova 3a, 79060 Lviv, Ukraine; slava.yaremchuk@gmail.com (Y.Y.); sophia_hryniv@ukr.net (S.H.)
[2] Polish Geological Institute—National Research Institute, Rakowiecka 4, 00-975 Warszawa, Poland
* Correspondence: tadeusz.peryt@pgi.gov.pl

Abstract: Clays deposited in marine evaporite sequences are strongly altered, and the most important factor determining their transformation is brine concentration. An X-ray diffraction study of clay minerals associated with the Lower and Middle Miocene evaporite formations of the Ukrainian Carpathian Foredeep indicated that the clay mineral assemblages in the gypsum facies are composed of smectite and illite, and, in some samples, mixed-layer chlorite–smectite and illite–smectite, as well as chlorite. In the halite facies, illite, chlorite, and mixed-layer illite–smectite occur in rock salt of Eggenburgian age (Vorotyshcha Suite); in addition to those minerals, smectite, corrensite, and mixed-layer chlorite–smectite occur in the Badenian rock salt (Tyras Suite); and in the potash facies, illite and chlorite were recorded. Such clay mineral assemblages resulted from the aggradational transformation of unstable and labile minerals and phases (kaolinite, smectite, and mixed-layer phases) that finally pass into illite and chlorite, minerals that are stable in an evaporite environment. In addition to brine concentration control, another important factor in the transformations of clay minerals was the sorption of organic components on the mineral structure, which slows the transformation processes. The assemblage of clay minerals in the weathering zone of the evaporite deposits, besides inherited illite and chlorite, also contains mixed-layer illite–smectite and kaolinite. The appearance of those clay minerals in hypergene deposits is the consequence of two processes: degradational transformation (illite–smectite) and neoformation (kaolinite) in conditions of decreased ionic concentrations during desalination.

Keywords: clay minerals; marine evaporites; brines; Miocene; caprock; X-ray diffraction

1. Introduction

The origin of clay minerals in various geological environments (weathering, sedimentary, and diagenetic-hydrothermal) are provided by three mechanisms of clay mineral formation: inheritance, neoformation, and transformation [1,2].

One of these, transformation has exceptional importance in the forming of clay mineral assemblages. The concept of transformation was introduced by [3] and includes degradation (negative transformation related to the weathering and removal of matter) and aggradation (positive transformation related to the growth of the crystalline structure and the incorporation of matter) [3]. Transformations of clay minerals by degradation are common in the soil-forming process, where leaching is intense. Aggradational transformations are characteristic of the depositional environment, where ions from concentrated solutions can participate in the building of well-structured lattices. Illite and chlorite are typical of the premetamorphic stage of aggradation; the mixed-layer minerals represent intermediate stages of the processes of degradation and aggradation [4,5].

One of main transformation factors of clay minerals in different geological contexts is the burial diagenesis of clay-rich deposits, for example, the illitisation of smectite ([6] and references therein). In evaporative conditions, the most important factor in the transformation of clay minerals is brine concentration [7,8].

Clay minerals of evaporite deposits are mostly regarded as authigenic [1,7,9–12]. The transformation or neoformation of clay minerals at the stage of sedimentation in an evaporite basin or at the stage of diagenesis within evaporite deposits are controlled by physico-chemical conditions, and, in a hypersaline environment, this depends on factors such as temperature [13], lateral salinity change and alkalinity [14], and salinity combined with higher burial-related temperatures [15,16].

Assemblages of clay minerals of the Lower and Middle Miocene evaporites of the Carpathian region are well studied (see [17], for a review). Considerable attention has been paid to the transformation processes of clay minerals under the influence of brines in salt basins or buried within deposits. Being unstable in conditions of increased salinity, allogenic minerals (kaolinite, smectite, and mixed-layer phases) are transformed, through a series of transitional phases, into illite and chlorite [5,18]. As the brine concentration increases further, the structure of those end-minerals becomes better ordered. Less attention has been paid to the degradational transformation processes of clay minerals of the Miocene evaporites that formed as brine concentrations decreased under the influence of surface waters. The mechanism of these transformations is related to the destabilisation of the interlayer spaces of clay structures and the formation of labile phases.

The present research considers factors controlling the transformational processes of clay minerals in the Miocene evaporite deposits of the Ukrainian Carpathian Foredeep.

Geological Setting

The Miocene Carpathian Foredeep is asymmetrical and developed as a peripheral foreland basin related to the moving Carpathian front [19]. The molasse, predominantly siliciclastic, deposits are up to 6 km thick in the Ukrainian Carpathian Foredeep, where three tectonic zones are distinguished: the outer (Bilche-Volytsya Zone), central (Sambir Nappe thrusted over the Bilche-Volytsya Zone), and inner (Boryslav-Pokuttya Nappe thrusted over the Sambir Nappe) [19,20] (Figures 1 and 2).

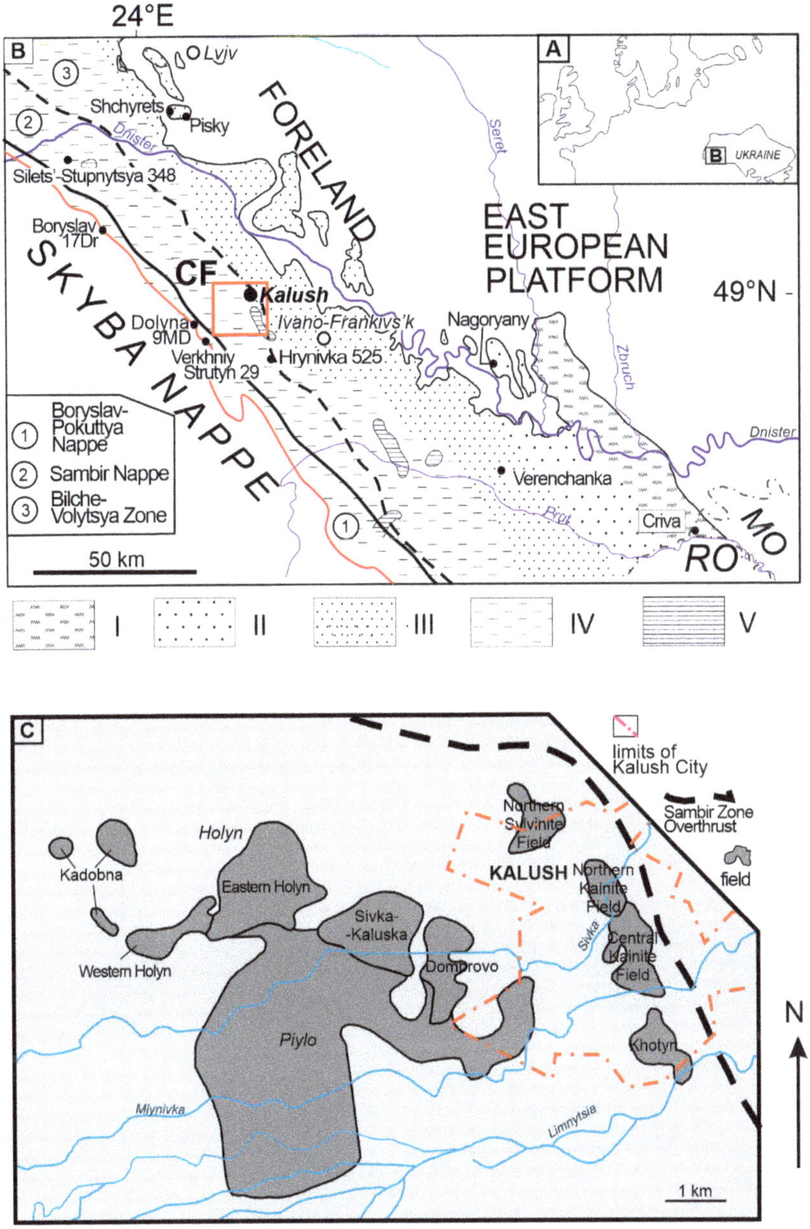

Figure 1. (**A**,**B**) Location of sample sites; tectonic zonation of the Carpathian Foredeep after [21]; CF—Carpathian Foredeep; MO—Moldova; RO—Romania; I–V—evaporite facies zones of the Tyras Suite for the Outer (Bilche-Volytsya) Zone and below the Sambir and Boryslav-Pokuttya nappes (after [22]): I–IV indicate gypsum facies (in the case of IV, gypsum and/or anhydrite facies), V—halite facies; the red box indicates the area of the Kalush-Holyn potash deposit with the distribution of exploited areas (after [23], simplified) shown in (**C**).

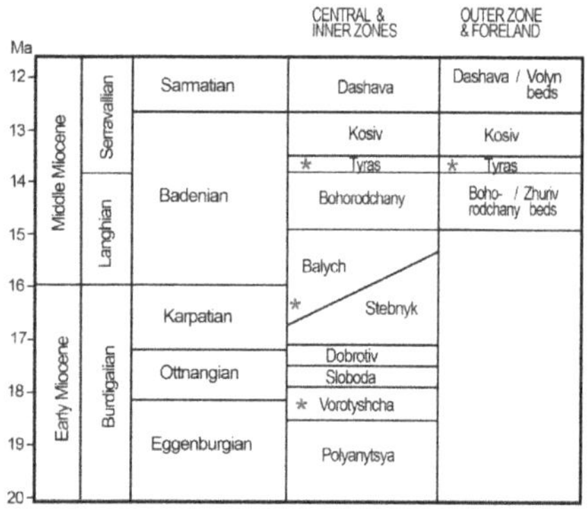

Figure 2. Regional stratigraphic scheme of Miocene strata in the Ukrainian Carpathian Foredeep (after [19], based on [19,24–26]). Note that the boundaries of formations are strongly diachronous according to [25], but this is here visualised only in the case of the boundary between the Stebnyk and Balych formations. The Miocene timescale is partly recalibrated and correlated to regional stages of the Central Paratethys (after [27]). Asterisks show the occurrence of evaporates in the stratigraphic section.

Evaporite deposits occurring in the Carpathian Foredeep fill (Figure 2) are involved in the thrusting of the Carpathian nappes, and, thus, their stratigraphic sections are likely multiplied [28–31]. For example, the Lower Miocene (Eggenburgian) Vorotyshcha Suite of the inner zone shows a total thickness more than 2000 m, but the primary section was 100–125 m thick (Figure 3; [31]). The stratigraphic position of another potash deposit that was recorded to be higher in the stratigraphic section, the Kalush-Holyn deposit, is controversial. A recent map synthesis [32] and the officially approved stratigraphical scheme [23] consider the Kalush-Holyn deposit to be related to the upper part of the Balych Suite of Karpatian age, and [33,34] correlate "the Kalush Beds" with the Upper Badenian salt of the Wieliczka and Bochnia region (Poland).

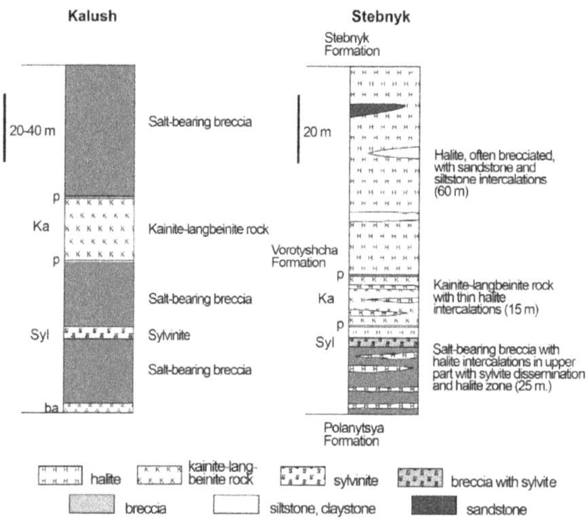

Figure 3. Presumed undisturbed lithological-stratigraphical sections of evaporites of the Holyn' fields group of the Kalysh-Holyn' deposit (after [31,35]); ba—basal anhydrite; Syl—sylvinite bed; Ka—kainite–langbeinite bed; p—polyhalite–anhydrite bed; and the Vorotyshcha Suite in the Stebnyk Mine (after [31,36]).

The potash-bearing sequence of the Kalush-Holyn deposit is up to 500 m thick and consists of interbedded salt claystones, salt-bearing breccias, potash, and rock salt, but it seems that only two beds of potash salts formed during sedimentation: the lower chloride and the upper sulphate [29–31] (Figure 3). Potash beds of the Kalush-Holyn deposit, in general, are composed of kainite, kainite–langbeinite (polymineralic), and langbeinite rocks; more rare is sylvinite. Clay material is dispersed and takes the form of halopelite layers (clays containing up to 30% of various salts), which are interbedded with beds of potash deposits or halite. Potash beds occur in salt breccia (or salt-bearing clay) composed of fragments of clays cemented by halite. Such clay–halite rocks with silt- and clay-rich siliciclastic rocks are commonly termed "zuber" [36–38].

Evaporites of the Kalush-Holyn deposit occur close to the surface in the Dombrowo field, where potash deposits are dissolved by an aquifer within Upper Pleistocene alluvial deposits, the main source of which is from rainfall [39]. A gypsum–clay caprock up to 21 m thick continues to form, as indicated by Late Pleistocene–Holocene radiometric ages of hypergene salt minerals [40].

The youngest evaporite deposits in the Carpathian Foredeep Basin are of late Badenian age (see [41]). These are included, in Ukraine, in the Tyras Suite, and consist of gypsum in the peripheral parts (e.g., [42–44]) laterally passing through anhydrite [45] into halite interbedded with siliciclastic deposits (e.g., [46,47]). In Poland, the latter are included in the Wieliczka Formation ([48], with references therein), and in Romania in the Cosmina Formation ([49], with references therein).

2. Materials and Methods

2.1. Materials

For this research, we applied our previously published mineralogical data on clay minerals [17,50–55], considered here in more detail and from a different perspective. Eleven samples of the pelitic fraction of water-insoluble residue contained in the Tyras gypsum in five gypsum quarries have been studied: four quarries are in Ukraine (Shchyrets, Pisky, Verenchanka, and Nagoryany), and one is in Moldova, at the boundary with Ukraine (Criva) (Figure 1). These samples represent various parts of the gypsum stratigraphic section and various lithological varieties of gypsum (stromatolitic, bedded, laminated, sabre, massive gypsum and gypsum breccia).

Twenty-three samples of the pelitic fraction of water-insoluble residue contained in the Tyras rock salt were from the Silets-Stupnytsya (boreholes 348 and 671, six samples) and Hrynivka (borehole 525, seventeen samples) localities. Ten samples of the pelitic fraction of water-insoluble residue of rock salt from the Vorotyshcha Suite (Eggenburgian) have been studied; they include samples from the following boreholes: Verkhniy Strutyn (borehole 29, three samples), Boryslav (borehole 17Dr, four samples), and Dolyna (borehole 9MD, three samples). The sample set from the Kalush-Holyn deposit (Holyn, Dombrovo, Northern Kainite and Khotyn fields) included twelve samples from the potash rocks (kainite, langbeinite, kainite-langbeinite, polyhalite, sylvinite), twelve from halite and halopelite layers, and four from salt-bearing breccia. From the Dombrovo quarry, ten samples of clays were taken from the gypsum-clay caprock: four of them from the northern escarpment of the quarry at the level of +265 m, and six of them from the eastern escarpment of the quarry at the level of +277 m in boreholes 714 and 747. The thicknesses of the gypsum–clay caprock were there 4.0 and 7.5 m, respectively (Figure 4). Six samples of clays from the gypsum–clay caprock were taken from the upper, middle, and lower parts of the gypsum–clay caprock above potash rocks (samples 2308, 2309, 2310) and above the salt-bearing breccia (samples 2305, 2306, 2307).

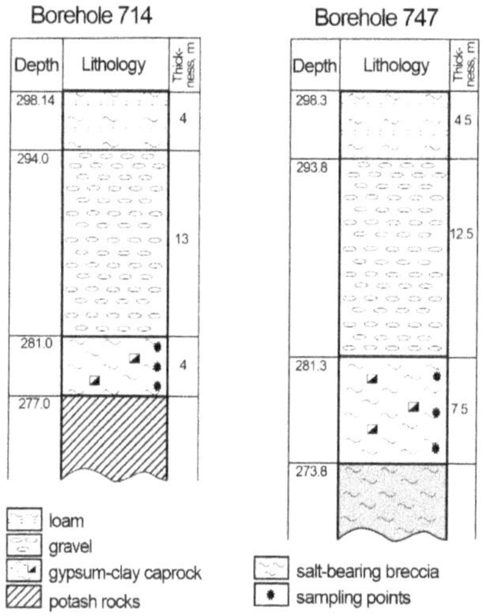

Figure 4. Sections above the potash rocks and salt-bearing breccia, Dombrovo Quarry, Kalush-Holyn deposit.

2.2. Methods

Assemblages of clay minerals have been studied in the pelitic fraction by X-ray diffraction (XRD), differential thermal analysis (DTA), derivative thermogravimetry (DTG), thermogravimetric analysis (TG), and scanning electron microscopy (SEM).

The pelitic fraction was separated by decantation. First, evaporite samples were dissolved in distilled water and washed to complete elimination of salt minerals. From the resultant insoluble residue, the sand fraction was separated, and a finer fraction was washed in several steps until a stable suspension was obtained. According to the present methodology, particles smaller than 0.005 mm occur in stable suspension 40 min after shaking in a 5 cm column [56]).

XRD patterns were obtained using two different powder diffractometers: DRON-0.5 (Bourevestnik, St. Petersburg, Russia) and ADP-2.0 (Anatek Services PVT. LTD., Mumbai, India). The operating conditions for DRON-0.5 were: 34 kV, 4 mA, Ni-filtered Cu-radiation, and a speed of counter movement of 2°/min. For ADP-2.0, the operating conditions were 30–36 kV, 9–15 mA, with different wavelengths of radiation used. These included Fe-filtered Co-radiation at 0.025 2θ/step with a counting time of 1.5 s and Mn-filtered Fe-radiation with a counting time of 0.75 s. For some samples in the 6–14° 2θ interval, counting times were 1.5 or 3.0 s.

Identification of clay minerals using XRD was based on the methodology of recognition of aluminosilicates in mineral assemblages [56,57]. To define clay minerals, oriented preparations were obtained by applying several drops of suspension of pelitic particles on a glass plate and drying at room temperature.

Oriented preparations that were air-dried, ethylene-glycolated, and heated at $T = 550$ °C for 1 h were studied. For determination of the kaolinite content in the presence of chlorite, samples were processed by a 15% solution of hydrochloric acid during a heating time of 2.5 h ($T = 80$ °C). Chlorite minerals under such circumstances decompose completely, and kaolinite remains stable even when subjected to concentrated hydrochloric acid. Randomly oriented aggregates of pelitic particles were studied to determine the structural type of the clay minerals (060 peak position, interval 74–98° 2θ).

Differential thermal and thermogravimetric analysis was performed on a Q-1500 D derivatograph on a Paulik-Paulik-Erdey system (MOM Szerviz Kft., Budapest, Hungary).

The analysis was carried out in a temperature range of 20–1000 °C with a heating rate of 10 °C/min in air. Aluminium oxide served as a reference substance. Endothermic and exothermic effects recorded during this thermal analysis indicated the structural features of the clay minerals and thermal destruction by oxidation of organic matter [58].

Scanning electron microscopy (SEM) analysis was carried out with a Jeol-JSM 6490 instrument (JEOL Ltd., Tokyo, Japan).

3. Results

3.1. Gypsum Facies

The pelitic fraction of the water-insoluble residue of the Badenian gypsum samples studied contains two main clay minerals: a considerable amount of smectite, and a small amount of illite [51]. In addition, in some samples, mixed-layer chlorite–smectite and a small amount of illite–smectite as well as chlorite were found (Table 1, Figure 5).

Table 1. Mineral composition of the pelitic fraction of water-insoluble residue of the Badenian gypsum in the Ukrainian and Moldavian Carpathian Foredeep basin.

Locality	Sample Number	Gypsum Type	Clay Minerals					Other Minerals
			Smectite	Chlorite-Smectite	Illite-Smectite	Chlorite	Illite	
Shchyrets	2313	Stromatolitic gypsum	++	–	+	–	+	Ca +
	2316	Bedded gypsum	++	(+)	–	(+)	++	Q +
	2317	Gypsum breccia	++	–	+	–	+	Ca ++
Pisky	2319	Sabre gypsum	++	+	+	–	+	Q (+)
	2320	Grass-like gypsum	++	–	(+)	+	+	Q +,
	2321	Laminated gypsum with salin spar	+	++	+	–	+	Q (+), Fs (+)
Verenchanka (borehole 20)	822	Grass-like gypsum	++	+	(+)	–	+	Fs (+), Ca +
	20	Sabre gypsum	++	+	+	–	+	Fs (+), Do ++, Ca +
Nagoryany	818	Sabre gypsum	++	–	–	+	+	Q (+), Fs (+)
Criva (borehole 84)	102	Massive gypsum	++	(+)	+	–	+?	–
	113	Massive gypsum	+	+	+	–	–	Ca ++

Other minerals: Q—quartz; Fs—feldspar; Do—dolomite; Ca—calcite. Content in a sample: ++ considerable; + small; (+) admixture; +? presence in doubt; – mineral lacking.

Smectite was recorded in all samples studied and is defined by a broad basal reflection 001 of high intensity. In some samples, this diffractional maximum is split at the apex, with interlayer spacings at 1.51–1.58 and 1.63 nm, that, when ethylene-glycolated, are shifted to 1.68–1.71 and 1.73 nm, respectively. The broadening of the 001 reflection with a release of the 1.63 nm line on the diffraction picture of the air-dried preparation is an indication of two minerals of the smectite group (Figure 5). The position of the 060 reflection of preparations of randomly oriented pelitic fraction at 0.149–0.150 nm indicates a dioctahedral type of smectite structure (that occurs in a considerable amount), although, in that interval, a weakly expressed 0.153 nm reflection also occurs that can be indicative of an admixture of trioctahedral smectite. A trioctahedral structure, similarly, may indicate chlorite, but this reflection was recorded also in samples were chlorite is absent. Thus, smectite in the pelitic fraction is heterogeneous: there occurs a mixture of Al-Fe dioctahedral allogenic, and a small quantity of Mg trioctahedral authigenic, material [51]. In thermally treated preparations, reflections at 1.51–1.63 nm are shifted to 0.98 nm, and, in some preparations, in the area of low angles, there are additional reflections of small intensity and poor separability which correspond to chlorite and chlorite–smectite (Figure 5). Illite was recorded in most samples, but its content is small. It was diagnosed based on 0.98, 0.49, and 0.332 nm reflections that do not change their location in ethylene-glycolated conditions and during thermal treatment. Illite corresponds to a dioctahedral structure type, but the spacing of the 060 reflection at 0.149 nm is superimposed on the same reflection of smectite.

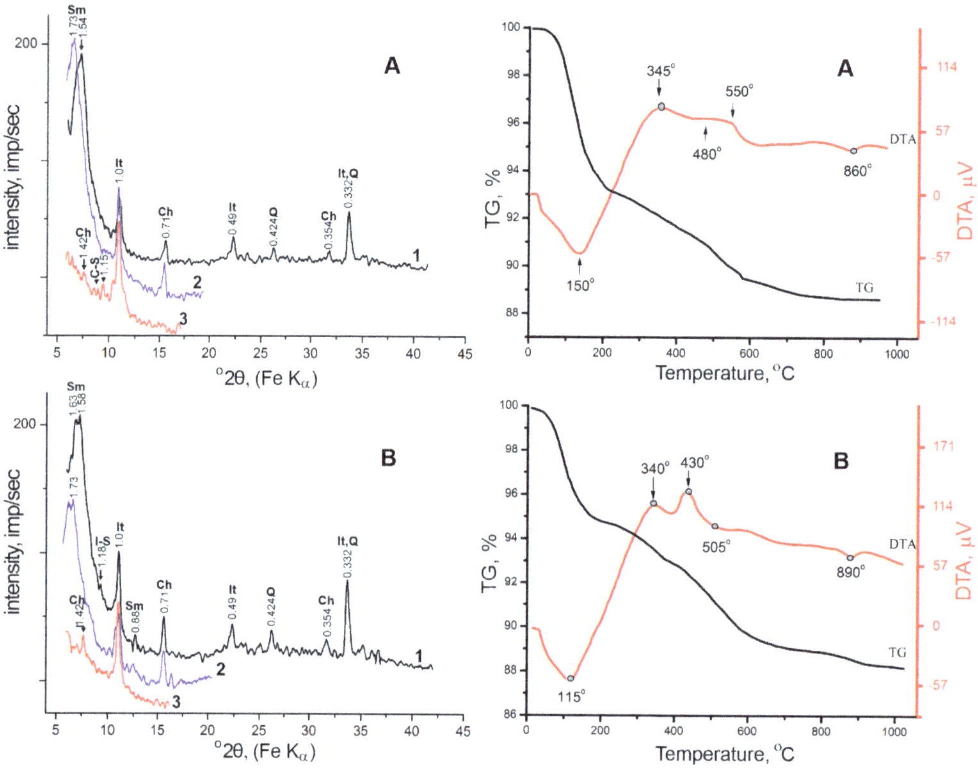

Figure 5. XRD data (**left side**) and DTA and TG data (**right side**) of the pelitic fraction of the water-insoluble residue of Badenian gypsum deposits; (**A**)—Shchyrets Quarry, sample 2316; (**B**)—Pisky Quarry, samples 2320. **Left side**: The smectite 001 reflection of 1.54–1.58 nm is shifted to 1.73 nm in preparations saturated with ethylene glycol. Mixed-layer illite–smectite is defined by a small peak of 1.18 nm, which after saturation moved to the low-angle side. After heating, the labile packets shrank: for illite–smectite to 1.0 nm, and peaks of 1.26, 1.15 nm (within 8.8–9.6° 2θ) indicate a chlorite-smectite admixture. Oriented preparations: 1—air-dried, 2—ethylene-glycolated; 3—heated at T = 550 °C. Sm—smectite; Ch—chlorite; It—illite; mixed-layer: I-S—illite–smectite, C-S—chlorite–smectite; Q—quartz. **Right side**: Endothermic peak of the loss of interlayer water a by smectite and illite admixture (115, 150 °C DTA pattern), with a corresponding mass loss of about 6% (TG pattern). Exothermic peaks with maxima of 340, 345, and 430 °C (DTA pattern) characterise the combustion of organic matter with corresponding mass loss of 1.3%–1.9% (TG pattern). Curves: DTA—differential thermal, TG—thermogravimetric analysis.

On the DTA curve, a profound low-temperature endothermic effect of the loss of interlayer water is noted in the temperature range of 115–150 °C. It corresponds to a mass loss (5.1%–6.9%, Figure 5 according to the TG curve) by smectite, which dominates in the samples, and by illite as well as mixed-layer chlorite–smectite and illite–smectite, which are present in small amounts. Illite on the DTA curves shows a low-temperature endo-effect that corresponds to the loss of interlayer water and is superimposed on the endo-effect of smectite across the whole temperature range. Since the interlayer spaces of illite are able to include a smaller amount of water molecules and release them at somewhat lower temperature (compared to smectite), on the DTA curve, a weak endo-effect in the range of 500–700 °C corresponds to the release of constitutionally bound water from the smectite structure (1.5%–2.7%) as well as from illite and mixed-layer phases. The small endo-effect, in the temperature range of 800–900 °C, corresponds to the release of the residue of constitutionally bound water and the destruction of the structure. In the case of subsequent temperature increase from the decomposition products, a new mineral crystallises, and this is expressed by the appearance of an exo-effect on the DTA curve (Figure 5) [51]. Exothermic peaks in the region of 300–480 °C (DTA curve) characterise the combustion of organic matter with corresponding mass loss of 1.3%–1.9% (TG curve).

A SEM microphotograph (Figure 6A) shows the crystal shape of smectite from the pelitic fraction of water-insoluble residue of the Badenian gypsum.

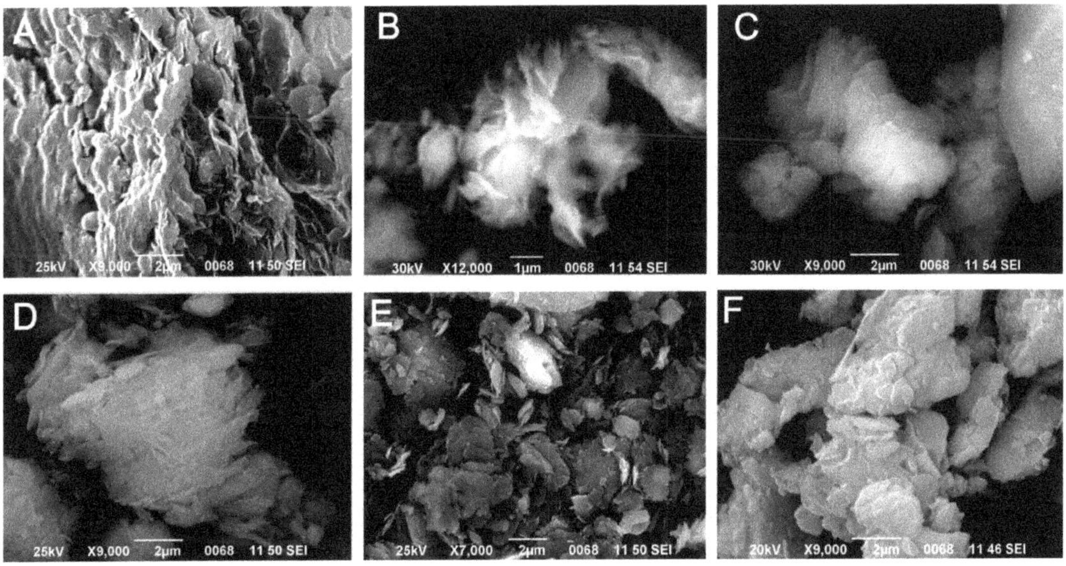

Figure 6. SEM microphotographs of clay minerals of the pelitic fraction of water-insoluble residue of Badenian gypsum. ((**A**) Pisky Quarry, sample 2320) and Badenian rock salt. ((**B**,**C**) Hrynivka 525 borehole, sample 1357), as well as SEM microphotographs of illite in the pelitic fraction of water-insoluble residue of Eggenburgian rock salt. (**D**,**E**) Boryslav 17 Dr borehole, sample 2570) and Badenian rock salt. (**F**) Hrynivka 525 borehole, sample 1350). (**A**) Smectite—leaf-like curved plates with bright rims; (**B**) smectite—semi-transparent twisted and bent plates; (**C**) corrensite—aggregates of ordered pseudohexagonal plates; (**D**) illite—elongated plates with sharp, locally split ends within an aggregate; (**E**) illite—isometric plates with uneven contours; (**F**) illite—aggregates of isometric plates.

3.2. Halite Facies

Illite, chlorite, and mixed-layer illite–smectite occur in rock salt of Eggenburgian age (Vorotyshcha Suite). In addition to those minerals, trioctahedral smectite and mixed-layer chlorite–smectite occur in Badenian rock salt (Tyras Suite), and, at the Hrynivka and Silets-Stupnytsya sites, corrensite was also recorded (Table 2, Figure 7).

Table 2. Mineral composition of the pelitic fraction of water-insoluble residue, rock salt of the Ukrainian Carpathian Foredeep.

Sample Number	Depth [m]	Clay Minerals						Other Minerals
		Smectite	Corrensite	Chlorite-Smectite	Illite-Smectite	Chlorite	Illite	
Borehole 29, Verkhniy Strutyn, Vorotyshcha Suite								
2573	562	–	–	–	+	++	++	Q (+), Fs –, Do –, Ca –, Ma –
2574	626	–	–	–	+	++	++	Q (+), Fs (+), Do +, Ca –, Ma –
2575	698	–	–	–	++	++	++	Q (+), Fs +, Do –, Ca –, Ma –
Borehole 17Dr, Boryslav, Vorotyshcha Suite								
2567	352	–	–	–	–	++	++	Q +, Fs +, Do –, Ca –, Ma ++
2566	353	–	–	–	–	++	++	Q ++, Fs +, Do –, Ca –, Ma +
2569	377	–	–	(+)	+	++	++	Q +, Fs (+), Do (+), Ca –, Ma –
2570	409	–	–	–	+	++	++	Q (+), Fs –, Do (+), Ca –, Ma –
Borehole 9MD, Dolyna, Vorotyshcha Suite								
858	38	–	–	+?	+	++	++	Q –, Fs (+), Do (+), Ca –, Ma –
859	73	–	–	–	+	++	++	Q +, Fs (+), Do –, Ca –, Ma –
863	152	–	–	–	+	++	++	Q (+), Fs (+), Do –, Ca –, Ma –
Boreholes 348 and 671, Silets-Stupnytsya, Tyras Suite								
902	132	+	+	+	+	+	+	Q (+), Fs–, Do –, Ca –, Ma –
907	172.5	–	–	++	+	+	+	Q +, Fs –, Do +, Ca –, Ma –
58	272	+	–	++	+	(+)	+	Q +, Fs –, Do +, Ca –, Ma –
54	302	–	–	++	(+)	+	+	Q (+), Pt –, Do +, Ca –, Ma –
47	449	+?	–	++	+	(+)	+	Q (+), Pt –, Do +, Ca –, Ma –
48	464	+	+	+	(+)	(+)	+	Q +, Fs Do –, Ca –, Ma –
Borehole 525, Hrynivka, Tyras Suite								
1342	304	+	++	+	(+)	+	+	Q +, Fs (+), Do –, Ca –, Ma –
1344	310.1	+	+	+	+	+	+	Q +, Fs (+), Do –, Ca –, Ma –
1348	330	+?	–	++	(+)	(+)	+	Q +, Fs –, Do –, Ca –, Ma –
1350	342	+	+?	++	+	(+)	+	Q (+), Fs (+), Do +, Ca +, Ma –
1352	362	+	+?	++	(+)	(+)	+	Q (+), Fs –, Do ++, Ca –, Ma –
1354	376	–	–	++	+	+	++	Q +, Fs (+), Do +, Ca +, Ma –
1356	395–399	+	++	(+)	–	(+)	++	Q +, Fs –, Do +, Ca –, Ma –
1357	410–411	+	++	+		+	+	Q–, Fs–, Do +, Ca –, Ma –
1361	444–448	+	++	(+)	(+)	(+)	+	Q (+), Fs (+), Do +, Ca –, Ma –
1362	454–456	++	+?	(+)	+	+	+	Q (+), Fs –, Do +, Ca –, Ma –
1364	474	+	++	+	–	(+)	+	Q (+), Fs –, Do +, Ca –, Ma –
1365	480	+?	++	+	+	(+)	+	Q –, Fs –, Do +, Ca –, Ma –
1366	489	–	+?	++	–	+	++	Q (+), Fs –, Do +, Ca –, Ma –
1369	518	++	++	(+)	–	(+)	+	Q (+), Fs (+), Do +, Ca –, Ma –
1370	521–528	+	–	++	+	(+)	++	Q +, Fs (+), Do +, Ca +, Ma –
1371	532	+	+?	++	(+)	+	+	Q +, Fs (+), Do –, Ca –, Ma –
1372	536	+	–	++	+	+	++	Q +, Fs (+), Do –, Ca –, Ma –

Other minerals: Q—quartz, Fs—feldspar, Do—dolomite, Ca—calcite, Ma—magnesite; Content in a sample: ++ considerable; + small; (+) admixture; +? presence in doubt; – mineral lacking.

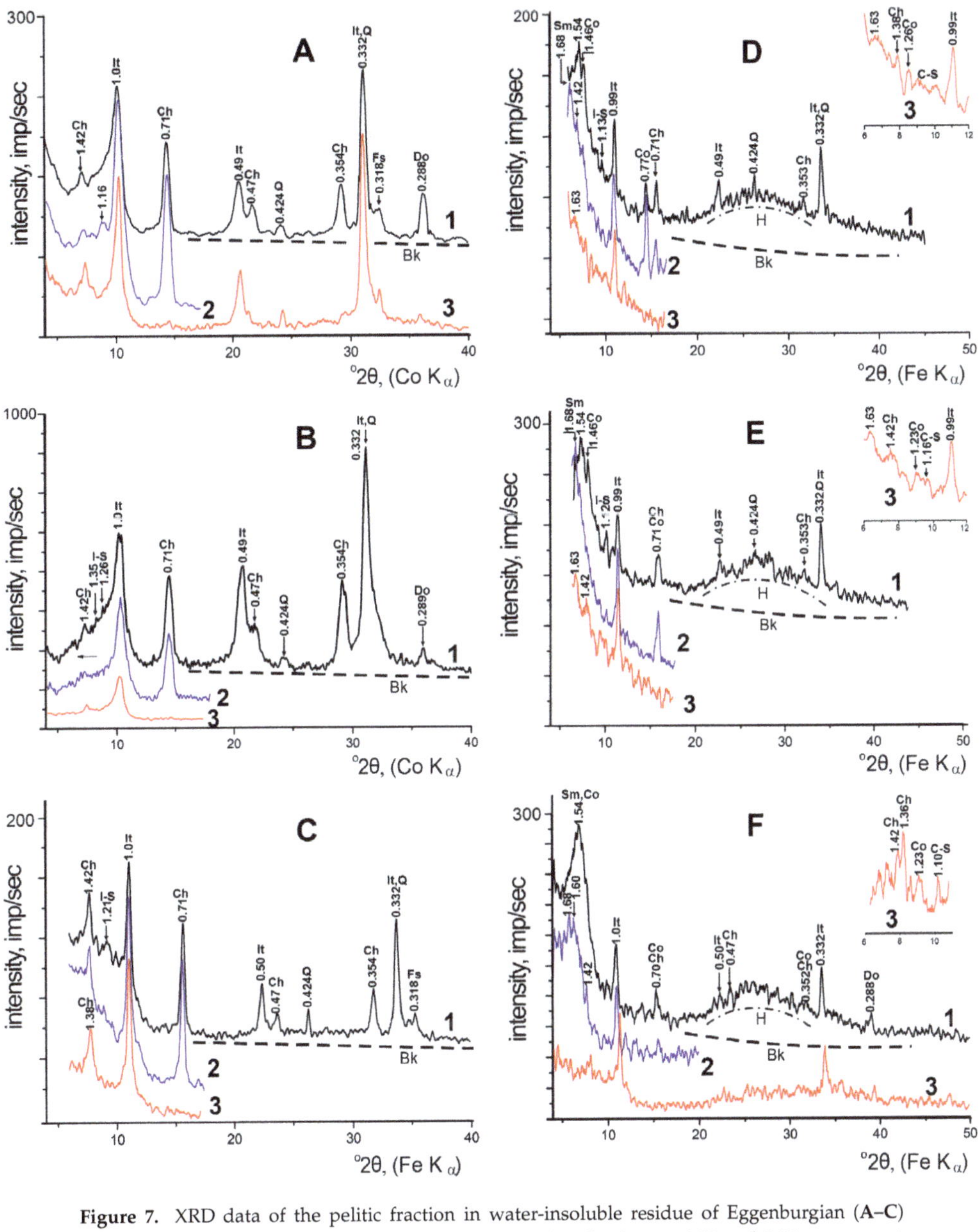

Figure 7. XRD data of the pelitic fraction in water-insoluble residue of Eggenburgian (**A–C**) and Badenian rock salt (**D–F**). After heating (curves 3), reflections at 1.23 and 1.26 nm suggest the presence of corrensite, and in the range of 1.23–0.99 nm, chlorite-smectite with predominantly smectite packets; (**D–F**) the convex raised background (halo) can be related to existing organic matter. (**A**) Verkhniy Strutyn 29 borehole, sample 2574; (**B**) Boryslav 17Dr borehole, sample 2570; (**C**) Dolyna 9MD borehole, sample 863; (**D**) Silets-Stupnytsya 671 borehole, sample 48; (**E**) Silets-Stupnytsya 348 borehole, sample 902; (**F**) Hrynivka 525 borehole, sample 1357. Oriented preparations: 1—air-dried; 2—ethylene-glycolated; 3—heated at $T = 550$ °C; Sm—smectite; Co—corrensite; Ch—chlorite; It—illite; mixed-layer: I-S—illite-smectite, C-S—chlorite-smectite; Q—quartz; Do—dolomite; Fs—feldspar; Bk—machine background, H—halo.

Clay minerals of both suites studied contain dioctahedral illite and trioctahedral chlorite, which are indicated by reflections 060 at 0.149 and 0.153 nm; for both of them, Fe-Mg-chlorite is recorded. Its composition is indicated by smaller intensities of the first and third basal reflections by comparison with the second one. Illite and chlorite from the Eggenburgian rock salt has intense, distinct basal reflections (Figure 7A–C), indicating that these are well crystallised, while in the Badenian rock salt, illite is characterised by a considerable asymmetry of the first basal reflection at 1.0 nm in the range of small angles that could have resulted from poor crystallisation, a content of interlayer water, or swelling packets (Figure 7D–F). Clay minerals of the Badenian rock salt contain more swelling minerals than the Eggenburgian rock does. Mixed-layer illite–smectite was diagnosed by small reflections in the field of 1.12–1.28 nm that, in ethylene-glycolated preparations, was shifted to 1.46 nm. Chlorite–smectite was diagnosed by a series of 1.42–1.51 nm reflections shifted to 1.60–1.71 nm.

Smectite is diagnosed by the behaviour of 001 reflections with a spacing of 1.51–1.55 nm in air-dried preparations that, in ethylene-glycolated, are shifted to 1.71 nm and, in diffractograms of thermally treated samples, decreased to 0.98 nm. This diffraction maximum has various forms, and poorly resolved spacings can be attributed to chlorite–smectite mixed layers [59]. Corrensite is established by reflections 1.42–1.46, 0.70–0.77, 0.472, and 0.353 nm on diffractograms of air-dried preparations; in ethylene-glycolated ones, they acquire values of 1.60, 0.78, and 0.355 nm, and on the diffractograms of preparations heated at 550 °C, a reflection in the field 1.23–1.26 nm is observed. A reflection of 2.88 nm is reported for corrensite in the literature; on the diffractograms of the samples studied, it has the form of an inflexion or is completely absent, which may indicate some disorder in the corrensite structure [59].

On the X-ray curves of the pelitic fraction of the Eggenburgian and Badenian rock salt, there are a series of reflections in the range of small angles. These reflections do not change their position in ethylene-glycolated conditions, and part of them remains on diffractograms after thermal processing (Figure 8) [54].

All diffraction spectra of the pelitic fraction of the Badenian rock salt show an increased position of the line of background and a halo with a large number lines of low intensity and poor separability in the area of the 22–36° 2θ, Fe Kα radiation (Figure 7D–F). In addition, the first two basal reflections 001 and 002 are extended at the base, and their intensity is too low for the main lines of minerals which prevail in a sample. In the range of low angles of diffraction, distinct peaks of low intensity with spacing 2.91–1.54 nm are recorded that, however, do not coincide with the reflection 001 of corrensite—2.84 nm. Thus, on diffractograms of air-dried and ethylene-glycolated preparations, distinct lines 2.70, 2.61, 2.51, 2.40–2.37, 2.21, 2.05, and 1.91 nm are recorded, and among these, the reflections 2.73–2.70, 2.51, 2.21, and 2.05 nm are particularly frequent; at the Hrynivka site, they occur on all diffraction curves (air-dried or ethylene-glycolated). After the thermal treatment (T = 550 °C), distinct peaks remain: 2.76–2.63; 2.40–2.30; 2.05; and 1.78–1.58 nm (Figure 8).

Diffractometric curves of oriented preparations of the pelitic fraction from the Eggenburgian rock salt showed that these sites lack a marked halo and increased background (which is characteristic of the Badenian rock salt) (Figure 7A–C). On the DTA curve of the pelitic fraction of water-insoluble residue of the Eggenburgian and Badenian rock salt, a distinct low-temperature endo-effect is noted with a maximum (110–140 °C), which corresponds to the loss of interlayer water (1.8%–3.4% as the mass loss on the TG curve). In Figure 9, for the Eggenburgian rock salt, sample 2570, which contains illite, chlorite, and a small amount of illite–smectite, the loss of interlayer water is 3.4% (Figure 9A), while for Badenian rock salt, sample 902, in which small amounts of illite, chlorite, as well as swelling minerals and phases—smectite, corrensite, chlorite–smectite, and illite–smectite—

are present, the loss of interlayer water is smaller at 2.4% (Figure 9B). We assume that the small loss of interlayer water can be caused by the presence of captured organic compounds in interlayer spaces of those labile structures. The exothermic effect in the temperature interval of 230–480 °C corresponds to a process of thermooxidative destruction of organic matter related to clay particle surfaces. Endothermic effects at the DTA curves within 480–810 °C correspond to dehydroxilation, i.e., discharge of constitutional water of clay minerals and structural destruction. Also, organic compounds sorbed by interlayer spaces are lost in this temperature interval (Figure 9).

According to our SEM data, the pelitic fraction of the halite facies mostly consists of particles 2–4 μm in size, although some tabular and bulky particles as small as 0.5 μm are present. SEM microphotographs show smectite, corrensite (Figure 6B,C), and illite (Figure 6D–F).

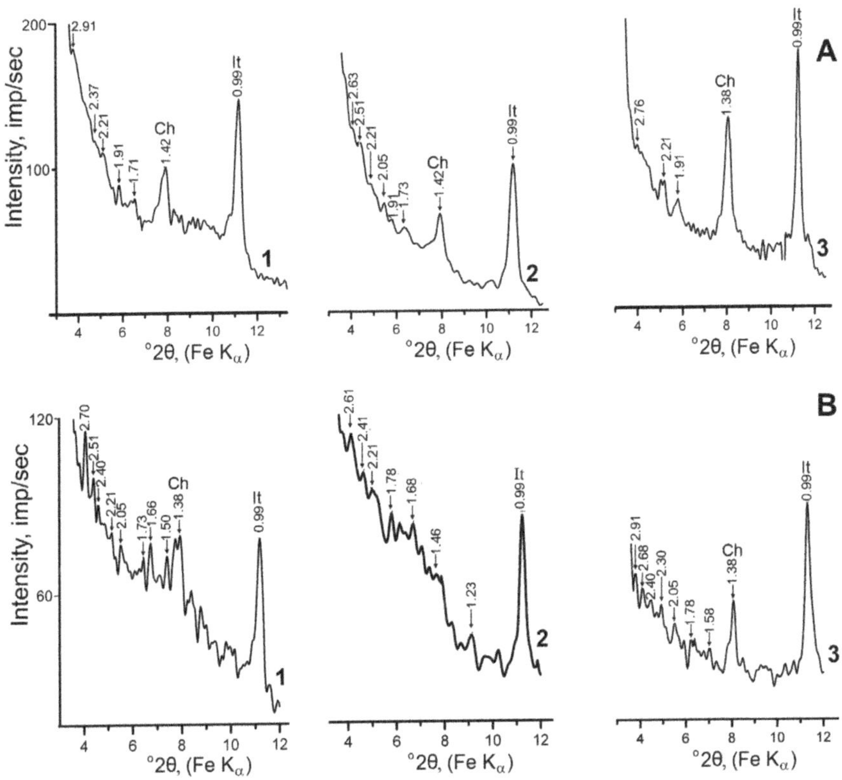

Figure 8. XRD data in the range of small angles of the pelitic fraction in water-insoluble residue of the Eggenburgian ((**A**) Dolyna 9MD borehole, sample 863) and Badenian ((**B**) Hrynivka 525 borehole, sample 1354) rock salt. Scattered lines of low intensity with d 1.73–2.91 nm indicate the existence of interlayer spaces of clay structures filled with organic compounds (curves 1, 2). Organic matter trapped in the structure, resistant to high-temperature processing (spacings from 2.91 to 1.58 nm appearing without displacement in air-dried, ethylene-glycolated; and heated aggregates were related to organic matter presence [54], curves 3). Oriented preparations: 1—air-dried; 2 –ethylene-glycolated; 3—heated at T = 550 °C. It—illite, Ch—chlorite.

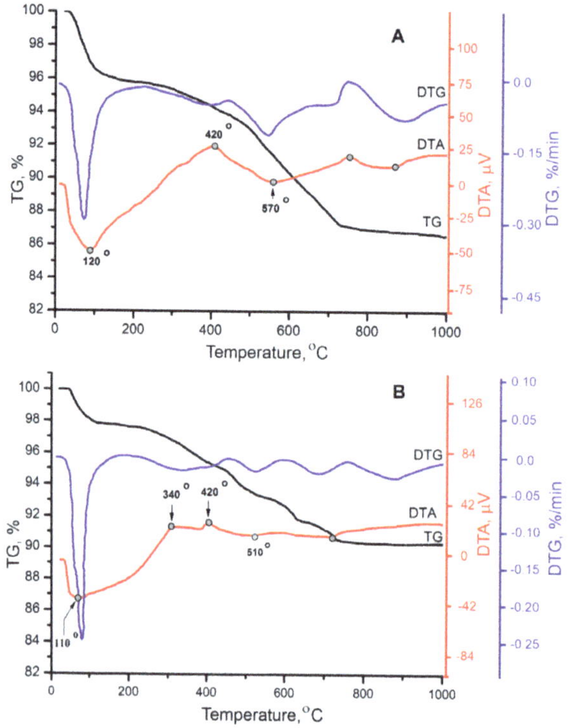

Figure 9. DTA-TG data for the pelitic fraction of water-insoluble residue of the Eggenburgian ((**A**) Boryslav 17Dr borehole, sample 2570) and Badenian ((**B**) Silets-Stupnytsya 348 borehole, sample 902) rock salt. (**A**) Endothermic peak is due to the loss of interlayer water by an illite and illite-smectite admixture at 120 °C (DTA pattern), with corresponding mass loss of about 3.4% (TG pattern); (**B**) endothermic peak of the loss of interlayer water by clay minerals at 110 °C (DTA pattern), with corresponding mass loss of about 2.4% (TG pattern); the insignificant mass loss is due to replacement of part of the interlayer water by organic compounds in the interlayer spaces of labile minerals. Curves: DTA—differential thermal, DTG—differential thermogravimetric, TG—thermogravimetric analysis.

3.3. Potash Facies

The assemblage of clay minerals of the Miocene potash facies of the Carpathian Foredeep is composed of illite and chlorite (Table 3, Figure 10).

Chlorite in the Kalush-Holyn deposit is a trioctahedral Fe-Mg (more rarely a Mg-Fe) mineral. Dioctahedral illite in these deposits is characterised by structural ordering that is expressed by the form of the first basal reflection and its location on the diffractogram. The 001 reflections of illite from the salt-bearing breccia are widened on their low-angle side, while those from the potash rocks are sharp and symmetrical; for some of them (kainite-langbeinite rocks), that reflection is relatively narrow (of small width) and enabled diagnosis of that mineral as mica [53]. It is not fully explained to what extent those were structurally ordered. In the potash rocks, illite is transformed into mica either fully [60] or partially. By mica, we mean a stable and ordered form of illite with a small K^+ deficit in the interlayer spaces. For such mica-like clay minerals, the following terms are commonly applied: illite $2M_1$ or muscovite $2M_1$. We use the term "mica" because the determination of polytype structures was not applied. Our study of diffractograms of pelitic fraction of the potash-bearing rocks of both air-dried and ethylene-glycolated preparations show that the basal 001 reflections are bifurcated at the top, with (d) 0.99 and 1.0 nm (Figure 10A,B, curve 4). Our results suggest that illite, as well as mica, is present in the pelitic fraction.

When scanning in slow-motion at angles of 8–13° 2θ in diffractograms, clear splits are seen (see Figure 10A,B, curve 4), which indicates the occurrence of the structurally similar minerals, mica and illite.

Table 3. Mineral composition of the pelitic fraction of water-insoluble residue of potash rocks and salt-bearing breccia from the potash facies of the Kalush-Holyn potash salt deposit.

Locality	Sample Number	Lithology	Clay Minerals		Other Minerals
			Illite	Chlorite	
Dombrovo	3	Clayey kainite rock	++	++	–
Northern Kainite field	1006b	Kainite rock	++	++	–
Northern Kainite field	1029	Kainite rock	++	++	–
Dombrovo	2251	Kainite rock	++	++	Ma +
Dombrovo	2	Langbeinite rock	++	++	*
Dombrovo	1105	Langbeinite rock	++	++	*
Dombrovo	2250	Langbeinite rock	++	++	Ma +
Dombrovo	1	Kainite–langbeinite rock	++	++	Ma +
Dombrovo	1097a	Kainite–langbeinite rock	++	++	Ma +
Dombrovo	2249	Kainite–langbeinite rock	++	++	Ma +
Khotyn field	1017	Sylvinite	++	++	Ma +
Dombrovo	5	Clayey polyhalite rock	++	++	–
Northern Kainite field	1006a	Salt-bearing breccia	++	++	Ma +
Khotyn field	1019	Salt-bearing breccia	++	++	–
Dombrovo	1117	Salt-bearing breccia	++	++	–
Holyn	2219	Halite from salt-bearing breccia	++	++	Ma +
Holyn	1035	Greenish-grey halite from the potash rocks	++	+	–
Holyn	1039	Greenish-grey halite from the potash rocks	++	+	Ma +
Dombrovo	1097	Greenish-grey halite from the potash rocks	++	+	Ma +
Dombrovo	14	Halite from alternated halite and halopelite layers	++	+	Ma +
Dombrovo	15	Halite from alternation of halite and halopelite layers	++	+	Ma +
Holyn	2185	Halite from alternated halite and halopelite layers	++	++	Ma +
Northern Kainite field	1026	Halite from alternated halite and halopelite layers	++	+	–
Holyn	1060	Greenish-grey halite from alternated halite and halopelite layers	++	+	–
Holyn	1062	Halite from alternated halite and halopelite layers	++	+	*
Dombrovo	25	Halopelite from alternated halite and halopelite layers	++	+	Ma +
Dombrovo	1096	Halopelite from alternated halite and halopelite layers	++	+	*
Dombrovo	1104	Halopelite from alternated halite and halopelite layers	++	+	–

Other minerals: Ma—magnesite, *—carbonate not determined. Content in a sample: ++ considerable; + small; – mineral lacking.

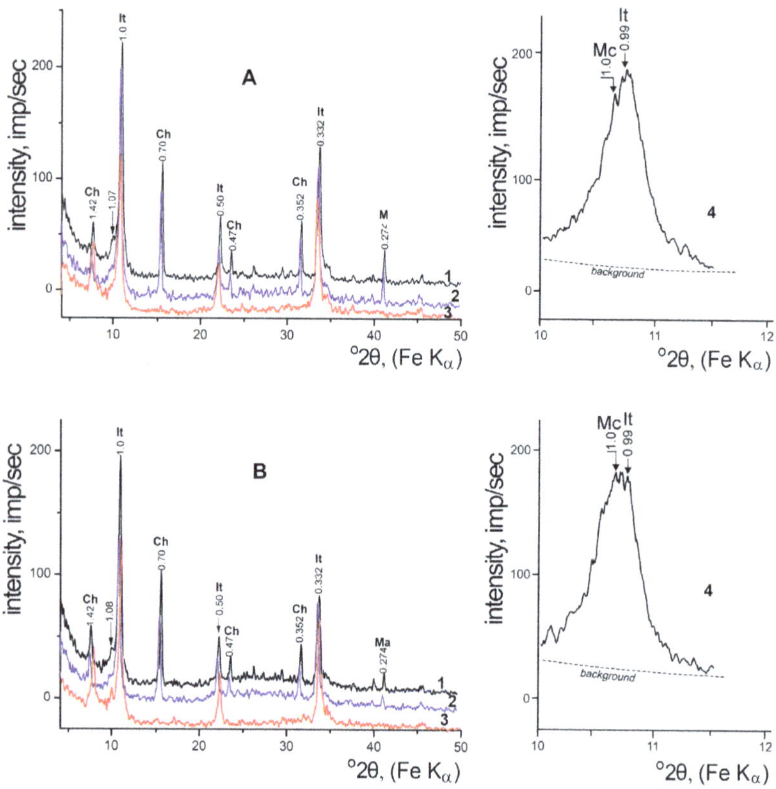

Figure 10. XRD data of the pelitic fraction of water-insoluble residue from the potash facies of the Kalush-Holyn potash deposit. Intense, narrow basal reflections of illite and chlorite indicate that these are well crystallised. The reflections at 1.07 nm (at angles of 10.4° 2θ) are the basal reflections of mica in the interlayer space of which relic hydrated K^+ cations are still present. A split 001 reflection (scanning in slow-motion—4) indicates the occurrence of the structurally similar minerals, mica and illite. (**A**) Halite from alternation of halite and halopelite layers, Holyn, sample 2185; (**B**) langbeinite rock, Dombrovo, sample 2250. Oriented preparations: 1—air-dried; 2—ethylene-glycolated; 3—heated at T = 550 °C; 4—part of the diffraction curve scanned in slow motion (right side). It—illite; Ch—chlorite; Ma—magnesite, Mc—mica.

Thermogravimetric analysis of the pelitic fraction of potash rock at low temperature (up to 200 °C) showed an endothermic effect due to the loss of interlayer water by illite, indicating a mass loss of almost 1% (Figure 11B). Such a small loss of interlayer water at a significant content of illite in the sample corroborates the XRD data indicating that the mica-like mineral with a reflection at 1.0 nm is composed of illite and mica. A high-temperature endothermic effect with a maximum at 575 °C in the DTA curve, which is accompanied by a 9.4% mass loss in the TG curve, is responsible for the dehydroxilation of illite and chlorite, as well as magnesite dissociation (Figure 11B).

On the SEM microphotographs the pelitic fraction of the potash facies (Figure 12) are observed particles of both clay and pelitic dimension and, in places, the aggregates of particles. The pseudohexagonal particles in Figure 12D may be $2M_1$ mica, and isometric particles with smoothened outlines may be illite.

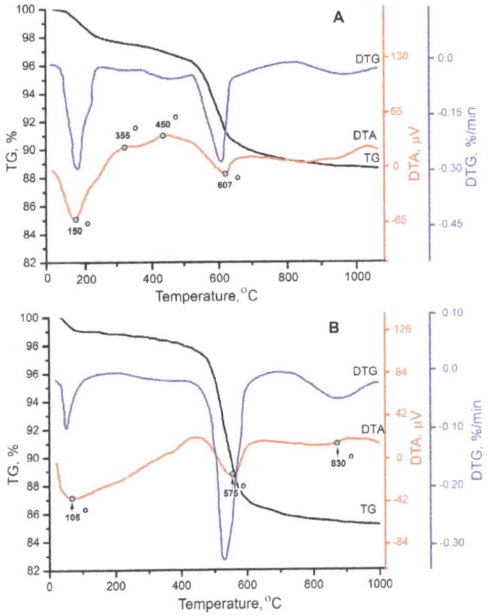

Figure 11. DTA-TG data for the pelitic fraction of water-insoluble residue from the potash facies of the Kalush-Holyn potash deposit. (**A**) Endothermic peak due to the loss of interlayer water by illite at 150 °C (DTA pattern), with corresponding mass loss of about 3.2% (TG pattern); halite from salt-bearing breccia, Holyn, sample 2219; (**B**) the endothermic effect of the loss of interlayer water by illite (105 °C, DTA pattern) corresponds to a mass loss of about 1% (TG pattern), and the effect of dehydroxylation of clay minerals and of magnesite dissociation corresponds to a loss of mass of 9.4% (TG pattern); kainite rock, Dombrovo, sample 2251. Curves: DTA—differential thermal, DTG—differential thermogravimetric, TG—thermogravimetric analysis.

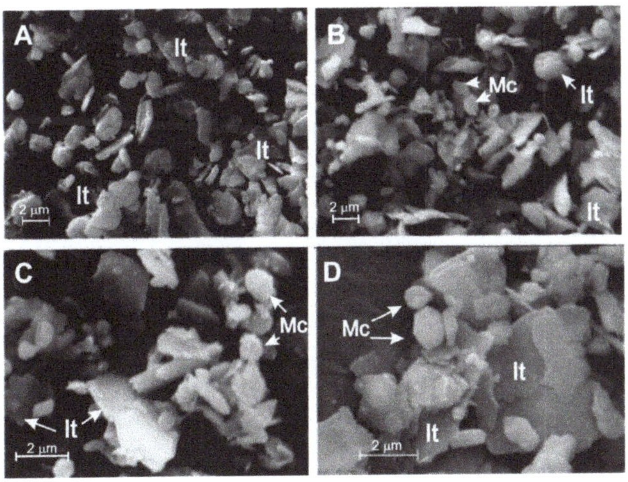

Figure 12. SEM microphotographs of clay minerals in the pelitic fraction of water-insoluble residue of the potash facies of the Kalush-Holyn potash deposit. Morphology of illite (It) and mica (Mc): illite particles are of isometric shape; the pseudo-hexagonal-shaped particles are supposedly mica. (**A**) Halite of salt-bearing breccia (Holyn, sample 2219), (**B**) kainite-langbeinite rock (Dombrovo, sample 2249), (**C**) kainite rock (Dombrovo, sample 2251), (**D**) halite from alternation of halite and halopelite layers (Holyn, sample 2185).

3.4. Weathering Zone of the Kalush-Holyn Deposit

Our data indicate that the assemblage of clay minerals in the hypergene deposits—in gypsum–clay caprock, above potash rocks, and salt-bearing breccia—contain, in addition to illite and chlorite, consistently mixed-layer illite–smectite, and, in five samples, kaolinite was additionally found that was not recorded in the evaporite deposits before dissolution (Table 4).

Table 4. Mineral composition of the pelitic fraction of water-insoluble residue of the weathering zone of the Kalush-Holyn potash salt deposit, Dombrovo quarry.

Sample Number	Lithology	Clay Minerals				Other Minerals
		Illite	Chlorite	Illite-Smectite	Kaolinite	
Clays from the gypsum-clay caprock above potash rocks						
2253	Clay with gypsum	++	++	+	–	Ma +
2255	Clay with syngenite	++	++	+	–	–
2257	Clay with mirabilite	++	++	+	–	Ma +
2300	Clay with syngenite	++	++	+	–	Ma +
2308	Grey clay, upper part of the gypsum–clay caprock	++	++	+	+	–
2309	Grey clay, middle part of the gypsum–clay caprock	++	++	+	+	–
2310	Grey clay, lower part of the gypsum–clay caprock	++	++	+	+	–
Clays from the gypsum–clay caprock above the salt-bearing breccia						
2305	Grey clay, upper part of the gypsum–clay caprock	++	++	+	+	–
2306	Grey clay with gypsum and syngenite, middle part of the gypsum–clay caprock	++	++	+	–	–
2307	Grey clay, lower part of the gypsum–clay caprock	++	++	+	+	Ma +

Other minerals: Ma—magnesite. Content in a sample: ++ considerable; + small; – mineral lacking.

The mixed-layer phase shows a considerable asymmetry in the first basal reflection of 1.0 nm in the range of low angles and small reflections at 1.04–1.28 nm. In ethylene-glycolated mounts, the width of the basal line 001 of illite diminishes, and the additional reflections decrease on its low-angles side to 1.51 nm and can be traced between reflections at 1.0 and 1.7 nm. All this gives an opportunity to diagnose such a mixed-layer phase as illite–smectite, mainly of illite composition with a small proportion of smectite [61].

The first basal reflection of illite is shifted from its low-angles side by an admixture of the mixed-layer phase. The location of line 060 (0.149–0.150 nm) indicates that illite in the gypsum–clay caprock is dioctahedral.

Reflection 1.38–1.40 nm does not change the location on the diffractograms of ethylene-glycolated preparations; therefore, this line is the basal reflection 001 of chlorite. The location of line 060 on diffractograms of preparations of randomly oriented pelitic particles (0.153–0.155 nm) indicates that chlorite in the gypsum–clay caprock is trioctahedral.

A clearly defined line 0.70 nm, available on diffractograms of samples subject to solution by hot HCl, indicates the presence of kaolinite. It was recorded in half of the samples of gypsum–clay caprock studied.

4. Interpretation

4.1. Brine Concentration Control

We consider that the most important factor in the transformation of clay minerals is the brine concentration [7,8], and the higher the water mineralization, the more advanced are the transformation processes [1]. An increased brine concentration causes the aggradational transformation of clay minerals, including changes involving structural ordering through the capture of cations and a decrease in molecular volume. This is clearly supported by the example of transformation of unstable and labile minerals and phases (kaolinite, smectite, and mixed-layer phases) that pass into illite and chlorite, these minerals being stable in an evaporite environment. Kaolinite is stable in acid conditions at pH ~5 [62]. In an evaporite basin, pH conditions depend on brine concentration: the pH value at the sulphate–carbonate stage is ~8.6 and decreases to 7.0–7.5 at the halite stage [63]. In evaporite deposits, kaolinite is destroyed in the middle of the halite stage; in deposits of higher stages, it occurs very rarely [64]. As well as kaolinite destruction, its illitisation is possible [65], but this process has not been proven in marine conditions [5]. The clay mineral assemblage of the Badenian gypsum is represented mainly by allogenic clay minerals (dioctahedral smectite, illite, and rarely chlorite), and it indicates a relatively low brine concentration in the evaporite basin. The presence of a small content of authigenic minerals (trioctahedral smectite and illite) and mixed-layer chlorite–smectite and illite–smectite indicates the beginning of aggradational transformation processes [51]. Smectite in evaporite basins has two modes of provenance, allogenic and dioctahedral smectite from the land, while trioctahedral smectite forms through the decomposition of volcanic glass of pyroclastic material, where volcanism is coeval with salt accumulation. Thus, the transformation of smectite occurred through mixed-layer illite–smectite and chlorite–smectite to dioctahedral illite and trioctahedral chlorite [5,59]. At the sulphate-carbonate stage of halogenesis, the content of potassium ion increases from 1.5 to 3.9 g/L [66], so, at this stage, the transformation of unstable allogenic clay minerals was only initiated. However, the concentration of brines from which the Badenian gypsum precipitated was considerably lower due to an important inflow of continental water into the Badenian evaporite basin [67–71]. The low concentration of K^+ and Mg^{2+} ions caused a slow transformation of unstable clay minerals.

The clay mineral assemblage in the Badenian rock salt (Hrynivka and Silets-Stupnytsya sites) contains minerals with labile (swelling) structures: smectite, corrensite, and mixed-layer chlorite–smectite with an admixture of illite, chlorite, and illite–smectite. Such an assemblage of clay minerals is consistent with the influence of contemporaneous volcanism [17,50,54]. During the decomposition of volcanic glass of pyroclastic material that entered the salt basin, trioctahedral smectite was formed that was subject to further transformation through mixed-layer phases into chlorite, this being the end-product of a number of transformations [59]. This interpretation explains the presence of a considerable number of swelling clay minerals, but it does not answer why they did not transform in the Badenian salt-forming basin under the influence of brines.

One of the reasons may be low brine concentrations. Geochemical conditions in the Miocene evaporite basins of the Ukrainian Carpathian Foredeep were influenced by substantial inflow of continental water [67–73]. Low concentrations of K^+ and Mg^{2+} ions caused the slow transformation of unstable clay minerals. Authigenic trioctahedral Mg-smectite is the initial clay mineral at the sulphate-carbonate stage of halogenesis (e.g., [10,59], and, due to its transformation through intermediate mixed-layer phases, a number of clay minerals originated.

In the Eggenburgian and Badenian rock salt, the brine concentration during halite precipitation was determined through the use of an ultramicrochemical method ([69] and references therein) that indicated, at the Hrynivka site, a low brine concentration from which the halite crystallised [50]. In the Silets-Stupnytsya site, although halite crystallised from brines of high concentration, the assemblage of clay minerals is the same as in the rock salt of the Hrynivka site. So, it is insufficient to explain the considerable content of swelling clay minerals in the Badenian rock salt only by the influence of brine concentration on the transformation of clay minerals. In addition to brine concentration, another significant factor must have influenced the transformation of layered aluminosilicates, slowing their aggradational transformation. This influence likely involved organic compounds, sorbed by interlayer spaces of labile clay minerals (smectite, corrensite, and mixed-layer phases) that prevented the entrance of inorganic cations (potassium, magnesium, and others) and increased the resistance of those minerals to transformation in the conditions of a salt-forming basin. Diffractograms of clay minerals of the Hrynivka and Silets-Stupnytsya sites show the presence of sorbed organic compounds (see below).

An increased brine concentration causes aggradational transformation of clay minerals—changes targeted to the structural ordering through the capture of cations and the decrease in molecular volume. This process is clearly supported by the example of the transformation of unstable and labile minerals and phases (kaolinite, smectite, and mixed-layer phases) that were supplied to Miocene evaporite basins of the Carpathian Foredeep from the continent and through a number of transformations, transformed into illite and chlorite—minerals stable in an evaporite environment. Those two minerals compose the assemblage of clay minerals in the Kalush-Holyn potash deposit.

Potassium enters the interlayer spaces of smectite and mixed-layer phases due to electrostatic forces of attraction of cations to negatively charged siloxane sheets, where the excess of negative charge is created by Al moving into tetrahedral positions in the place of Si and Al replacing Mg, Fe^{2+} in octahedral positions [74]. The illitisation of smectite as a pure phase, as well as packages in mixed-layer phases, is through a heterogeneous reaction that is described by the following equation [75]:

$$\text{montmorillonite} + K^+ \rightarrow \text{illite} + (Fe^{2+}, Mg^{2+}) + SiO_2$$

In evaporitic conditions, smectite is magnesium-rich; during its aggradational transformation, the entrance of potassium in the mineral structure leads to the liberation of silicon and magnesium, which cause the crystallisation of magnesium-rich chlorite and authigenic quartz. The process of the illitisation of smectite in hypersaline conditions is accompanied by the further development of the illite structure. The generally accepted crystallisation order during progressive illitisation is as follows (e.g., [76]):

$$\text{smectite} \rightarrow \text{I-S, random} \rightarrow \text{I-S, ordered} \rightarrow \text{illite, } 1M_d \rightarrow \text{muscovite, } 2M_1$$

Such transformation is noted during the diagenetic change in smectite at a temperature of ~360 °C [77]. In an evaporitic environment, we infer that, rather than temperature, the main factor is high salinity, which causes a further aggradation of illite. This is why micas of $2M_1$ polytype are the only stable minerals among the entire series of aggradational transformation of smectite at progressively higher salinities. Better ordering of the illite structure and its transformation to mica is observed in clay minerals of the Kalush-Holyn potash deposit.

Thin intercalations of tuffs and tuffites occurring in the Kalush-Holyn deposit [78,79] indicate that pyroclastic material was also supplied to the evaporite basin. During the decomposition of volcanic glass, trioctahedral smectite was formed that, through chlorite-smectite, was finally transformed into trioctahedral chlorite. No intermediate products of this transformation were recorded, however. In clay minerals related to the rock salt occurring in the Sambir zone of the Ukrainian Carpathian Foredeep, an excess of mixed-layer phases was observed [50,54], and they are clearly fixed products of decomposition of pyroclastic material. A more complex picture emerges from study of the Tyras gypsum. Ref. [51] concluded that the composition of clay minerals in that gypsum does not indicate volcanogenic input into the evaporite basin, possibly because the decomposition of pyroclastic material that leads to the formation of trioctahedral smectite had just begun due to low brine concentration in the basin, this conclusion being supported by an admixture of trioctahedral smectite in the pelitic fraction. However, ref. [80] recorded a bed 5–20 cm thick of bentonite clay that is a local marker of the Badenian gypsum in the region between the Seret and Zbruch rivers.

4.2. Organic Matter Control

Previous research indicated that the low reactivity of clay minerals may be attributed to the presence of organic matter in the samples [81], and that this anomalous lack of evolution (i.e., transformations) of clay minerals is due to specific interactions between organic matter and clay minerals [82].

Clay mineral particles actively sorb organic compounds both onto outer surfaces and, in the mineral structure, in interlayer spaces. In the first case, their surface defects and sides perpendicular to the silicate sheets, broken chemical bonds of which actively contribute to sorption of organic ions [83], and, to a lesser extent, onto cleavage surfaces. Due to the high dispersion of clay particles, the total surface is large and can adsorb a significant amount of organic matter. Organic matter (including scattered and epigenetic), as well as others amorphous materials in diffractograms, appears in the form of a halo, which looks like a broad hump of low intensity. The position of the amorphous halo (°2θ) on the diffractograms is correlated with the material of the X-ray tube anode and varies in the d/n range of 0.88–0.225 nm [84]. The halos with maxima of 0.45 and 0.34–0.37 nm characterise the sapropel and aromatic components of organic matter, respectively [85]. The presence of a halo on diffractograms of clay minerals in the field 22–36° 2θ may be caused by the adsorption of amorphous organic matter onto clay surfaces. The adsorption of organic compounds is characteristic on external surfaces of all types of clay particles, although the intercalation of organic compounds into interlayer spaces is typical only for swelling structures [56,86,87]. In the labile interlayer spaces of clay minerals, organic molecules may be preferentially intercalated from a solution relative to cations of alkaline or alkaline earth elements, and can also replace those cations in these spaces [83]. At the same time, the structure of smectite and mixed-layer phases becomes disordered, extending along the C axis [86], leading to the appearance of additional low-intensity reflections in the 3–7° 2θ area of the diffractograms. When a mount with swelling mineral (e.g., smectite) is ethylene-glycol saturated, this organic compound (ethylene-glycol) is sorbed by all interlayer spaces of the mineral, and we observe one peak often more intensive than that in the air-dried mount with a d-spacing of 1.68 nm. In our case, probably many different organic compounds occur, and this is reflected by many narrow peaks. In the case of the line with d-spacing more than 1.68 nm, we can assume the occurrence, in interlayer spaces, of molecule bigger than ethylene–glycol or the penetration of several organic compounds [87]. When labile (smectite) packages are interacting with organic matter compounds, the nature of organic matter, its quantity and ways of arrangement in interlayer spaces are of importance [87].

Organic ions or compounds that, in the clay mineral structure, occupy exchange positions, are linked by Van der Waals bonds. An additional factor in retaining aliphatic organic compounds is the similarity of the structure of hydrocarbon chains, composed of hydrogen tetrahedra, to certain directed elements in the layered aluminosilicate (clay mineral) sheets, represented by siloxane tetrahedra [56].

The diffraction spectra of clay minerals associated with rock salt of the Carpathian Foredeep Basin are interpreted both in terms of the sorption of organic compounds on the faces and edges of the clay crystals. All spectra of the basal reflections of the Badenian rock salt show an increased position of the line of background and halo with numerous lines of low intensity in the area of 22–36° 2θ. The pelitic fraction of both the Badenian and Eggenburgian rock salt in the range of low angles (4–7° 2θ) is fixed by a series of reflections that do not cover the diffraction from the basal planes of the clay structures. These reflections do not change their position in ethylene-glycolated conditions, and a part of them remains on diffractograms after thermal treatment (T = 550 °C) [54].

The possible influence of epigenetic organic matter (bitumen) on the transformation of clay minerals remains controversial, especially as regards the duration of the processes involved. At the stages of sedimentation and early diagenesis, syndepositional, biochemically decomposed organic matter in a brine-saturated deposit in an evaporite basin, adsorbed onto labile clay minerals, became intercalated into the interlayer spaces, hindering the entrance of inorganic cations and slowing their transformation. Epigenetic organic matter got into evaporite deposits considerably later at the late diagenetic stage. In the Carpathian Foredeep, the formation of regional overthrusts created pathways of bitumen migration that are geochemically recorded in the Miocene evaporite deposits of that region (e.g., [88]). Bitumen compounds are undoubtedly adsorbed onto clay minerals; there remains an unresolved question as to how this influenced the transformation of the clay minerals. We commonly assume that, in a lithified evaporite succession, the transformation of clay minerals continued, although at a slower rate, and intercrystal brines in the deposit provided a supply of cations or organic compounds. That is, epigenetic organic matter also participated in the transformation of clay minerals in the Badenian evaporite deposits.

Adsorbed organic matter on surfaces and edges prevent the entrance of inorganic cations in interlayer spaces. This stopped or slowed the transformation of clay minerals and increased their resistance to changes in physico-chemical conditions of the environment. In evaporite deposits, as salt concentration increases, the clay minerals transform, ordering their structure and decreasing the number of swelling layers [53]. The final products of such transformation during the progressive evaporite process are illite and chlorite. Smectite and mixed-layer chlorite–smectite are characteristic only of the initial sulphate-carbonate stage of evaporation. At the halite stage, only a small content of mixed-layer phases is present, and the main minerals are illite and chlorite. The same is the case for the assemblage of clay minerals in the Eggenburgian rock salt of the Vorotyshcha Suite; this is represented by illite and chlorite with a small content of mixed-layer illite–smectite, rarely (only in two samples out of ten) with an insignificant admixture of mixed-layer chlorite–smectite.

4.3. *Fresh Water Control During Hypergenesis*

Clay minerals of the hypergenesis zone deposits of the Carpathian Foredeep have been studied only rarely: ref. [89] found illite and mixed-layer illite-smectite in the gypsum-clay caprock of Dombrovo Quarry, the Kalush-Holyn deposit, and ref. [90] recorded kaolinite and illite in the gypsum–clay caprock of the Stebnyk deposit.

The hypergene deposits of the Kalush-Holyn potash salt accumulation are exposed in Dombrovo Quarry, where, in the gypsum–clay caprock, the assemblage of clay minerals also contains, in addition to illite and chlorite, mixed-layer illite–smectite, and, in half of the samples, additionally, kaolinite was found that was not recorded in the evaporite rocks before they suffered dissolution by fresh water. Assemblage of clay minerals above the potash rocks, as well as above salt-bearing breccia, do not differ (see Table 4).

In the weathering zone of the Kalush-Holyn deposit, fresh surface waters come into contact with evaporites and dissolve them. A considerable decrease in brine concentration caused degradational transformation of the clay minerals. Our data indicate that the assemblages of clay minerals of the weathering zone, in addition to illite and chlorite, inherited from evaporite deposits, also contain mixed-layer illite–smectite and kaolinite—minerals appearing in deposits of the hypergene zone. The appearance of these clay minerals in hypergene deposits is a consequence of the transformation (degradation) of clay structures and neoformation when ionic concentrations decrease under the influence of fresh surface waters. The degradational transformation of mica and illite is the reverse process to aggradation and acts to leach potassium from the structure with water molecules penetrating into interlayer spaces. To restore electrostatic equilibrium, Fe and Mg ions migrate into the structure from octahedral sheets to interlayer spaces, and Al moves from tetrahedral to octahedral positions. Degradation starts from the peripheries of clay particles. As size decreases, the specific surface area increases, which promotes migration and element liberation. Structural defects accelerate the degradation process [4].

During degradational transformation, mica is completely converted into illite that, in interlayer spaces, contain partly hydrated potassium ions [77,91]. With the continued formation of labile interlayer spaces, mixed-layer illite–smectite appears. Ionic bonds are moderately strong and are difficult to break, which explains the small content of mixed-layer illite–smectite. The capture of potassium by mixed-layer illite–smectite and its transition into illite (aggradational transformation) happens more easily than the reverse process—the leaching of potassium and transformation of illite into mixed-layer illite–smectite (degradational transformation). We infer that kaolinite in hypergene deposits of the Kalush-Holyn accumulation is authigenic newly formed mineral. This is indicated by an absence of clay minerals in the evaporite deposits that could be transformed in kaolinite. Kaolinite forms during the reaction of Al-(hydr)oxides that are released during intense leaching, with silica in weakly acidic conditions (pH ~5), when silica activity is low and the concentration of basic cations is low [62].

Let us consider if such conditions occurred in the gypsum–clay caprock. During dissolution of potash deposits from the lower part of the gypsum–clay caprock, conditions in the brine were weakly acidic, as they also were during the sedimentation of the potash deposits. Thus, in the precipitation field of K-Mg salts, pH values decreased to 5.7 [63,66]. Similar results were obtained in a fluid inclusion study of Miocene evaporites of the Carpathian Foredeep [92]: pH ranged from 4.5 to 6.6, averaging 5.6. Kaolinite formed only in a part of the gypsum–clay caprock adjacent to the salt, and, due to continual caprock growth, it now occurs in the whole caprock. Its presence indicates that conditions existed here that favoured its preservation. Oxides and hydroxides of aluminium and silicon, necessary for the crystallisation of kaolinite, were released during the degradation of illite to mixed-layer illite–smectite. All of this indicates that, during the formation of the gypsum–clay caprock, there existed the physico-chemical conditions necessary for kaolinite crystallisation.

4.4. Burial Depth and Geothermal Regime Control

The Eggenburgian and Badenian evaporites studied occur at depths of 600–700, rarely 1500 m [23]. The thermobarimetric analysis of fluid inclusions in the Miocene evaporite deposits of the Ukrainian Carpathian Foredeep indicates that the temperature at the diagenesis stage in the halite of the salt-bearing breccia was 35–40 °C, 60–80, and even 110 °C in the potash facies [93,94]. However, those higher values are explained by [94] due to various exothermic reactions related to evaporite minerals, such as the conversion of metastable mineral phases and minerals into stable ones, or the radioactive decay of potassium.

Surface heat flow within the Ukrainian Carpathian fold–thrust belt and adjacent East European platform varies from 35 to 130 mW/m^2, with lower values dominating the East European platform [95]. Heat flow has not been elevated in the nappes adjacent to the East European platform since the Oligocene [96].

A geothermal gradient of ~37 °C/km is assumed to be the maximum value for the Ukrainian Carpathians (including the Carpathian Foredeep) [97]. Based on an analysis of fluid inclusions, ref. [98] (and references therein) proposed slightly higher (up to 3–5 °C/km) gradient estimates than those proposed by [96].

Ref. [97] concluded that the outermost and innermost Carpathian thrust sheets were heated to less than 60 °C and less than ~120 °C, respectively; the heating depended entirely on the burial, and the cooling occurred in two main phases (15–30 °C/m.y. between ca. 12 and ca. 5 Ma, 3–6 °C/m.y. from ca. 5 Ma to the present) and was induced by exhumation. The pressure and temperature in the Miocene evaporite deposits of the Ukrainian Carpathian Foredeep did not reach values characteristic of the late diagenesis and anchizone, and thus the transformation processes of clay minerals in those deposits are controlled by other causes than pressure and temperature.

5. Conclusions

Brine concentration control is the most important factor determining the clay mineral transformation of marine evaporite sequences. Clay minerals associated with the gypsum, halite, and potash facies of the Miocene evaporite formations of the Ukrainian Carpathian Foredeep reflect the increased concentration of brines that promoted the aggradational transformation of labile clay minerals into illite and chlorite that are stable in hypersaline conditions. Further ordering of the structure led to the transformation of illite into mica.

Brine concentration decrease (inflow of fresh water) led to a slowing of the transformation of clay minerals, and, as a result, in the Badenian gypsum and rock salt, an assemblage untypical for the gypsum and halite facies originated: only a small content of mixed-layer phases in the gypsum facies and, in the halite facies, a considerable amount of labile swelling minerals that did not transform into illite and chlorite.

Organic matter control is another important (but local) factor of the transformation of clay minerals of the Miocene evaporites. Organic compounds, sorbed by labile clay minerals and mixed-layer phases, prevented the entrance of inorganic cations into their interlayer spaces, suspending or delaying their aggradational transformation. A large number of labile clay minerals in the Badenian rock salt (smectite, corrensite, and mixed-layer phases) are still present, owing to organic compounds captured on the faces and edges of the labile clay minerals.

Clay minerals of the Badenian rock salt interacted with epigenetic organic matter entering the Miocene evaporite deposits during the origin of regional overthrusts in the Carpathian Foredeep that created pathways of bitumen migration.

Freshwater control in hypergene conditions during the dissolution of evaporite deposits by fresh surface waters is the common factor determining the clay minerals' degradational transformation that involves partial leaching of potassium from the interlayer spaces of illite and the formation of illite–smectite (degradational transformation) and kaolinite (neoformation) in the gypsum–clay caprock of Dombrovo Quarry (the Kalush-Holyn potash deposit).

Author Contributions: Conceptualization: Y.Y., S.H. and T.P.; methodology: Y.Y. and S.H.; investigation: Y.Y. and S.H.; data curation: Y.Y. and S.H.; formal analysis: Y.Y., S.H. and T.P.; visualisation: Y.Y., S.H. and T.P.; writing—original draft preparation: Y.Y. and S.H.; writing—review and editing: T.P., Y.Y. and S.H. All authors have read and agreed to the published version of the manuscript.

Funding: This research was supported by the National Academy of Sciences of Ukraine statutory funds.

Data Availability Statement: Research data are available on request.

Acknowledgments: We are very grateful to the late V.M. Kovalevych for consultations and general guidance during the initial stage of this research (2004–2013). We also thank J. Zalasiewicz for his helpful suggestions on the manuscript and to the journal referees for their comments.

Conflicts of Interest: The authors declare no conflicts of interest.

References

1. Millot, G. *Geology of Clays: Weathering, Sedimentology, Geochemistry*; Springer: New York, NY, USA; Berlin, Germany, 1970; p. 429.
2. Eberl, D.D.; Farmer, V.C.; Barrer, R.M. Clay Mineral Formation and Transformation in Rocks and Soils [and Discussion]. *Philos. Trans. R. Soc. Lond.* **1984**, *311*, 241–257.
3. Lucas, J. La transformation des minéraux argileux dans la sédimentation. Etudes sur les argiles du Trias. *Mém. Serv. Carte Géol. d'Alsace Lorraine* **1962**, *23*, 1–202.
4. Millot, G.; Lucas, J.; Paquet, H. Evolution géochimique par dégradation et agradation des minéraux argileux dans l'hydrosphère. *Geol. Rundsch.* **1966**, *55*, 1–20. [CrossRef]
5. Dunoyer de Segonzac, G. The transformation of clay minerals during diagenesis and low-grade metamorphism: A review. *Sedimentology* **1970**, *15*, 281–346. [CrossRef]
6. Lanson, B.; Sakharov, B.A.; Claret, F.; Drits, V.A. Diagenetic smectite-to-illite transition in clay-rich sediments: A reappraisal of X-ray diffraction results using the multi-specimen method. *Am. J. Sci.* **2009**, *309*, 476–516. [CrossRef]
7. Bodine, M.W., Jr. Trioctahedral Clay Mineral Assemblages in Paleozoic Marine Evaporite Rocks. In *Sixth International Symposium on Salt*; Salt Institute: Alexandria, VA, USA, 1985; Volume 1, pp. 267–284.
8. Warren, J.K. *Evaporites: A Compendium*; Springer: Berlin/Heidelberg, Germany, 2016; p. 1854.
9. Kossovskaya, A.G.; Drits, V.A. Kristallokhimiya dioktaedricheskikh slyud, khloritov i korrensitov kak indikatorov geologicheskikh obstanovok. In *Problemy Litologii i Geokhimii Osadochnykh Porod i Rud*; Nauka: Moskva, Russia, 1975; pp. 60–69. (In Russian)
10. Sokolova, T.N. Autigennoe silikatnoe mineraloobrazovanie raznykh stadiy osolonennya. *Tr. GIN* **1982**, *361*, 1–164. (In Russian)
11. Bilonizhka, P.M. The clay minerals as indicators of salt deposition conditions in Precarpathian Foredeep. *Geol. Geochem. Comb. Min.* **1992**, *78*, 95–102. (In Ukrainian with English abstract)
12. Bilonizhka, P.M. The transformation of terrigenous clay minerals in the saliferous process. *Miner. Zb.* **1992**, *45*, 51–56. (In Ukrainian with English abstract)
13. Calvo, J.P.; Blanc-Valleron, M.M.; Rodriguez Arandia, J.P.; Rouchy, J.M.; Sanz, M.E. Authigenic clay minerals in continental evaporitic environments. *Palaeoweathering Palaeosurfaces Relat. Cont. Depos.* **1999**, *27*, 129–151.
14. Turner, C.E.; Fishman, N.S. Jurassic Lake T'oo'dichi: A large alkaline, saline lake, Morison Formation, eastern Colorado Plateau. *Geol. Soc. Am. Bull.* **1991**, *103*, 538–558. [CrossRef]
15. Uhlík, P.; Honty, M.; Šucha, V.; Franců, J.; Biroň, A.; Clauer, N.; Hanzelyová, Z.; Majzlan, J. Influence of salt-bearing environment to illitization. In Proceedings of the XVII Congress of CBGA, Bratislava, Slovakia, 1–4 September 2002.
16. Honty, M.; Uhlík, P.; Šucha, V.; Čaplovičova, M.; Franců, J.; Clauer, N.; Biroň, A. Smectite-to-illite alteration in salt-bearing bentonites (East Slovak Basin). *Clay Clay Miner.* **2004**, *52*, 533–551. [CrossRef]
17. Bilonizhka, P.; Iaremchuk, I.; Hryniv, S.; Vovnyuk, S. Clay minerals of Miocene evaporites of the Carpathian Region, Ukraine. *Biul. Państwowego Inst. Geol.* **2012**, *449*, 137–146.
18. Velde, B. *Clay Minerals. A Physico-Chemical Explanation of Their Occurrence*; Elsevier: Amsterdam, The Netherlands, 1985; p. 427.
19. Oszczypko, N.; Krzywiec, P.; Popadyuk, I.; Peryt, T. Carpathian Foredeep Basin (Poland and Ukraine): Its sedimentary, structural, and geodynamic evolution. *AAPG Mem.* **2006**, *84*, 293–350.
20. Oszczypko, N.; Uchman, A.; Bubniak, I. The Dobrotiv Formation (Miocene) in the Boryslav-Pokuttya and Sambir nappes of the Ukrainian Carpathians: A record of sedimentary environmental change in the development of the Carpathian Foredeep Basin. *Geol. Quart.* **2014**, *58*, 393–408. [CrossRef]
21. Hnylko, O.M. Tectonic zoning of the Carpathians in terms of the terrane tectonics Article 2. The Flysch Carpathian—Ancient accretionary prism. *Geodynamics* **2012**, *12*, 67–78. (In Ukrainian with English summary) [CrossRef]
22. Peryt, T.M. The beginning, development and termination of the Middle Miocene Badenian salinity crisis in Central Paratethys. *Sediment. Geol.* **2006**, *188–189*, 379–396. [CrossRef]
23. Rudko, H.I.; Petryshyn, V.Y. *Soliani Resursy Peredkarpattia ta Perspektyvy Yikh Vykorystannia*; Bukrek: Kyiv, Chernivtsi, Ukraine, 2017. (In Ukrainian)
24. Andreyeva-Grigorovich, A.S.; Oszczypko, N.; Ślączka, A.; Oszczypko-Clowes, M.; Savitskaya, N.; Trofimovicz, N. New data on the stratigraphy of the folded Miocene Zone at the front of the Ukrainian Outer Carpathians. *Acta Geol. Pol.* **2008**, *58*, 325–353.
25. Vashchenko, V.O.; Hnylko, O.M. About stratigraphy and sedimentary features of the Neogene molasse of the Boryslav-Pokuttya and Sambir Nappes of the Ukrainian Fore-Carpathians. *Heol. Heokh. Hor. Kop.* **2003**, *106*, 87–101. (In Ukrainian with English summary)
26. Vashchenko, V.O.; Hnylko, O.M. About stratigraphy of the salt-bearing molasse of the Ukrainian Fore-Carpathians *Zbirn. Nauk. Prats UkrDGRI* **2013**, *2*, 71–77. (In Ukrainian with English summary)
27. Piller, W.E.; Harzhauser, M.; Mandic, O. Miocene Central Paratethys stratigraphy—Current status and future directions. *Stratigraphy* **2007**, *4*, 151–168. [CrossRef]

28. Vyalov, O.S. *Stratigrafiya Neogenovykh Molass Predkarpatskogo Progiba*; Naukova Dumka: Kiev, Ukraine, 1965; p. 192. (In Russian)
29. Dzhinoridze, N.M. Tertiary potassium basins. In *Deposits of Potassium Salts of the USSR*; Raievskiy, V.I., Fiveg, M.P., Eds.; Nedra: Leningrad, Russia, 1973; pp. 183–234. (In Russian)
30. Dzhinoridze, N.M. Karpatskiy kalienosnyy region. In *Zakonomernosti Razmeshchenia i Kriterii Poiskov Kaliynykh Soley SSSR*; Izdatelstvo "Mecniereba": Tbilisi, Russia, 1980; pp. 73–159. (In Russian)
31. Koriń, S.S. Geology of the Miocene salt-bearing formations of the Ukrainian Fore-Carpathians). *Przegląd Geol.* **1994**, *42*, 744–747. (In Polish)
32. Vashchenko, V.O.; Turchynova, S.M.; Turchynov, I.I.; Polikha, G.G. *Derzhavna Geologichna Karta Ukrainy 1:200,000. Karpatska Seriya. Arkush M-35-XXV (Ivano-Frankivsk)*; Poyasnyu valna zapiska; UkrDGI: Kyiv, Ukraine, 2007. (In Ukrainian)
33. Andreyeva-Grigorovich, A.S.; Vashchenko, V.O.; Hnylko, O.M.; Trofimovich, N.A. Stratigraphy of Neogene deposits of the Ukrainian Carpathians and Carpathian Foreland. *Tekton. Strat.* **2011**, *28*, 67–77. (In Ukrainian with English summary) [CrossRef]
34. Hnylko, O. Olistostromes in the Miocene salt-bearing folded deposits at the front of the Ukrainian Carpathian orogen. *Geol. Quart.* **2014**, *58*, 381–392. [CrossRef]
35. Korin, S.S. Tektonicheskiye usloviya formirovaniya struktury kaliynykh mestorozhdeniy v Boryslavsko-Pokutskom pokrove. *Otech. Geol.* **1992**, 20–26. (In Russian)
36. Peryt, T.M.; Kovalevich, V.M. Association of redeposited salt breccias and potash evaporites in the Lower Miocene of Stebnyk (Carpathian Foredeep, West Ukraine). *J. Sediment. Res.* **1997**, *67*, 913–922.
37. Korenevskiy, S.M.; Donchenko, K.B. Geologiya i usloviya formirovaniya kaliynykh otlozheniy Sovetskogo Predkarpatia. *Tr. VSEGEI N.S.* **1963**, *99*, 3–153. (In Russian)
38. Czapowski, G.; Bukowski, K.; Tomassi-Morawiec, H.; Toboła, T. Geochemistry and mineralogy of Miocene Zuber rocks in the Carpathian Foredeep (southern Poland). *Przegląd Solny* **2023**, *17*, 39–50.
39. Shestopalov, M.; Liutyi, H.; Sanina, I. Suchasni pidkhody do hidroheolohichnoho raionuvannia Ukrainy. *Miner. Resur. Ukr.* **2019**, *2*, 3–12. (In Ukrainian)
40. Wójtowicz, A.; Hryniv, S.P.; Peryt, T.M.; Bubniak, A.; Bubniak, I.; Bilonizhka, P.M. K-Ar dating of the Miocene potash salts of the Carpathian Foredeep (West Ukraine): Application to dating of tectonic events. *Geol. Carp.* **2003**, *54*, 243–249.
41. de Leeuw, A.; Bukowski, K.; Krijgsman, W.; Kuiper, K.F. Age of the Badenian salinity crisis; impact of Miocene climate variability on the circum-Mediterranean region. *Geology* **2010**, *38*, 715–718. [CrossRef]
42. Peryt, T.M. Sedimentology of Badenian (middle Miocene) gypsum in eastern Galicia, Podolia and Bukovina (West Ukraine). *Sediment* **1996**, *43*, 571–588. [CrossRef]
43. Bąbel, M. Badenian evaporite basin of the northern Carpathian Foredeep as a drawdown salina basin. *Acta Geol. Pol.* **2004**, *54*, 317–337.
44. Bąbel, M. Selenite-gypsum microbialite facies and sedimentary evolution of the Badenian evaporitic basin of the northern Carpathian Foredeep. *Acta Geol. Pol.* **2005**, *55*, 187–210.
45. Kasprzyk, A.; Orti, F. Paleogeographic and Burial Controls on Anhydrite Genesis: The Badenian Basin in the Carpthian Foredeep (Southern Poland, Western Ukraine). *Sediment* **1998**, *45*, 889–907. [CrossRef]
46. Hryniv, S.P.; Dolishniy, B.V.; Khmelevska, O.V.; Poberezhskyy, A.V.; Vovnyuk, S.V. Evaporites of Ukraine: A review. *Geol. Soc. Spec. Publ.* **2007**, *285*, 309–334. [CrossRef]
47. Galamay, A.R.; Bukowski, K.; Zinchuk, I.M.; Meng, F. The Temperature of Halite Crystallization in the Badenian Saline Basins, in the Context of Paleoclimate Reconstruction of the Carpathian Area. *Minerals* **2021**, *11*, 831. [CrossRef]
48. Bukowski, K. *Badenian Saline Sedimentation Between Rybnik and Dębica Based on Geochemical, Isotopic, and Radiometric Research; Dissertation Monographs 236*; AGH: Kraków, Poland, 2011; p. 184. (In Polish with English summary)
49. Palcu, D.V.; Mariș, I.; de Leeuw, A.; Melinte-Dobrinescu, M.; Anton, E.; Frunzescu, D.; Popov, S.; Stoica, M.; Jovane, L.; Krijgsman, W. The legacy of the Tethys Ocean: Anoxic seas, evaporitic basins, and megalakes in the Cenozoic of Central Europe. *Earth-Sci. Rev.* **2023**, *246*, 104594. [CrossRef]
50. Iaremchuk, I.; Galamay, A. Mineral composition of the water insoluble residue from rock salt Hrynivka area of the Carpathian Foredeep. *Heol. Heokh. Hor. Kop.* **2009**, *146*, 79–90. (In Ukrainian with English abstract)
51. Iaremchuk, I.; Poberezhsky, A. Clay mineral composition of Badenian gypsum of the Dnister-River region. *Miner. Zb.* **2009**, *59*, 116–127. (In Ukrainian with English summary)
52. Iaremchuk, I. *Clay Minerals of Phanerozoic Evaporates and Their Dependence Upon Brine Concentration Stage and Seawater Chemical Type*; Avtoref. Dis. Kand. Heol. Nauk: Lviv, Ukraine, 2010; p. 24. (In Ukrainian with English summary)
53. Iaremchuk, I. Dependence of clay mineral associations of the Neogene evaporates from the Carpathian region upon the salt brines concentration in salt-producing basins. *Heol. Heokh. Hor. Kop.* **2012**, *160–161*, 119–130. (In Ukrainian with English summary)
54. Iaremchuk, I.; Hryniv, S. Organic matter effect on composition and genesis of clay minerals of rock salt deposits of the Carpathian Foredeep. *Heol. Heokh. Hor. Kop.* **2013**, *162–163*, 60–70. (In Ukrainian with English abstract)

55. Hryniv, S.; Yaremchuk, Y.; Radkovets, N. The influence of marine and continental waters on the clay minerals transformation processes of evaporite deposits (on the example of the Kalush-Holin' deposit, Carpathian Foredeep). *Geol. Geochem. Comb. Min.* **2023**, *191–192*, 122–134. (In Ukrainian with English summary) [CrossRef]
56. Moore, D.M.; Reynolds, R.C., Jr. *X-ray Diffraction and the Identification and Analysis of Clay Minerals*; Oxford University Press: New York, NY, USA, 1997; p. 376.
57. Brown, G.; Brindley, G.W. X-Ray Diffraction Procedures for Clay Mineral Identification. *Mineral. Soc. Monogr.* **1980**, *5*, 305–356.
58. Bish, D.L.; Duffy, C.J. Thermometric analysis of minerals. *Clay Miner. Soc. Workshop Lect.* **1990**, *3*, 96–157.
59. Drits, V.A.; Kossovskaya, A.G. *Glinistyye Mineraly: Smektity, Smeshanosloynyye Obrazovaniya*; Nauka: Moskva, Russia, 1990; p. 214. (In Russian)
60. Bilonizhka, P.M. Pryroda mizhsharovoi vody v hidrosliudakh. *Miner. Zb.* **2001**, *51*, 142–148. (In Ukrainian)
61. Drits, V.A.; Sakharov, B.A. *X-Ray Structure Analysis of Interstratified Minerals*; Nauka: Moskva, Russia, 1976; p. 256. (In Russian)
62. Galán, E. Genesis of Clay Minerals. In *Developments in Clay Science: Vol. 1. Handbook of Clay Science*; Bergaya, F., Theng, B.K.G., Lagaly, G., Eds.; Elsevier: Amsterdam, The Netherlands, 2006; Chapter 14; pp. 1129–1162.
63. Bąbel, M.; Schreiber, B.C. Geochemistry of Evaporites and Evolution of Seawater. *Treatise Geochem.* **2014**, *9*, 483–560.
64. Yaremchuk, Y.; Hryniv, S.; Peryt, T.; Vovnyuk, S.; Meng, F. Controls on Associations of Clay Minerals in Phanerozoic Evaporite Formations: An Overview. *Minerals* **2020**, *10*, 974. [CrossRef]
65. Lanson, B.; Beaufort, D.; Berger, G.; Bauer, A.; Cassagnabere, A.; Meunier, A. Authigenic kaolin and illitic minerals during burial diagenesis of sandstones: A review. *Clay Miner.* **2002**, *37*, 1–22. [CrossRef]
66. McCaffrey, M.A.; Lazar, B.; Holland, H.D. The evaporation path of seawater and the coprecipitation of Br and K with halite. *J. Sediment. Res.* **1987**, *57*, 928–937.
67. Dopieralska, J.; Belka, Z.; Zieliński, M.; Górka, M.; Poberezhskyy, A.; Stupka, O.; Walczak, A.; Wysocka, A. Neodymium and strontium isotopes track the origin of parent brines of primary gypsum deposits (Miocene, Fore-Carpathian Basin). *Chem. Geol.* **2024**, *648*, 121963. [CrossRef]
68. Petrichenko, O.I.; Peryt, T.M.; Poberegsky, A.V. Pecularities of gypsum sedimentation in the Middle Miocene Badenian evaporite basin of Carpathian Foredeep. *Slovak Geol. Mag.* **1997**, *3*, 91–104.
69. Kovalevich, V.M.; Petrichenko, O.I. Chemical composition of brines in Miocene evaporite basins of Carpathian region. *Slovak Geol. Mag.* **1997**, *3*, 173–180.
70. Peryt, T.M.; Szaran, J.; Jasionowski, M.; Halas, S.; Peryt, D.; Poberezhskyy, A.; Karoli, S.; Wójtowicz, A. S and O isotope composition of the Badenian (Middle Miocene) sulphates in the Carpathian Foredeep. *Geol. Carp.* **2002**, *53*, 391–398.
71. Peryt, T.M.; Hryniv, S.P.; Anczkiewicz, R. Strontium isotope composition of Badenian (Middle Miocene) Ca-sulphate deposits in West Ukraine: A preliminary study. *Geol. Quart.* **2010**, *54*, 465–476.
72. Galamay, A.R. Physico-chemical conditions of the formation of the Badenian salt deposits of the Ukrainian Forecarpathians (Grynivka area). *Heol. Heokh. Hor. Kop.* **2010**, *151*, 64–77. (In Ukrainian with English summary)
73. Galamay, A.R. Influence of continental run-off on the composition of marine brines of Badenian salt basin central part (Ukrainian Carpathian Foredeep). *Miner. Zb.* **2012**, *62*, 228–235. (In Ukrainian with English summary)
74. Środoń, J. Illite group clay minerals. In *Encyclopedia of Sediments and Sedimentary Rocks*; Middleton, G.V., Church, M.J., Coniglio, M., Hardie, L.A., Eds.; Springer: Dordrecht, The Netherlands, 1978; p. 115.
75. Meunier, A. *Clays*; Springer: Berlin/Heidelberg, Germany, 2005; 472p.
76. Zhao, G.; Peacor, D.R.; McDowell, S.D. "Retrograde diagenesis" of clay minerals in the Precambrian Freda Sandstone, Wisconsin. *Clays Clay Miner.* **1999**, *47*, 119–130. [CrossRef]
77. Rosenberg, P.E. The nature, formation, and stability of end-member illite: A hypothesis. *Am. Miner.* **2002**, *87*, 103–107. [CrossRef]
78. Korenevskiy, S.M. Miotsenovyye vulkanicheskiye tufy Predkarpatia. *Tr. VNIIGalurgii* **1954**, *29*, 176–196. (In Russian)
79. Dzhinoridze, N.M.; Rogova, M.S.; Telegin, V.P. Vulkanogennyye porody Kalush-Golynskogo mestorozhdeniya kaliynykh soley. *Tr. VNIIGalurgii* **1974**, *71*, 36–56. (In Russian)
80. Turchinov, I.I. The Badenian (Middle Miocene) gypsum section in Kryvche (Podolia, West Ukraine). *Biul. Państw. Inst. Geol.* **1999**, *387*, 70–74.
81. Claret, F.; Bauer, A.; Schäfer, T.; Griffault, L.; Lanson, B. Experimental investigation of the interaction of clays with high pH solutions: A case study from the Callovo-Oxfordian formation, Meuse-Haute Marne underground laboratory (France). *Clays Clay Miner.* **2002**, *50*, 633–646. [CrossRef]
82. Claret, F.; Sakharov, B.A.; Drits, V.A.; Velde, B.; Meunier, A.; Griffault, L.; Lanson, B. Clay minerals in the Meuse-Haute Marne underground laboratory (France): Possible influence of organic matter on clay mineral evolution. *Clays Clay Miner.* **2004**, *52*, 515–532. [CrossRef]
83. Klubova, T.T. *Glinistye Mineraly i Ikh Rol v Genezise, Migratsii i Akkumulatsii Nefti*; Nedra: Moskva, Russia, 1973; p. 255. (In Russian)
84. Rowe, M.C.; Brewer, B.J. AMORPH: A statistical program for characterizing amorphous materials by X-ray diffraction. *Comput. Geosci.* **2018**, *230*, 21–31. [CrossRef]

85. Korchagina, Y.I.; Chetverikova, O.P. *Metody Issledovaniya Rasseyannogo Organicheskogo Veshchestva Osadochnych Porod*; Nedra: Moskva, Russia, 1976; p. 229. (In Russian)
86. Grim, R.E. *Clay Mineralogy*; McGraw-Hill Book Company: New York, NY, USA, 1969; p. 422.
87. Lagaly, G.; Ogawa, M.; De'ka'ny, I. Clay Mineral Organic Interactions. In *Developments in Clay Science: Vol. 1. Handbook of Clay Science*; Bergaya, F., Theng, B.K.G., Lagaly, G., Eds.; Elsevier: Amsterdam, The Netherlands, 2006; pp. 309–378.
88. Więcław, D.; Lytvyniuk, S.; Kovalevych, V.; Peryt, T.M. Fluid inclusions in halite and bitumens in rock salt from Miocene evaporites in the Ukrainian Fore-Carpathian region: Evidence for hydrocarbon accumulations in the underlying strata. *Prz. Geol.* **2008**, *56*, 837–841. (In Polish with English summary)
89. Semchuk, Y.M. *Naukovi ta Metodychni Osnovy Okhorony Heolohichnoho Seredovyshcha v Raionakh Rozrobky Kaliinykh Rodovyshch (na Prykladi Peredkarpattia)*; Vasyl Stefanyk Precarpathian National University: Ivano-Frankivsk, Ukraine, 1995. (In Ukrainian)
90. Lipnitskiy, V.K. Litologicheskiye osobennosti i solevoy kompleks chetvertichnykh otlozheniy i porod gipsovo-glinistoy shlyapy Stebnikskogo mestorozhdeniya kaliynykh soley. In *Materialy po Gidrogeologii i Geologicheskoy Roli Podzemnykh Vod*; Izdatelstvo Leningradskogo Universiteta: Leningrad, Russia, 1971; pp. 98–108. (In Russian)
91. Środoń, J.; Elsass, F.; McHardy, W.J.; Morgan, D.J. Chemistry of illite-smectite inferred from TEM measurements of fundamental particles. *Clay Miner.* **1992**, *27*, 137–158. [CrossRef]
92. Petrichenko, O.Y. *Fiziko-Khimicheskiye Usloviya Osadkoobrazovaniya v Drevnikh Solerodnykh Basseynakh*; Naukova Dumka: Kiev, Russia, 1988; p. 128. (In Russian)
93. Kovalevich, V.M. *Fiziko-Khimicheskie Uslovia Formirovania Soley Stebnikskogo Kaliynogo Mestorozhdenia*; Naukova Dumka: Kiev, Russia, 1978; p. 100. (In Russian)
94. Petrichenko, O.I. *Epigenez Evaporitov*; Naukova Dumka: Kiev, Russia, 1989; p. 64. (In Russian)
95. Kutas, R.I. Heat flow and geothermal crustal model of the Ukrainian Carpathians. *Geophys. J.* **2014**, *6*, 3–25.
96. Kotarba, M.J.; Koltun, Y.V. The origin and habitat of hydrocarbons of the Polish and Ukrainian parts of the Carpathian Province. *Am. Ass. Petrol. Geol. Mem.* **2006**, *84*, 395–442.
97. Andreucci, B.; Castelluccio, A.; Corrado, S.; Jankowski, L.; Mazzoli, S.; Szaniawski, R.; Zattin, M. Interplay between the thermal evolution of an orogenic wedge and its retro-wedge basin: An example from the Ukrainian Carpathians. *Geol. Soc. Am. Bull.* **2015**, *127*, 410–427. [CrossRef]
98. Vityk, M.O.; Bodnar, R.J.; Dudok, I.V. Fluid inclusions in "Marmarosh Diamonds": Evidence for tectonic history of the Folded Carpathian Mountains, Ukraine. *Tectonophysics* **1996**, *255*, 163–174. [CrossRef]

minerals

Article

The Burela Kaolin Deposit (NW Spain): Genesis, Composition and Micro- and Nanotexture

Blanca Bauluz [1,*], Alfonso Yuste [1], Sergio Alvira [2] and Andrea García-Vicente [3]

1 Instituto Universitario de Ciencias Ambientales de Aragón-Universidad de Zaragoza, Campus San Francisco, 50009 Zaragoza, Spain; alfon@unizar.es
2 Departamento de Ciencias de la Tierra, Universidad de Zaragoza, Campus San Francisco, 50009 Zaragoza, Spain; 735526@unizar.es
3 Geology Department, Universidad de Salamanca, 37008 Salamanca, Spain; andrea_gar@usal.es
* Correspondence: bauluz@unizar.es

Abstract: A set of samples from the Monte Castelo kaolin deposits (Burela, NW Spain), corresponding to igneous acidic rocks affected by chemical weathering with variable intensity have been investigated in order to establish the mineralogical and chemical changes with weathering, and the micro- and nano-scale textures developed. For the study, XRD, FESEM, HRTEM and chemical analyses have been used. The more intense the weathering, the more the dissolution of aluminosilicates (albite, K feldspars and K micas) and the crystallization of kaolinite is favored. Kaolinite grows, forming booklets and generating a fine-grained matrix and, along the cleavages of muscovite, forming mica–kaolinite intergrowths. Bidimensional crystallographic continuity between mica and kaolinite has been observed and no intermediate phases have been identified as a consequence of the high W/R ratio. Kaolin mainly contains kaolinite with high crystallinity; however, when there are quartz impurities, they interfere with the 'optimal' reflections for the calculation of the Hinckley index. In this case, the use of the AGFI index almost eliminates the effect that the relative intensities of the quartz and feldspar impurities may have on those of kaolinite. With weathering, there is a progressive decrease in the contents of most chemical elements, except Al, TiO_2, HREEs, Ta, Hf, Th, U, V, Cr, S, Zr, Mo and Sn.

Keywords: kaolin; genesis; SEM; TEM; geochemistry

1. Introduction

After the closure of the Sargadelos factories in 1875, the extraction of local kaolin continued in Burela, which was subsequently exported to other locations in the Iberian Peninsula. At the beginning of the 20th century, a foreign industrial bourgeoisie settled in Burela and took over the exploitation of kaolin. Kaolin is an industrially relevant mineral. It has been used for centuries for numerous purposes, such as the manufacture of white ceramics, which gained great fame in the 16th century. The main uses for which it is currently used are paper and ceramic manufacturing, production of refractories, paints and varnishes, generation of polymers and fiberglass, and the manufacture of pesticides, fertilizers, and cosmetics.

The genesis of primary kaolin may be a consequence of different geological processes. The primary environments of kaolinization are (1) vein-type deposits, (2) (sub) volcanic and pyroclastic deposits, (3) skarn to epithermal deposits, (4) granitic rocks and their affiliated rocks (pegmatites and greisen). By contrast, the secondary kaolinization are (1) kaolin and soils, (2) layered residual kaolin deposits (mixed-type residual kaolin-bauxite deposits, exposed residual kaolin, hidden residual kaolin), (3) vein-like kaolin, (4) alluvial-fluvial environments, fluvial meandering streams, (5) prograding fluvial deltas into a marine or lacustrine basin, (6) lakes and ponds (permanent and ephemeral lakes), (7) coal-bearing environments (suspended load deposits in coal swamps, underclays, composite residual and hydrothermal kaolinization in coal swamps, kaolin tonsteins), (8) marine terrigenous shoreline deposits. As tertiary environments of kaolinization, the following can be cited: (1) burial diagenesis, (2) very-low-grade regional dynamo metamorphism and (3) contact metamorphism [1]. It is also possible to cite a combination of several processes.

In the Variscan belt of Galicia (NW Spain), the kaolinitization of crystalline rocks, metapelites and volcanic rocks are widely spread. Large deposits of kaolin, and also minor occurrences, are hosted by feldspathic rocks, mainly granites and rhyolites. Kaolinitization processes occurred during post-Alpine weathering events that obliterated most of the earlier hydrothermal features [2–6]. The area of Burela (NW Spain) has the largest kaolin deposits in Spain.

The aim of this study is to investigate the deposits of Montecastelo (Burela, NW Spain) that are formed by large-scale weathering of Variscan granitoids. The kaolin is formed from two mica leucogranites and alkali feldspars granites belonging to a complex and heterogeneous magmatic body (San Ciprian pluton) that intruded into an early Cambrian quartzite sequence during the latest stages of the Varican orogeny [5,7].

2. Geological Setting

In the Variscan belt of Galicia (NW Spain), the kaolinitization of crystalline rocks, metapelites and volcanic rocks are widely spread (Figure 1). Large deposits of kaolin and also minor occurrences are hosted by feldspathic rocks, mainly granites and rhyolites. Kaolinitization processes occurred during post-Alpine weathering events that obliterated most of the earlier hydrothermal features [2–6].

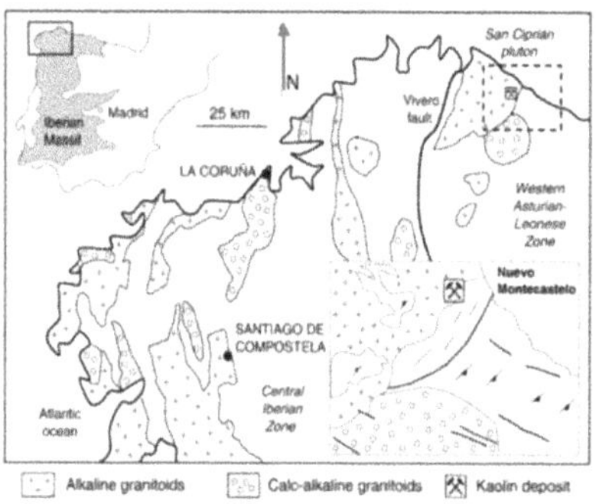

Figure 1. Geological sketch of the NW Iberian massif showing the location of the Nuevo Montecastelo kaolin deposits (modified from [5]).

According to [6], temperature ranges from 15 to 35 °C, with an average of approximately 28 °C, which was calculated on the basis of the isotopic signatures ($\delta^{18}O$, δD) for the kaolin materials. This scatter suggests that, if continental weathering was involved in the kaolin formation on the lower side of the temperatures, it was not the only process, especially for the kaolin associated with felsites and metapelites. The higher temperatures are indicative of a hydrothermal auto-metamorphic alteration. The area of Burela (NW Spain) has the largest kaolin deposits in Spain. They have been mined intermittently since the 19th century, and they are still being mined today.

These deposits of kaolin are characterized by a white to light beige color which clearly differentiates them from other types of deposits such as those in Georgia (USA) with pink and cream colors [1] and from the deposits of ball clays in the NE of the Iberian Peninsula, which are characterized by gray colors due to the presence of organic matter [8].

3. Materials and Methods

Eight samples were extracted from boreholes that cross granitic facies with different degrees of alteration. Samples BU-G1 and BU-G2 correspond to fresh granite, BU-GA1 and BU-GA2 correspond to altered granites, samples BU-C1 and BU-C2 correspond to highly altered granite and samples BU-Kaol1 and BU-Kaol2 correspond to pure kaolin (Figure 2).

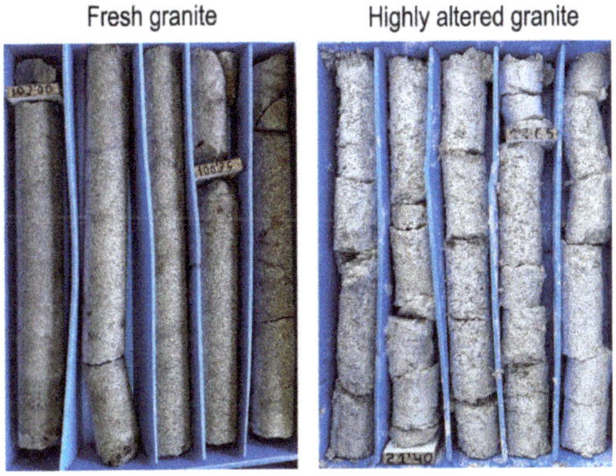

Figure 2. Boreholes of fresh and highly altered granites.

3.1. X-Ray Diffraction (XRD)

The mineralogy of the samples was analyzed by X-ray diffraction (XRD). To obtain the diffraction patterns, a Philips 1710 diffractometer was used at the University of Zaragoza (Zaragoza, Spain), with 40 kV voltage, 30 mA current, CuKα radiation, an automatic slit, and a graphite monochromator. The XRD patterns were acquired from 3 to 60 ° 2θ for the whole rock samples. The goniometer velocity was 0.1°2θ/s and the integration time was 0.45 s. The recording was made with XPowderX software [9]. The relative proportions of the minerals were calculated using Reference Intensity Ratio (RIR) values from the literature [10–12]. These RIR values were calculated in accordance with [13]. The X-ray diffraction measurements in capillary were acquired using an X'Pert PRO MPD PANanalytical diffractometer (Eindhoven, The Netherlands) configured in 2theta with molybdenum radiation, Mo $K_{\alpha 1}$ (λ = 0.70929 Å). The X-ray diffraction patterns were obtained at the Central Research Support Services (SCAI) of the University of Malaga. The experimental conditions used for XRD data collection were step-size 0.02°2θ and a counting time of 1.33 s/step. To achieve high-resolution X-ray diffraction data for distinguishing kaolinite polytypes and calculating crystallinity indices, we employed Mo Kα radiation in a capillary geometry setup. Mo radiation, with its shorter wavelength than Cu radiation, improves angular resolution and reduces sample absorption. The capillary setup allows for a uniform and random orientation of crystallites, minimizing preferred orientation artifacts. This configuration provides strictly monochromatic radiation and enhanced data quality, essential for resolving subtle structural differences among kaolinite polytypes. Table 1 clarifies the use of Mo and Cu radiation.

Table 1. Equivalence of Miller indexes and 2θ position of kaolinite peaks for both copper and molybdenum anodes.

d (Å)	2θ (Mo Kα)	2θ (Cu Kα)	hkl
7.15	5.69	12.37	001
4.472	9.10	19.84	020
4.363	9.32	20.34	$1\bar{1}0$
4.182	9.73	21.23	$\bar{1}\bar{1}1$
4.137	9.84	21.46	111
3.849	10.57	23.09	021
3.732	10.91	23.82	$0\bar{2}1$
3.58	11.37	24.85	002
2.56	15.93	35.02	$\bar{1}30$
2.53	16.12	35.45	$\bar{1}30$
2.49	16.38	36.04	$\bar{1}31$
2.38	17.14	37.77	003
2.34	17.43	38.44	$1\bar{3}1$
2.29	17.82	39.31	131
1.49	27.54	62.26	060

Kaolinite crystallinity has been extensively examined through the analysis and modeling of X-ray diffraction patterns, leading to the proposition of various crystallinity indices [14–17]. For this study, three commonly used indices in the literature—Hinckley index, AGFI index and Liétard index—were selected for assessing kaolinite crystallinity.

The Hinckley index (HI) is calculated as the ratio of the sum of A and B parameters, corresponding to the intensity of the ($1\bar{1}0$) and ($1\bar{1}1$) X-ray diffraction peaks, respectively (Figure 3). These measurements are taken from a background traced between the ($020/1\bar{1}0$) inter-peak base and the ($110, 1\bar{1}1$)/(111) base-peak. This index is computed when the percentage of kaolinite is sufficiently high to accurately measure the intensities described. As outlined by [15], the Hinckley index (HI) is sensitive to the proportion of well-ordered versus high-defect kaolinite in a given sample. Higher HI values indicate a greater abundance of well-ordered kaolinite domains. In general, Hinckley index values can vary from <0.5 for poorly ordered kaolinite to 1.5 for well-ordered kaolinite [18].

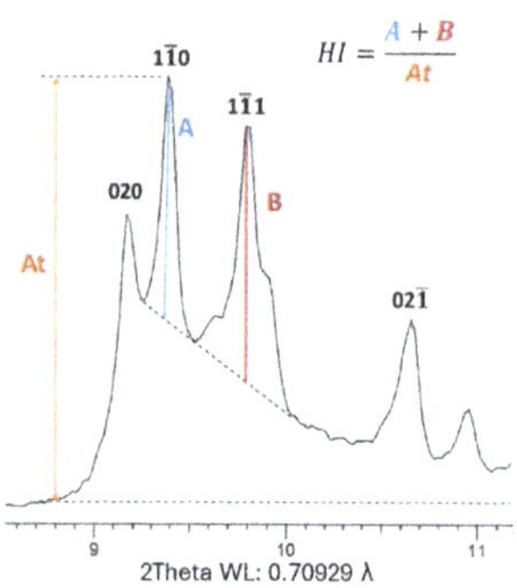

Figure 3. Hinckley index measurement.

The AGFI index ([17]) is calculated for samples containing more than 10 wt.% of kaolinite. It is determined as the sum of deconvoluted intensities of the $1\bar{1}0$ and 111 reflections of kaolinite, divided by twice the intensity of the 020 reflection. This index, known as the Aparicio–Galán–Ferrell index, is measured on 02l and 11*l* reflections after decomposing individual peaks within the complex diffraction band to minimize interferences. The AGFI index exhibits a strong correlation with the percentage of low-defect kaolinite and the Hinckley Index. Notably, it is less susceptible to interference from associated minerals and X-ray amorphous phases compared to other indices. The AGFI index can be effectively utilized for determining the order-disorder nature of kaolinite in a broad range of kaolins and kaolinitic rocks, with the condition that the kaolinite content exceeds 10 wt.% for reproducible results.

The Liétard index (R2) is computed as the ratio between half of the sum of the K2 and K1 parameters, corresponding to the intensity of the (132, $1\bar{3}1$) and (131) diffraction peaks, respectively. This is subtracted from the intensity of the inter-peak between these two reflections, denoted as the parameter K, over the general XRD background. The denominator consists of a third of the sum of the K, K2 and K1 parameters (refer to Figure 4). The Liétard index is designed to assess the presence of random defects in the (ab) plane, which have been interpreted as stacking faults, according to [19]. According to [18], the Liétard index ranges from <0.4 for high-defect kaolinite to 1.2 for low-defect kaolinite.

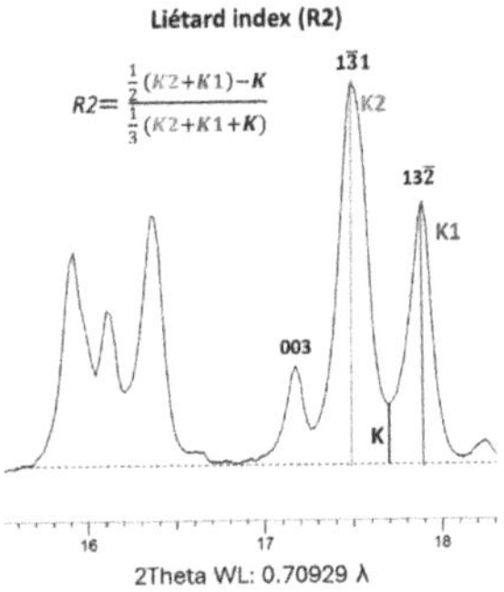

Figure 4. Liétard index measurement.

The Hinckley and AGFI indices differ in their interaction with impurities in samples. The Hinckley index is influenced by quartz, as its 100 reflection overlaps with the 111 reflection of kaolinite. This presence of quartz and feldspar raises the Hinckley index, while the AGFI index remains stable. Additionally, the AGFI index seems to be not affected by other minerals or amorphous phases. It can be applied in all cases except when the amount of halloysite exceeds 50 wt.%. The AGFI index aligns with other methods for estimating the degree of structural order/disorder in a sample, and its primary advantage lies in its applicability to complex samples, allowing the study of kaolinite crystallinity without the need for sample purification.

3.2. Field Emission Electron Microscopy (FESEM)

Twelve thin sections of the granite samples with variable degrees of alteration were studied using a Carl Zeiss Merlin field emission scanning electron microscope (FESEM) equipped with an Oxford energy-dispersive X-ray (EDS) detector at the University of Zaragoza (Zaragoza, Spain). For this, the thin sections were previously carbon-coated. Compositional images of the samples were obtained using two types of backscattered electron detectors: angular selective (AsB). The accelerating voltage used for the former and for the EDS was 15 kV with a beam current of 400 pA. In addition, morphological images were obtained from fragments of the rocks using a secondary electron (SE) detector (Inlens). In this case, the accelerating voltage was 5 kV with a beam current of 800 pA. Semi-quantitative analyses were carried out using the EDS detector, which has a detection limit of 0.1%. From these analyses, the mica formula was calculated on the basis of $O_{10}(OH)_2$.

3.3. Transmission Electron Microscopy (TEM)

For Transmission Electron Microscopy (TEM) study, two lamellas were extracted perpendicularly to two thin carbon-coated thin sections using a Dual Beam equipment and working with backscattered electron images (Figure 5). Specifically, they were extracted from muscovite–kaolinite intergrowths, with the aim of determining the textural and compositional relationships between these two phases. One of the lamellas was analyzed to achieve high-resolution transmission electron images (HRTEM) using a Tecnai F30 TEM (Eindhoven, The Netherlands), with a HADDF detector and equipped with an energy-dispersive spectroscopy (EDS) system in the Advanced Microscopy Laboratory of the University of Zaragoza, Spain. The accelerating voltage was 300 kV with a beam current of 92 μA. The second lamella was analyzed using a HRTEM working in Scanning TEM mode at 300 Kv;this equipment is equipped with a high-brightness XFEG field emission gun, a monochromator and a 2 k × 2 k Gatan camera for high-resolution image acquisition.

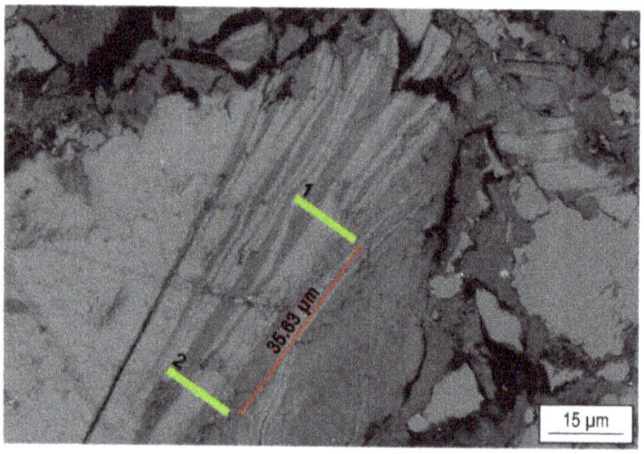

Figure 5. BSE image showing with two green lines the areas from which lamellas 1 and 2 were extracted on the mica–kaolinite intergrowths.

3.4. Methodology of Chemical Analysis

Chemical analyses of major and trace elements in bulk samples were performed at Actlabs Laboratories (Canada). Major elements and some trace elements (V, Sr, Ba, Sc, Be) were determined by inductively coupled plasma optical emission spectroscopy (ICP-OES). Fused samples (lithium metaborate/tetraborate fusion) were rapidly digested in a weak nitric acid and run on a combination of simultaneous/sequential Thermo Jarrell-Ash ENVIRO II ICP or a Varian Vista 735 ICP. Calibration was performed using seven prepared USGS- and CANMET-certified reference materials. One of the seven standards was used during the analysis for every group of ten samples. The detection limit for the major elements was 0.01% (except for MnO and TiO_2 with detection limits of 0.001%). The rest of the trace elements were analyzed by inductively coupled plasma mass spectrometry (ICP-MS). Fused samples were diluted and analyzed by a Perkin Elmer Sciex ELAN 6000, 6100, or 9000 ICP-MS. Three blanks and five controls were analyzed per group of samples. Duplicates were fused and analyzed every 15 samples. The detection limit for each trace element is indicated in brackets: Cr and Ni (20 ppm), Co, Rb, and Zr (1 ppm), Y (0.5 ppm), Nb (0.2 ppm), Cs and Hf (0.1 ppm), Th, La, Ce, and Nd (0.05 ppm), U, Pr, Sm, Gd, Tb, Dy, Ho, Er, and Yb (0.01 ppm), Lu (0.002 ppm), and Eu and Tm (0.005 ppm).

4. Results

4.1. Mineralogy

XRD patterns show the presence of quartz, albite and K feldspar, and micas, kaolinite and calcite in minor proportions (Figure 6). With the progress of the alteration, albite and K feldspar contents decrease notably producing and increase in the contents of quartz (up to 82%) and kaolinite (up to 8%). All the samples have calcite and kaolinite in low proportions. Additionally, two samples of pure kaolin were available (BU-Kaol1 and BU-Kaol2).

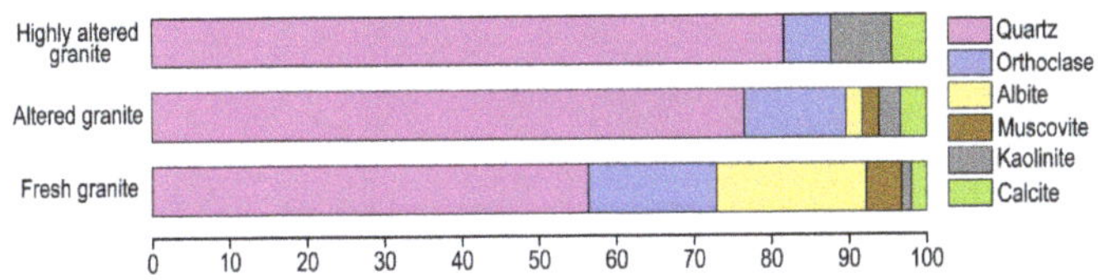

Figure 6. Average mineral composition (XRD) of unaltered rocks (BU-G), altered rocks (BU-GA) and highly altered rocks (BU-C).

Kaolin: X-Ray Characterization

All studied kaolin samples (BU-Kaol1 and BU-Kaol2) show the typical profile of a kaolinite at first glance, with the 001 reflection at ~7.15 Å, a similar sequence of peaks, and the 060-reflection located at ~1.49 Å, characteristic of dioctahedral minerals. The main identified phase is kaolinite, with the subordinate components (in minor quantities) being illite and quartz. A subdivision as a function of disorder degree into ordered kaolinite (BU-Kaol1) and medium-ordered kaolinite (BU-Kaol2) can be performed by estimating crystallinity indices. The difference between ordered kaolinite and medium-ordered kaolinite is noticed by a variation in hkl intensities, as well as the influence and increasing amount of quartz impurities affecting *hk0* and *hkl* reflections. Quartz impurities interfering with $\overline{1}\,\overline{1}1$ (4.168 Å) reflection of kaolinite lead to overestimating the Hinckley and Liétard indices in the case of the BU-Kaol2 sample. In the BU-Kaol1 sample, four reflections characteristic of kaolinite, 020 (4.472 Å), 110 (4.363 Å), 111 (4.182 Å) and (4.137Å) appear in the region between 19 and 24°2θ (λ = 1.54 Å). Several minor reflections assigned to kaolinite appear in this region, 3.849 Å and 3.732 Å. The region comprised between 34 and 40°2θ (λ = 1.54 Å) is affected by stacking disorder and the presence of stacking defects/faults of kaolinite. It is worth estimating the kaolinite crystallinity indices in both regions, that is, the Hinckley and Liétard index. The Hinckley index (HI) of the BU-Kaol1 sample is 1.11, whereas the Liétard index is 1.19. Little influence of quartz impurities is noticed in the case of the BU-Kaol1. Moreover, in the BU-Kaol2 sample, quartz impurities increase and lead to higher values of KCI (kaolinite crystallinity index). Hence, the AGFI index is a suitable method to "estimate" the degree of disorder in those reflections not biased by the presence of quartz. Pearson VII function deconvolution was used following methodology by [18]. In BU-Kaol 1, the AGFI index is about 1.05 and, in BU-Kaol2, the AGFI index is about 1.14, whereas the Hinckley index and Liétard index reached values up to 1.74 and 1.23, respectively. As the presence of quartz increases, hk0 and 131 relative intensities decrease.

4.2. FESEM

Backscattered electron images (BSE) of fresh granites (Figure 7A) show the typical textures of plutonic acidid rocks with scarce evidence of alteration. In some areas of these rocks, mica–kaolinite intergrows are observed (Figure 7B). Micas show light color in contrast with the dark gray of kaolinite, as a consequence of the different molecular weight of these phases. Figure 7C corresponds to altered rocks that contain frayed micas immersed on a dark matrix. The EDS mapping (Figure 7D) shows an Al-rich matrix indicating that the matrix is mainly composed of kaolinite. Figure 7E shows micas with fuzzy edges and irregular outlines; cleavage is not distinguished, indicating that the alteration process is destroying the mineral structure. On the other hand, the replacement of micas by Fe oxides can be observed (Figure 7F).

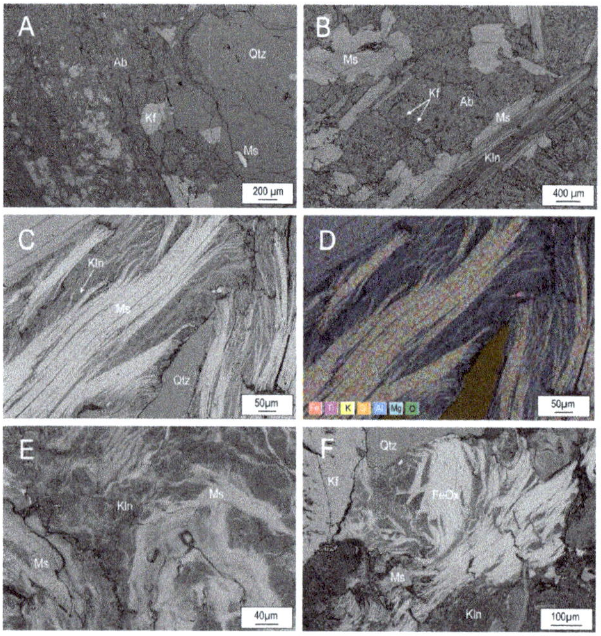

Figure 7. (**A**) Scarcely granite formed by quartz, K feldspar, albite and micas. (**B**) Granite with incipient alteration and formation of mica–kaolinite intergrowths. (**C**) Frayed micas immersed on a kaolinitic matrix. (**D**) EDS mapping on figure (**C**) showing the Al-rich matrix mainly composed of kaolinite. (**E**,**F**): Highly altered rocks. Kf = K feldspar, Ab = Albite, Qtz = Quartz, Ms= Mica, Kln = Kaolinite, FeOx = Fe oxides.

Figure 8 shows three compositional groups of micas: (1) those having high K content in the interlayer (0.84–0.93 K apfu) corresponding to fresh granites, (2) those with K in interlayer ranging from 0.50 to 0.80 K apfu included in altered granites, and (3) a third group of micas with very low K content in the interlayer (0.25–0.03 apfu), that are present in highly altered granites.

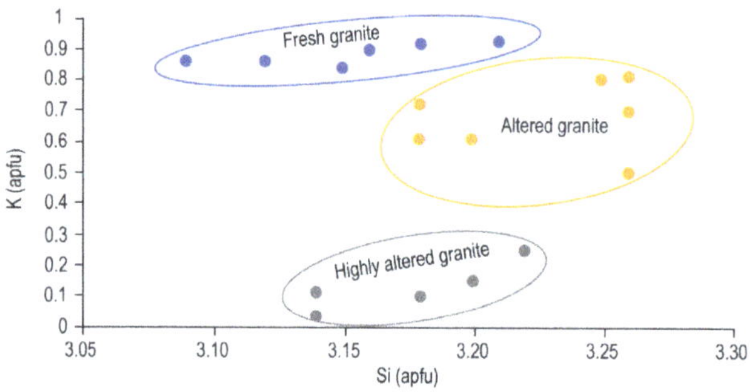

Figure 8. Si vs. K plot of the analyzed micas (EDS).

The aluminum distribution through the mica–kaolinite intergrowth shows sharp limits between these phases (Figure 9A), and Figure 9C,D shows booklets of kaolinites forming the matrix.

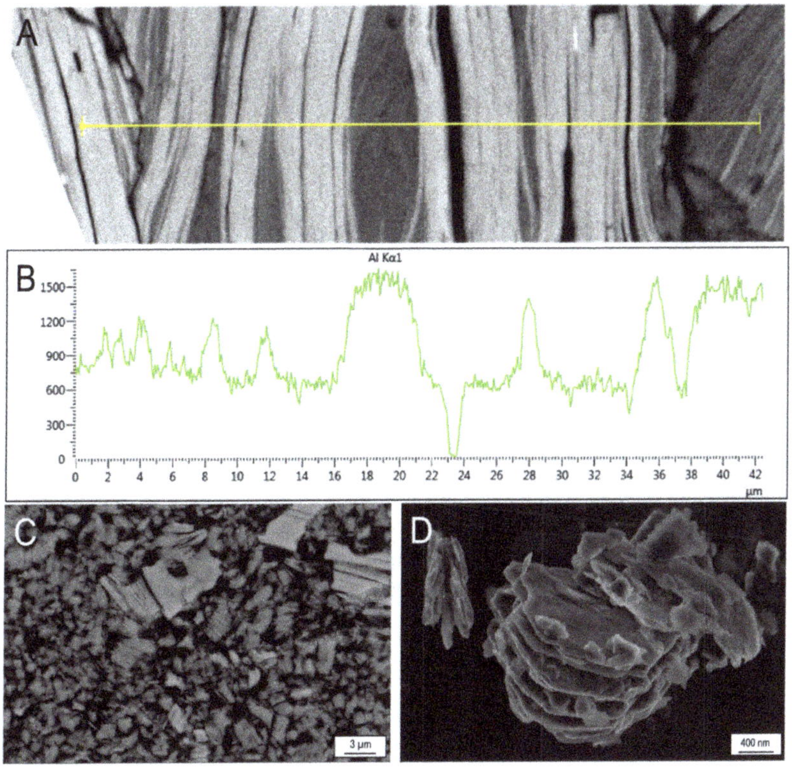

Figure 9. (**A**) The image shows an enlarged image of a mica–kaolinite intergrowth. (**B**) Image shows the aluminum distribution (EDS analysis) along a line perpendicular to the exfoliation planes of micas and kaolinite. The aluminum line shows sharp limits between mica and kaolinite. This image also shows an intense altered granite in which abundant crystals and kaolinite aggregates were formed. (**C**) It shows an SE image of euhedral kaolinites forming booklets, which reach up to 3 µm of longitudinal development in the direction of the c crystallographic axis, and mica relicts. (**D**) Image shows an SE image of purified kaolin.

Heavy minerals such as monazite apatite, zircon and Fe oxides have been detected by FESEM.

4.3. HRTEM

High resolution TEM images from muscovite–kaolinite intergrowths (Figure 10) show that lattices fringes of muscovite and kaolinite are parallel to each other. Phase boundaries between both minerals show bidimensional crystallographic continuity, and no intermediate phases have been identified. Kaolinite and muscovite occur as thick packets free of interlayering.

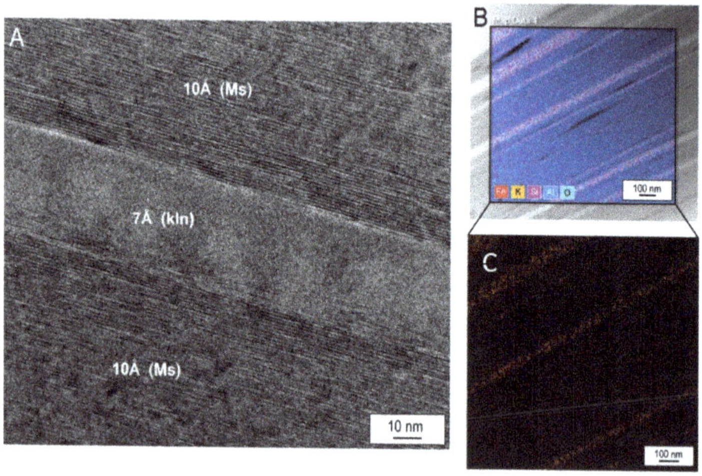

Figure 10. (**A**) TEM image from muscovite–kaolinite intergrows in sample BU-G2, (**B**) EDS mapping at the nanoscale showing the distribution of Fe, K, Si, Al and O in sample BU-G2, (**C**) EDS mapping showing the K distribution at the nanoscale.

4.4. Chemical Analysis

Fresh granite is characterized by high amounts of SiO_2 (71.67%) and Al_2O_3 (14.61%), and lower proportion of K_2O (5.41%) and Na_2O (2.21%). Much lower proportion of MgO (0.42%), CaO (0.29%), P_2O_5 (0.19%), TiO_2 (0.18%), MnO (0.02) and LOI (1.88%) has been detected. Data in Table S1 (Supplementary Material). Figure 11A shows that, from weathered to highly weathered granite, there is a decrease in K_2O contents, in accordance with the mineralogical variations described. In contrast, there is an increase in MnO and TiO_2 in weathered granite and in highly weathered granite, indicating that these elements have low mobility in these conditions. The high LOI values indicate the presence of hydrated phases as phyllosilicates and clays.

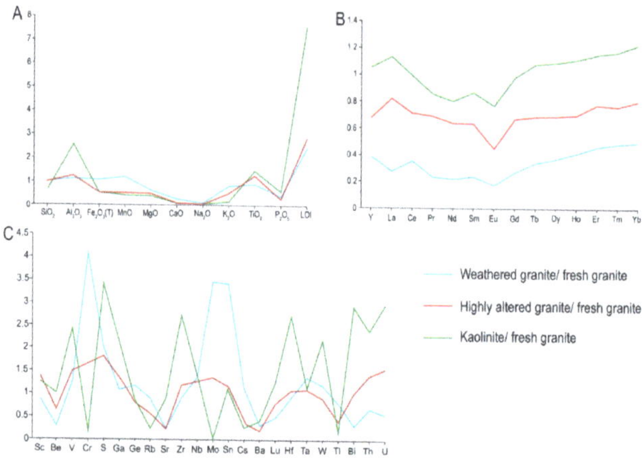

Figure 11. Major (**A**), REE + Y (**B**) and trace elements (**C**) contents in weathered granites, highly altered granites and pure kaolinite, normalized to fresh granite.

On the other hand, pure kaolinite is enriched in Al_2O_3, TiO_2 and has higher LOI. Figure 11B reflects that weathered and highly altered granites are depleted in Y + REE, and only kaolin shows slight enrichments in these elements. Trace elements (Figure 11C) show strong variabilities in weathered granites: there is an increase in Cr, Mo and Sn, whereas in highly altered granite, there is an increase in V, S and Mo. In contrast, pure kaolinite concentrates V, S and Zr. REE + Y are mobile under weathering conditions and Hf, Ta, W, U, Ta, and W are relatively immobile probably because they are hosted by heavy minerals.

5. Discussion

Fresh and altered granites show similar mineralogy. They consist of high amounts of quartz, K feldspar, albite, micas and a very low proportion of kaolinite and calcite. The replacement of alkali and plagioclase feldspars by albite is a prevalent phenomenon in the Earth's upper crust [20]. The process is known as albitisation and takes place in a wide variety of rock types and tectonic settings, e.g., during the hydrothermal alteration of granites [21]. In the studied fresh granite, the presence of minor calcite could be related to a previous hydrothermal event, since the Ca^{2+} liberated during the transition from oligoclase to albite, could have been fixed by secondary calcite [22] and such as [6] suggest. In comparison with fresh granites, in weathered granites, the content in micas and K feldspar decreases, but the most significant difference between the fresh and the weathered granites is the significant decrease in albite (from 13 to 6%) and the relative increase in quartz. Highly altered granites are mainly formed by quartz (more than 80%) and low proportions of K feldspar and muscovite have been destabilized, and kaolinite increases up to 8%. Changes in the proportions of the minerals with weathering indicate that albite is the aluminosilicate that is more easily altered, followed by K feldespar and K micas.

Fresh and altered granites show the typical phaneritic texture of plutonic rocks (Figure 7A,B) and scarce evidence of alteration. However, the images suggest that, in these rocks begins the growth of kaolinite between the exfoliation planes (001) of micas (Figure 7B). In weathered rocks, micas show signs of deterioration: they are frayed and immersed in a kaolinite matrix which is formed simultaneously with the destabilization of aluminosilicates (Figure 7C). The matrix is formed by micron-sized booklets of kaolinite and crystallyzed as a consequence of the alteration process (Figure 7C). In highly weathered rocks, micas show a higher degree of deterioration, with fuzzy edges, irregular outlines and cleavage that is not distinguished, suggesting the deterioration of the structure (Figure 7E). The replacement of micas by Fe oxides suggests the complete destabilization of the micas (Figure 7F).

The progressive alteration of the micas has a direct relation, with the K contents of these minerals in the interlayer site (Figure 8). In fresh granites, micas have high K contents in the interlayer (0.84–0.93K apfu), whereas, in weathered granites, their K content ranges between 0.50 and 0.80 K apfu, suggesting large compositional variability from mica to illite. Finally, in highly altered granites, the K content is very low (0.25–0.03 apfu), and micas have compositions similar to smectite. The data show that, as weathering progresses, micas lose K from the interlayer.

The AGFI index applied to characterize kaolinite aligns with other methods for estimating the degree of structural order/disorder in a sample, and its primary advantage lies in its applicability to complex samples, allowing the study of kaolinite crystallinity without the need for sample purification.

TEM images and EDS mapping at nanoscale from muscovite–kaolinite intergrows (Figure 10) show that these phases are parallel to each other. Phase boundaries between both minerals show bidimensional crystallographic continuity, and no intermediate phases have been identified. Therefore, kaolinite and muscovite occur as thick packets free of interlayering. The images suggest that muscovite only provides suitable substrate for the epitactic crystallization of kaolinite. Additionally, kaolinite growth continues as long as Al is available in the environment due to the destabilization of aluminiumsilicates (K feldspars, albite and micas). The absence of intermediate interstratifications is most likely a consequence of the high water/rock ratio. In closed systems, interstratifications between these phases have been described [8]. Other authors [23] observed similar nanotextures in mica–kaolinite intergrowths in the sandstones of diagenetic environments and indicate that kaolinitization begins at grain edges, forming fanned-out textures, and propagates toward the interior along the cleavages of muscovite. In this study, we do not have evidence to support this process, probably because, in weathering environments, the water/rock ratio is higher than in diagenetic environments. The higher water/rock ratio promotes faster mineral reactions.

Chemical composition of weathered and highly weathered rocks and kaolinite support the alteration processes described, based on mineralogy. K and Na content decrease significantly, suggesting a high mobility of these elements under weathering conditions; in contrast, MnO and TiO_2 are relatively immobile in this environment. Higher LOI is a consequence of the formation of hydrated phases such as clays. Pure kaolinite is enriched in Al_2O_3 and TiO_2, reflecting the low mobility of these elements (Figure 10). Weathered and highly altered granites are depleted in Y + REE, and only kaolin shows slight enrichment in these elements. Therefore, REE + Y are mobile under weathering conditions and Hf, Ta, U and W are relatively immobile, probably because they are hosted by resistant heavy minerals (Figure 10), such as monazite, apatite and zircon, or also adsorbed on the surface of Fe oxides.

6. Conclusions

The clear colors of the kaolin investigated in this study indicate an effective fluid circulation altering the rocks and leaching various components of the original igneous rock. A previous hydrothermal event most likely took place, and subsequently, weathering produced the kaolin deposits from the igneous rocks.

Weathering produces significant changes in the mineralogical composition of granites. Albite and K feldspars destabilized most rapidly, followed by K micas. The availability of aluminum and its relatively immobile behavior in this environment favors the formation of kaolinite.

Kaolinite grows forming booklets generating a fine-grained matrix and also along the cleavages of muscovite generating muscovite–kaolinite intergrowths. The intergrowths show bidimensional crystallographic continuity, and no intermediate phases have been identified, suggesting that muscovite only provides the suitable substrate for the epitactic crystallization of kaolinite. Kaolinite shows high crystallinity but the presence of quartz impurities interferes with the 'optimal' reflections for the calculation of the Hinckley index. In these cases, the use of AGFI index almost eliminates the effect that the relative intensities of the quartz and feldspar impurities may have on those of kaolinite.

With weathering, there is a progressive decrease in the contents of most chemical elements, except for Al, TiO_2, HREEs, Ta, Hf, Th, U, V, Cr, S, Zr, Mo and Sn.

Supplementary Materials: The following supporting information can be downloaded at: https://www.mdpi.com/article/10.3390/min15040416/s1, Table S1: Chemical composition of analized samples.

Author Contributions: Conceptualization, B.B.; methodology, B.B.; investigation, B.B., S.A., A.G.-V. and A.Y.; resources, B.B. and AY; writing—original draft preparation, B.B.; writing—review and editing, A.Y. and A.G.-V.; supervision, B.B.; project administration, B.B.; funding acquisition, B.B: All authors have read and agreed to the published version of the manuscript.

Funding: This research was funded by the Spanish Ministry of Science, Innovation and Universities [grant number PID2021-123127OB-I00] and by the European Regional Development Fund and the Government of Aragon [Aragosaurus Group: Geological Resources and Palaeoenvironments, grant number E18_20R].

Data Availability Statement: Data are contained within the article and Supplementary Materials.

Acknowledgments: We are very grateful to the academic editor and the anonymous reviewers for their suggestions that have improved the manuscript. The authors would like to acknowledge the use of Servicio General de Apoyo a la Investigación-SAI, Universidad de Zaragoza.

Conflicts of Interest: The authors declare no conflicts of interest.

References

1. Dill, H.G. Kaolin: Soil, rock and ore from the mineral to the magmatic, sedimentary and metamorphic environments. *Earth Sci. Rev.* **2016**, *161*, 16–129. [CrossRef]
2. Martín Pozas, J.M.; Galan, E.; Miartin-Vivaldi, J.L. Ill Giaciemento di Caolino di Jove. Congresso Nazionale; I Congresso Nazionale, AIPEA ıt, Lugo, Book of abstracts, Spagna. AIPEA: Pavia, Italy, 1971; pp. 89–109.
3. Galán, E.; Martin-Vivaldi, J.L. Caolines españoles. Geología, mineralogía y génesis. Parte VII: Depósitos hidrotermales. *Bol. Soc. Esp. Cerám Vidr.* **1975**, *14*, 123–144.
4. Galán, E.; Martín-Vivaldi, J.L. Caolines españoles. Geología, mineralogía y génesis. Parte VIII: Depósitos residuales y volcánicos. *Bol. Soc. Esp. Cerám Vidr.* **1975**, *14*, 351–370.
5. Fernández-Caliani, J.C.; Galán, E.; Aparicio, P.; Miras, A.; Márquez, M.G. Origin and geochemical evolution of the Nuevo Montecastelo kaolin deposit (Galicia, NW Spain). *Appl. Clay Sci.* **2010**, *49*, 4991–4997.
6. Galán, E.; Aparicio, E.; Fernández-Caliani, J.C.; Miras, A.; Márquez, M.; Fallick, A.E.; Clauer, N. New insights on mineralogy and genesis of kaolin deposits: The Burela kaolin deposit (Northwestern Spain). *Appl. Clay Sci.* **2016**, *131*, 14–26. [CrossRef]
7. Fernández-Suárez, J.; Dunning, G.R.; Jenner, G.A.; Gutiérrez-Alonso, G. Variscan collisional magmatism and deformation in NW Iberia: Constraints from U–Pb geochronology of granioids. *J. Geol. Soc.* **2000**, *157*, 565–576. [CrossRef]
8. Bauluz, B.; Mayayo, M.J.; Laita, E.; Yuste, A. Micro- and Nanotexture and Genesis of Ball Clays in the Lower Cretaceous (SE Iberian Range, NE Spain). *Minerals* **2021**, *11*, 1339. [CrossRef]
9. Martín, J.D. A software package for powder X-ray diffraction analysis. In *Qualitative, Quantitative and Microtexture*; 2017; 121p. GR 1001/04 ISBN: 84-609-1497-6 (see 2004.01 CDROM) Register number: 4071204. Available online: https://www.xpowder.com/ (accessed on 10 April 2025).
10. Biscaye, P.E. Mineralogy and sedimentation of recent deep-sea clay in the Atlantic Ocean and adjacent seas and oceans. *Geol. Soc. Am. Bull.* **1965**, *76*, 803–832. [CrossRef]
11. Schultz, L.G. *Quantitative Interpretation of Mineralogical Composition from X-ray and Chemical Data for the Pierre Shale*; Professional Paper 391-C; U.S. Geological Survey: Reston, VA, USA, 1964; pp. 1–31.
12. Smith, D.K.; Johnson, G.G., Jr. Digitized database quantification, DDBQ, analysis of complex mixtures using fully digitized patterns. *Adv. X-Ray Anal.* **2000**, *42*, 276–286.
13. Hillier, S. Quantitative analysis of clay and other minerals in sandstones by X-ray powder diffraction (XRPD). *Int. Assoc. Sedi. Spe. Pub.* **2003**, *34*, 213–251.
14. Liétard, O. Contribution à L'étude des Propités Phisicochimiques, Crystallographiques et Morphologiques des Kaolins. Ph.D. Thesis, University of Nancy, Nancy, France, 1977; 345p.
15. Plançon, A.; Giese, R.F., Jr.; Snydier, R.; Drits, V.A.; Bookin, A.S. Stacking faults in the kaolin-group minerals: Defect structures of kaolinite. *Clays Clay Min.* **1989**, *37*, 203–210. [CrossRef]
16. Amigó, J.M.; Bastida, J.; Sanz, A.; Signes, M.; Serrano, J. Crystallinity of Lower Cretaceous kaolinites of Teruel (Spain). *Appl. Clay Sci.* **1994**, *9*, 51–69. [CrossRef]
17. Aparicio, P.; Galán, E.; Ferrell, R.E. A new kaolinite order index based on XRD profile fitting. *Clay Min.* **2006**, *41*, 81–817. [CrossRef]
18. Aparicio, P.; Galán, E. Mineralogical interference on kaolinite crystallinity index measurements. *Clays Clay Min.* **1999**, *47*, 12–27. [CrossRef]
19. Cases, J.M.; Lietard, O.; Yvon, J.; Delon, J.F. Etude des propietes cristallochimiques, morphologiques, superficielles de kaolinites désordonneés. *Bull. Mineral.* **1982**, *105*, 439–455.
20. Hövelmann, J.; Putnis, A.; Geisler, T.; Schmidt, B.C.; Golla-Schindler, U. The replacement of plagioclase feldspars by albite: Observations from hydrothermal experiments. *Contrib. Mineral. Petrol.* **2010**, *159*, 43–59. [CrossRef]

21. Plümper, O.; Putnis, A. The complex hydrothermal history of granitic rocks: Multiple feldspar replacement reactions under subsolidus conditions. *J. Petrol.* **2009**, *50*, 967–987. [CrossRef]
22. Kaur, P.; Chaudhri, N.; Hormann, A.W.; Raczek, I.; Okrusch, M.; Skora, S.; Baumgartner, L.P. Two-Stage, Extreme Albitization of A-type Granites from Rajasthan, NW India. *J. Petrol.* **2012**, *53*, 919–948. [CrossRef]
23. Arostegui, J.; Irabien, M.J.; Nieto, F.; Sangüesa, J.; Zuluaga, M.C. Microtextures and the origin of muscovite-kaolinite intergrowths in sandstones of the Utrillas formation, Basque Cantabrian basin, Spain. *Clays Clay Miner.* **2001**, *49*, 529–539. [CrossRef]

minerals

MDPI

Article

Mineralogical and Geochemical Characterization of the Benavila (Portugal) Bentonites

Javier García-Rivas [1,*], Maria Isabel Dias [2,3], Isabel Paiva [2,3], Paula G. Fernandes [2], Rosa Marques [2,3], Emilia García-Romero [1,4] and Mercedes Suárez [5]

[1] Department of Mineralogy and Petrology, Complutense University of Madrid, 28040 Madrid, Spain; mromero@ucm.es
[2] Center for Nuclear Sciences and Technologies (C2TN), Instituto Superior Técnico, University of Lisbon, Estrada Nacional 10, km 139.7, 2695-066 Bobadela, Portugal; isadias@ctn.tecnico.ulisboa.pt (M.I.D.); ipaiva@ctn.tecnico.ulisboa.pt (I.P.); paulagf@ctn.tecnico.ulisboa.pt (P.G.F.); rmarques@ctn.tecnico.ulisboa.pt (R.M.)
[3] Department of Nuclear Sciences and Engineering (DECN), Instituto Superior Técnico, University of Lisbon, Estrada Nacional 10, km 139.7, 2695-066 Bobadela, Portugal
[4] Institute of Geosciences (IGEO), Complutense University of Madrid–Consejo Superior de Investigaciones Científicas, 28040 Madrid, Spain
[5] Department of Geology, University of Salamanca, 37008 Salamanca, Spain; msuarez@usal.es
* Correspondence: javier.garcia.rivas@ucm.es

Abstract: This work aims to perform a detailed mineralogical, crystal-chemical, and geochemical characterization of bentonites from the Benavila outcrop, the largest known deposit of bentonites in continental Portugal. Bulk samples and different size fractions were characterized through X-Ray Diffraction (XRD). Structural formulae of the smectites were fitted from point analyses acquired by analytical electron microscopy (AEM) with transmission electron microscopy (TEM). Smectites are the major component with variable amounts of calcite and minor amounts of quartz, feldspar, illite, and chlorite. Occasionally, amphiboles and dolomite have also been identified. The high content of carbonates in different parts of the sampling area is related to the circulation of carbonate-rich fluids. The smectites present high-layer charge, are intermediate terms of the montmorillonite–beidellite series, and also show an intermediate cisvacant–transvacant configuration. Major and trace elements concentrations were determined by ICP-MS. The geochemical analysis of the samples indicates an enrichment in SiO_2 and Al_2O_3 and a depletion of the more clayey materials in REE, HFSE, and Y, among others. The calculation of the PIA and CIA alteration indices, along with other parameters observed, shows the possible alteration pathways of the Benavila deposit. Research to evaluate the ability of these bentonites to be used as engineering barrier systems (EBS) and sealing materials for radioactive waste repositories is ongoing.

Keywords: bentonite; smectite; geochemistry; clay minerals; mineralogy

1. Introduction

Bentonites are clayey industrial rocks of great economic interest whose main component is smectites. The most important physical–chemical properties of these minerals are their high specific surface area and cation exchange capacity, which are deeply related [1–4], along with their variable layer charge, allowing them to react with inorganic and organic reagents. Hence, they have swelling and rheological properties and high plasticity, which play a very important role in the industrial field [5,6]. Their applications are diverse,

https://doi.org/10.3390/min15080836

contingent on the aforementioned properties. While their most prominent use is as adabsorbents, their significance in other domains cannot be overstated. For instance, they are used in civil engineering, as binding agents, as additives in animal nutrition, as catalysts or support for catalysts [5–7], which also have to be highlighted. Apart from the typical fields in which these materials have been traditionally used, in recent years there has been a spike in the use of bentonites in the fields of nanocomposites [8–10], pharmaceutics [11,12], medicine [13–15], among many others. The study of bentonites as engineering barrier systems (EBS) and sealing materials in radioactive waste repositories has to be highlighted, since this research field has been one of the most prominent ones in the last four decades regarding these materials [16–22].

The largest known deposit of bentonites in continental Portugal is located in Benavila (in the Alentejo region) (Figure 1). This deposit was formed due to the weathering of granitoids from the Benavila igneous massif [23,24], mainly composed of quartz, plagioclase, and amphibole, along with other accessory minerals [24]. The composition of these plutonic rocks will affect the resulting weathering products, which are the bentonites.

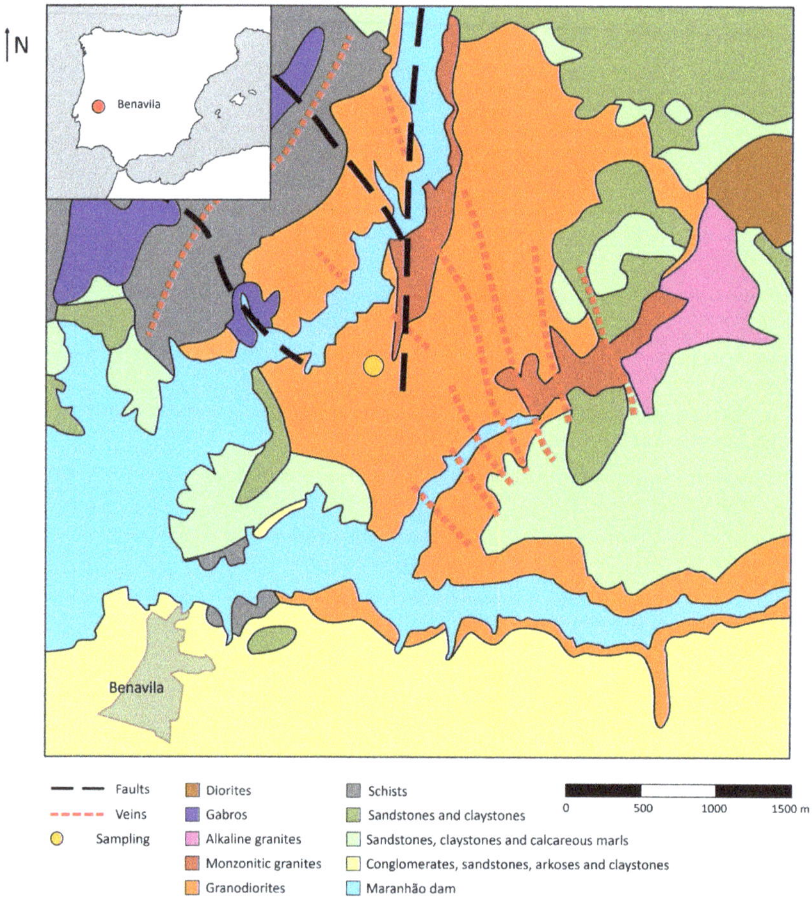

Figure 1. Geographical location of the samples and geological mapping of the area. Modified from [25]. All the plutonic and metamorphic rocks are dated as Silurian, while the sedimentary rocks are dated as Miocene [26,27].

The Benavila deposit reserves were estimated at around 5 M.t. of bentonite, with the highest purity levels reaching 2.0–2.5 Mt [28,29]. The mineralogy and applications of these bentonites have been studied by different authors, determining their mineralogy as being primarily constituted by dioctahedral smectites [29–34]. However, there are slight differences regarding their classification within the group of dioctahedral smectites, because some authors classify them as Tatatila-type montmorillonites [29] and others just as montmorillonites [31].

This work aims to attain a detailed mineralogical, crystal-chemical, and geochemical characterization of a representative outcrop of bentonites from the Benavila deposit. This work will serve as a basis for further research regarding the applications of these bentonites, mainly towards their use as lining barriers of disposed radioactive waste.

2. Geological Setting

The area under study, corresponding to the Benavila igneous massif, is located close to the Avis area, more precisely in Benavila, Portalegre District (Portugal). It is approximately 7 km long (E-W) and 4 km wide, occupying an area of about 40 km^2 (Figure 1). This area exemplifies the typical morphology of Alto Alentejo: a high plateau with scattered reliefs. The petrographic characteristics of the Benavila rocks lead them to be considered similar to those of the neighboring Ervedal, Fronteira, and Stª Eulália massifs. However, the rocks of this region, like those of Stª Eulália, are more diverse than those of the first two massifs mentioned, as the presence of gabbros is referred [35].

The Benavila igneous massif was formed during the Lower Silurian [26,27], and it is characterized by the diversity of petrologic types: granites, granodiorites, monzonites, quartz diorites, and, exceptionally, olivine gabbros. Granitoids, quartz diorites, and gabbros have a common characteristic: the presence of green-colored hornblende amphiboles. The massif is cut by numerous veins, running NW-SE, which are mostly constituted of granite. In addition to the aforementioned veins, there are also many enclaves of felsic nature, namely monzonites, always more basic than the surrounding granitoids. The entire massif is also affected by carbonated alterations of Paleogene–Quaternary age, due to the high content of carbonates in the groundwater [24]. Chemically, the global composition of these rocks located at the Benavila igneous massif is mainly constituted by SiO_2, Al_2O_3, Fe_2O_3, and CaO, representing at least 75% of the total oxide content [24].

The subject of study is bentonites of Miocene age, which have been previously described in the literature as "clayey and marly sandstones with a greenish hue, of continental facies, probably Miocene" [26]. These bentonites originated from the weathering of the preexisting granodiorites located within the massif, all of which present the same type of alteration that partially covers the massif, which seems to be favored by fractures in the NNE–SSW direction [26,27] and is dated in the Paleogene–Quaternary. The Geological and Mining Institute of Portugal refers to these clayey materials of residual origin as "very pure bentonites with relevant technological characteristics and difficult exposure and exploitation conditions" [36].

3. Materials and Methods

3.1. Samples

A bentonite outcrop (Universal Transverse Mercator coordinates: 29S 598,193.02 m E 4,330,882.19 m N) with a thickness of approximately 7 m and a NE–SW direction in the proximity of the locality of Benavila (Portalegre, Portugal) (Figure 1) was sampled. The outcrop is characterized by the greenish colors of the bentonites, as well as by the presence of sub-vertical carbonate-rich veins of whitish color (Figure 2a,d). The presence of these veins is the main difference that can be observed in the horizontal axis. In addition, it is possible to observe a higher degree of compaction of the bentonites towards the top of the outcrop, while the bottom shows more disaggregated materials.

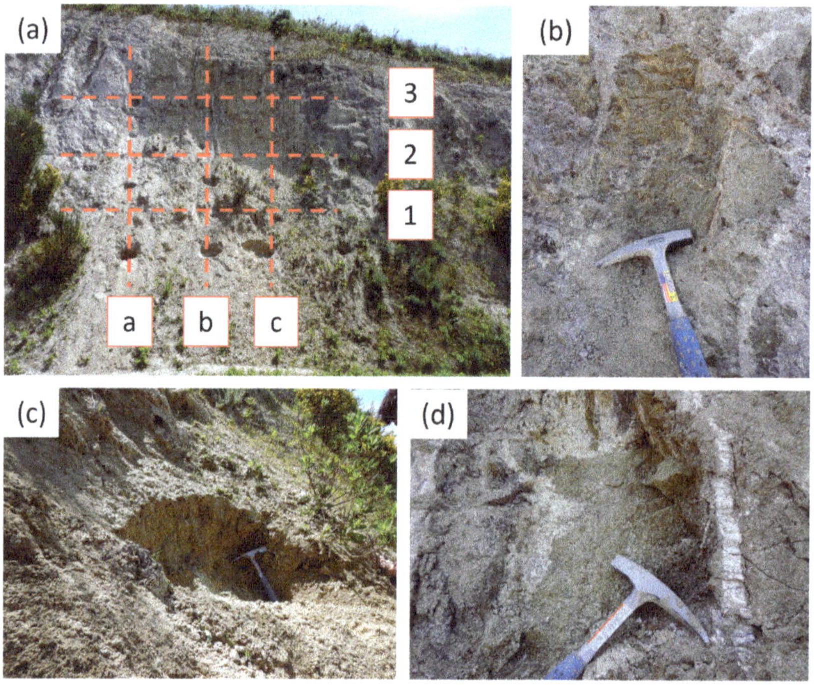

Figure 2. (**a**) General photograph of the outcrop and sampling grid, showing the horizontal profiles (1, 2, 3) and the vertical profiles (a, b, c). (**b**) Bentonite sample with part of the original texture preserved. (**c**) Bentonite sample without preserved original texture. (**d**) Carbonate vein.

As shown in Figure 2a, nine samples from the most representative and accessible parts of the deposit were collected in a grid, with the A profile located towards the NE and the C profile towards the SW. In order to obtain non-contaminated samples, the surface of the outcrop was cleaned at the places where these samples were collected. The samples from the top horizontal profile (BEN3) were interpreted in the field as the samples with the higher content of smectites. They are characterized by their high compaction, and, in some cases, they even preserve their original textures (especially towards the C profile) (Figure 2b), although they have been transformed almost entirely to clay minerals. According to the expected evolution of the weathering profile, which gave place to the formation of the deposit, the lower levels (BEN1 and BEN2) present a lower degree of compaction and higher disaggregation, not preserving the original textures (Figure 2c). Specially along the B line, the presence of veins of carbonates (Figure 2d) is usual.

3.2. Methodology

Different granulometric fractions of all the samples were obtained. Silt fractions ($\theta < 63$ μm and $\theta < 38$ μm) were separated from the bulk rock samples by wet sieving, while the clay size fraction ($\theta < 2$ μm) was separated by decantation following Stokes' law. The granulometric separation was performed not only to identify the variation in the relative proportion of the minerals within them, but also with a view to their possible future use in radionuclide sorption experiments to assess their applicability as lining barriers at radioactive waste disposal sites.

The mineralogical characterization of the samples was carried out by X-ray diffraction (XRD) using a Bruker D8 Advance and Cu kα radiation. All the granulometric fractions were powdered prior to their characterization, and, in addition, the clay fraction (θ < 2 μm) was also prepared as oriented air-dried aggregates (OA), solvated with ethylene glycol (OA + EG), and heated at 550 °C for 2 h (OA + TT). Powdered samples were scanned from 2° to 65° 2θ, and air-dried aggregates from 2° to 45° 2θ, using a speed of 0.05° 2θ/s. The clay size fraction was also measured from 59° to 63° 2θ at a speed of 0.03° 2θ/s to obtain a detailed diffractogram of the (060) reflection of clay minerals, to know the trioctahedral or dioctahedral character of the phyllosilicates present within the samples. Semi-quantification of the mineral phases observed at the diffractograms was performed through the reflecting powers method (RPM) [37], which is a standard in the study of clays, focusing on the ratio of peak areas to semi-quantify the minerals within the samples.

Structural formulae of the smectites were fitted from point analyses acquired by analytical electron microscopy (AEM) with transmission electron microscopy (TEM) at the National Electron Microscopy Center (Madrid, Spain) using a JEOL JEM-1400 microscope (JEOL Ltd., Tokyo, Japan), with an acceleration voltage of 100 kV. The microscope incorporates an OXFORD ISIS EDX spectrometer (Oxford Instruments, Oxfordshire, UK, 136 eV resolution at 5.39 keV), equipped with its own software for quantitative analysis. The TEM samples were prepared by depositing a drop of diluted suspension on a microscopic grid with collodion (cellulose acetate butyrate) and coated with Au. Homoionization was performed on the samples to obtain more accurate structural formulae that allow a more reliable classification of the smectites [38].

Thermogravimetric analysis (TGA) routine was performed using an SDT-Q600 from T.A. Instruments (T.A. Instruments, New Castle, DE, USA). Homoionized fractions < 2 μm were analyzed in an air atmosphere, with a heating rate of 10 °C/min, from room temperature up to 985 °C.

Geochemical analyses of major and trace elements were performed at Activation Laboratories Ltd. (ACTLABS, Ancaster, ON, Canada). For these analyses, 2 g of powdered sample was digested with aqua regia, along with appropriate international reference materials for the metals of interest, and diluted to 250 mL volumetrically. Major elements, along with Sc, Be, V, Sr, Zr, and Ba, were analyzed by lithium metaborate/tetraborate fusion inductively coupled plasma (ICP) and trace elements were analyzed by ICP-mass spectrometry (ICP-MS).

The chemical index of alteration (CIA) and plagioclase index of alteration (PIA) were calculated to elucidate the alteration rock degree [39–44]. These indices incorporate major bulk element oxide chemistry from the samples, giving information on the leaching of elements on each of the samples originating from homogeneous parent materials and indicating their relative degree of alteration.

Software package IBM SPSS version 29.0 was used for the statistical analyses.

4. Results

4.1. Mineralogical Characterization

4.1.1. XRD

Powdered samples from all the granulometric fractions were characterized by XRD, as well as the air-dried aggregates of <2 microns fractions, which allowed identifying the mineral assemblages within them (Figure 3). Powdered samples allow the identification of phyllosilicates, carbonates (mainly calcite, along with minor amounts of dolomite), and scarce quantities of tectosilicates (quartz, K-feldspar, and plagioclase) (Table 1). The rest of the minerals identified are always present, just at trace levels. These results agree with previous references [29,30]. Amphiboles were also identified, but only in samples belonging to the C vertical profile (Figure 2). The diffractograms corresponding to the finer granulometries (<63, <38, and <2 μm) show that the relative intensities of phyllosilicates increase compared with the other minerals identified, indicating a preferential concentration of phyllosilicates in the fine-grained fraction, as expected. The oriented aggregates (OA) and their treatments grant the capacity of studying clay minerals in more detail, being able to identify smectite as the main component along with traces of illite, chlorite, and possible kaolinite. Small amounts of a mixed-layer of chlorite and smectite (C/S) have also been identified due to the presence of a reflection at 12.4 Å that does not collapse in the AD + TT [45]. Also, in the OAs, there are small inflections at 12.3 Å (Figure 3b), which indicate a possible mixture of smectites with different predominant interlayer cations, ratified by their swelling in the OA + EG treatment. The dioctahedral character of the smectites was identified due to the d-spacings of the (060) reflections, whose values range between 1.499 and 1.507 Å, and in all the samples studied (Figure 3c).

Table 1. Semi-quantification of the mineral content of the samples from X-ray diffraction (XRD). (-): present at trace levels. Amphibole (Amp), calcite (Cal), chlorite (Chl), dolomite (Dol), phyllosilicates (Phyllo), plagioclase (Pl), potassium feldspar (Kfs), and quartz (Qz).

Sample	Amp	Cal	Dol	Kfs	Pl	Qz	Chl	Ilt	Sme	C/S
BEN1A		12	2	4	-	6	-	-	74	-
BEN1A < 63 μm		5	-	-	-	2	-	-	90	-
BEN1A < 38 μm		2	-	-	-	-	-	-	96	-
BEN1A < 2 μm		2					-	-	97	-
BEN1B		25	2	-	-	-	-	-	70	-
BEN1B < 63 μm		11	-	-	-	-	-	1	86	-
BEN1B <3 8 μm		9		-	-	-	-	1	89	-
BEN1B < 2 μm		4					-	-	95	-
BEN1C	-	16		-	-	-	-	-	82	-
BEN1C < 63 μm	-	4		-	-	-	-	-	94	-
BEN1C < 38 μm	-	5		-	-	-	-	-	92	-
BEN1C < 2 μm		2					-	-	97	-
BEN2A		17	-	-	-	-	-	-	81	
BEN2A < 63 μm		12		4	-	-	-	-	83	
BEN2A < 38 μm		5		-	-	-	-	-	93	
BEN2A < 2 μm		-					-	-	99	
BEN2B		27		-	-	-	-	-	71	-
BEN2B < 63 μm		9		3	-	-	-	-	86	-
BEN2B < 38 μm		15		4	-	-	-	-	80	-
BEN2B < 2 μm		2					-	-	97	-
BEN2C	9	26		2	7	-	-	-	55	
BEN2C < 63 μm	4	13					-	-	82	
BEN2C < 38 μm	3	5		-	-	-	-	-	90	
BEN2C < 2 μm		2					-	-	96	
BEN3A			-	-	-	2		-	95	1
BEN3A < 63 μm				-		-		-	97	1
BEN3A < 38 μm				-	-	-		-	96	1
BEN3A < 2 μm								-	98	1
BEN3B		34					-	-	64	1
BEN3B < 63 μm		27				-	-	-	71	1
BEN3B < 38 μm		28					-	-	70	1
BEN3B < 2 μm		1					-	-	96	2
BEN3C	4	38		4	-	2	-	-	51	
BEN3C < 63 μm	-	16		-	-	-	-	-	82	
BEN3C < 38 μm	2	11		-	-	-	-	-	86	
BEN3C < 2 μm		2					-	-	97	

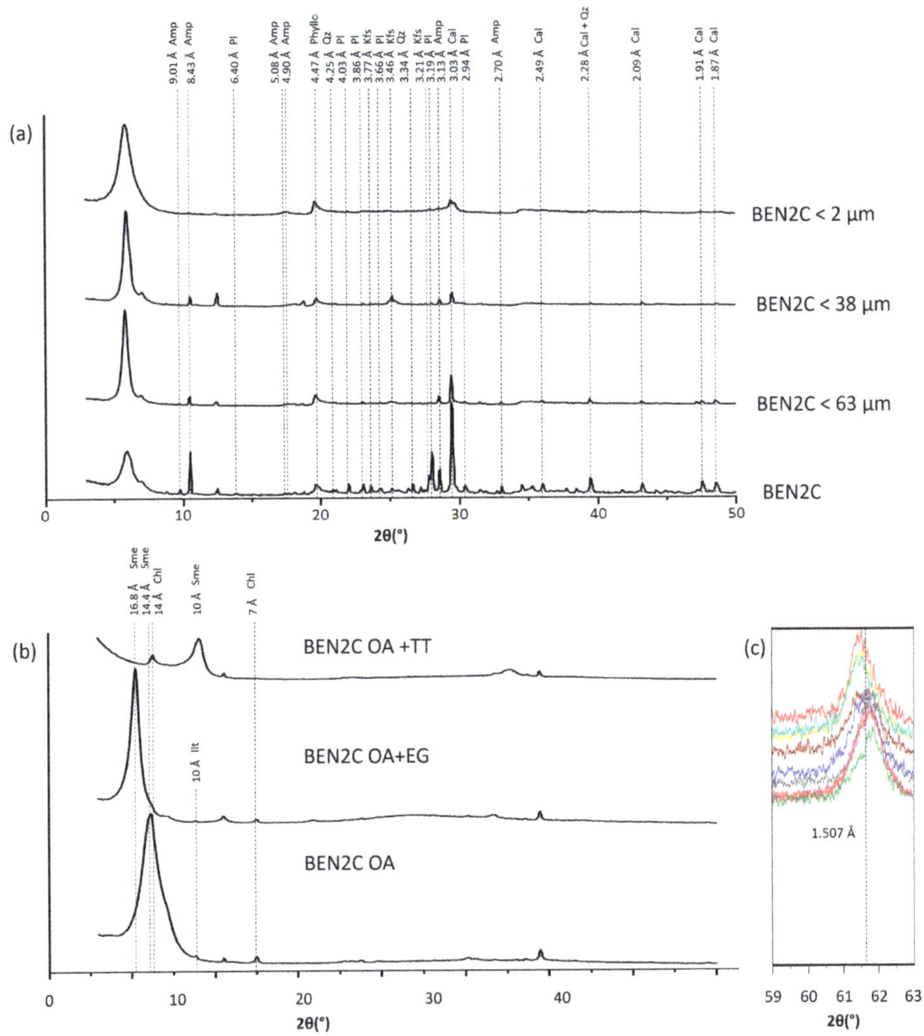

Figure 3. (**a**) Powder X-Ray diffractograms of sample BEN2C. (**b**) Air-dried oriented aggregates of sample BEN2C. (AD): air-dried oriented aggregate; (AD + EG): air-dried oriented aggregate solvated with ethylene glycol; (AD + TT): air-dried oriented aggregate calcinated at 550 °C. (**c**) Comparison of the (060) reflection of the fractions < 2 µm of all the samples. Amphibole (Amp), Calcite (Cal), chlorite (Chl), illite (Ilt), phyllosilicates (Phyllo), plagioclase (Pl), potassium feldspar (Kfs), smectite (Sme), and quartz (Qz).

The X-Ray diffractograms not only allow identifying minerals, but also semi-quantifying them (Table 1). In all the granulometric fractions, the main components are phyllosilicates (smectite plus quantities below 2% of illite, chlorite/smectite mixed-layers, and chlorite), followed by carbonates (calcite plus traces of dolomite) and scarce quantities of quartz and feldspars. In the bulk rock samples, phyllosilicates range from 97% (BEN3A) to 52% (BEN3C), with smectite being the main component (95% and 51%, respectively). Calcite is the major impurity and ranges from 0% (BEN3A) to 38% (BEN3C) with a mean content of 22%. It has to be pointed out that sample BEN3A is the richest sample in smectite because it is the only one that does not contain calcite, and a negative correlation between these two minerals can be observed. The <63 μm fraction shows percentages of phyllosilicates ranging from 99% (BEN3A) and 73% (BEN3B), with smectite being the major mineral, and of calcite ranging from 0% (BEN3A) and 27% (BEN3B). This tendency previously described is also seen in the <38 μm fraction, being the maximum content of phyllosilicates (mainly smectite) 98% (BEN3A) and the minimum, 72% (BEN3B), coinciding with the maximum and minimum contents in calcite (28% and 0% at samples BEN3B and BEN3A, respectively). Finally, in the <2 μm fraction, most of the minerals identified and quantified are essentially smectite with a minimum content of 95% and scarce impurities of chlorite, illite, and mixed-layer chlorite/smectite that can be identified but are difficult to quantify.

4.1.2. TEM-AEM

The top samples (BEN3A, BEN3B, and BEN3C) were used for the crystal-chemical study through TEM-AEM analyses. The TEM representative photographs (Figure 4) show platy particles with subhedral morphologies, which present some straight faces together with wavy and pointy ends (as usual in smectites) on the same particles. This resembles typical morphologies of micaceous materials (euhedral particles with straight faces), which, most likely, are the precursor phases whose alteration gave place to the smectites. The wavy and pointy ends indicate the transformation process and/or growth of smectites from these micaceous particles [46–49]. The photographs evidence an increase in the wavy and pointy-ended faces towards the A profile (Figure 4).

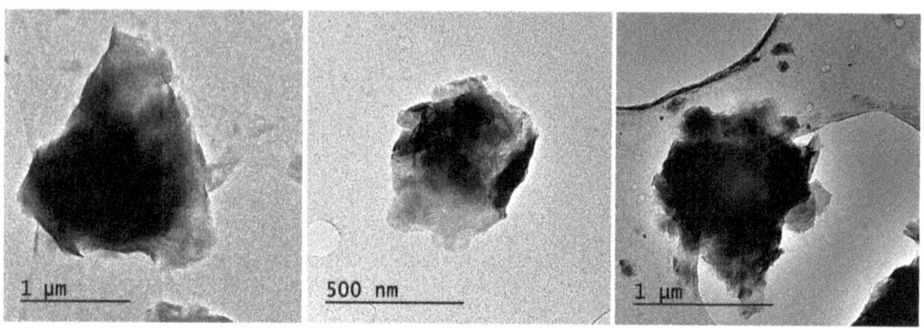

Figure 4. TEM representative photographs of smectie particles from samples BEN3A (**left**), BEN3B (**middle**), and BEN3C (**right**).

The point chemical data obtained by AEM analyses (Table 2) of isolated particles of smectite of the natural samples show that the main oxide (considering the mean values) is SiO_2, ranging from 63.98% (BEN3A) and 56.49% (BEN3B). Regarding the second most abundant oxide, there are two cases: sample BEN3A has more content in Al_2O_3 than Fe_2O_3 (19.90 and 7.89%), while samples BEN3B and BEN3C have higher contents in Fe_2O_3 (18.43 and 20.07%) than in Al_2O_3 (15.20 and 13.35%). Finally, the rest of the oxides analyzed follow the same trend in all the samples studied: MgO > CaO > K_2O > TiO_2 > NaO. It is also noteworthy that they show a broad compositional variability, as indicated by the differences between the maximum and minimum values, as well as the standard deviation, which is related to the continuous compositional variations in all the studied samples.

Table 2. Chemical analysis of major oxides (wt%) and structural formulae of the smectites obtained from point analyses by HRTEM–AEM. Max: maximum value; Min: minimum value; Mean: mean value; SD: standard deviation; ΣTC: sum of tetrahedral cations; ΣOC: sum of octahedral cations; TCh: tetrahedral charge; OCh: octahedral charge; LC: layer charge; n: number of analyses.

Samples		Chemical analysis of major oxides								Mean structural formulae															
										Tetrahedral			Octaheral					Interlayer							
		SiO_2	Al_2O_3	Fe_2O_3	MgO	CaO	NaO	K_2O	TiO_2	Si	Al	ΣTC	Al	Mg	Fe	Ti	ΣOC	Ca	Na	K	TCh	OCh	LC
BEN3A	Max	66.33	32.64	11.02	8.21	9.15	0.10	6.87	0.26														
	Min	54.81	18.33	0.00	2.57	0.25	0.00	0.00	0.00														
	Mean	63.98	19.90	7.89	6.84	0.92	0.00	0.40	0.06	7.67	0.33	8.00	2.48	1.22	0.71	0.01	4.42	0.12		0.06	−0.33	0.05	0.28
n = 26	SD	1.97	2.58	2.28	0.90	1.60	0.02	1.26	0.09														
BEN3B	Max	62.46	19.56	33.00	11.94	2.75	0.00	1.43	1.14														
	Min	48.07	11.58	10.40	4.60	0.46	0.00	0.00	0.00														
	Mean	56.49	15.20	18.43	8.03	1.20	0.00	0.54	0.11	7.12	0.88	8.00	1.39	1.51	1.75	0.01	4.66	0.16		0.09	−0.88	0.48	0.40
n = 29	SD	3.85	1.72	5.04	2.06	0.60	0.00	0.40	0.29														
BEN3C	Max	63.18	16.95	29.57	11.95	2.93	1.83	0.70	0.74														
	Min	49.74	11.85	11.77	4.20	1.41	0.00	0.00	0.00														
	Mean	56.80	13.35	20.07	7.08	1.89	0.12	0.38	0.12	7.22	0.78	8.00	1.23	1.34	1.92	0.01	4.50	0.26	0.03	0.06	−0.78	0.17	0..61
n = 18	SD	3.37	1.24	4.10	1.79	0.41	0.47	0.24	0.23														
BEN3ACa	Max	66.65	29.57	13.24	6.85	16.03	0.00	9.76	0.57														
	Min	54.10	15.68	2.04	1.13	2.22	0.00	0.04	0.00														
	Mean	61.53	19.19	7.64	4.71	5.49	0.00	1.40	0.06	7.54	0.46	8.00	2.31	0.86	0.70	0.01	3.88	0.72		0.22	−0.46	−1.20	1.66
n = 20	SD	3.42	3.74	3.17	1.41	2.68	0.00	2.88	0.20														
BEN3BCa	Max	63.19	17.80	19.73	13.66	6.48	0.00	2.77	1.38														
	Min	49.28	12.84	11.67	4.03	1.77	0.00	0.00	0.00														
	Mean	58.80	15.13	14.75	6.54	3.67	0.00	0.70	0.42	7.36	0.64	8.00	1.59	1.22	1.39	0.04	4.24	0.49		0.11	−0.64	−0.46	1.10
n = 26	SD	3.36	1.05	1.98	2.17	1.05	0.00	0.81	0.33														
BEN3CCa	Max	62.60	16.75	16.39	13.30	8.25	0.00	7.10	1.03														
	Min	53.74	12.86	10.97	6.04	2.18	0.00	0.31	0.02														
	Mean	57.10	14.73	13.27	8.99	3.74	0.00	1.67	0.50	7.20	0.80	8.00	1.38	1.69	1.26	0.05	4.38	0.48		0.34	−0.80	−0.50	1.30
n = 17	SD	2.49	1.11	1.77	2.30	1.39	0.00	1.94	0.25														

The structural formulae were calculated from the mean oxide contents of the point chemical analyses (Table 2). The resulting formulae from the natural samples correspond to dioctahedral smectites. Their average number of octahedral cations ranges from 4.42 to 4.66. These formulae exhibit positive octahedral charge ranging from 0.05 to 0.48 and a very low interlayer charge ranging from 0.30 to 0.61, indicating that the results are unreliable. To properly fit the smectite structural formulae, the samples were homoionized with Ca^{2+} [38]. The homoionized analyses show similar distributions as in non-homoionized samples (also considering the mean values): SiO_2 is the main oxide, ranging from 61.53% (BEN3ACa) and 57.10% (BEN3CCa). The second most abundant oxide is always Al_2O_3 (in opposition to the non-homoionized samples), which ranges from 19.19% (BEN3ACa) and 14.73% (BEN3CCa), while Fe_2O_3 is the third, ranging from 14.75% (BEN3B) and 7.64% (BEN3A). Samples BEN3BCa and BEN3CCa show MgO as the fourth oxide in terms of abundance (6.54 and 8.99% respectively) and CaO as the fifth (3.67 and 3.74% respectively), while this trend is inverted in sample BEN3ACa (5.49% of CaO and 4.71% of MgO). The rest of the oxides analyzed follow the same trend in all the samples studied: $K_2O > TiO_2 > NaO$. The structural formulae also correspond to dioctahedral smectites. The mean octahedral charges vary, being −1.66 (BEN3ACa), −0.46 (BEN3BCa), and −0.50 (BEN3CCa), and an increase in the interlayer charge. It is possible to see that the main difference between non-homoionized and homoionized samples is their content in Fe_2O_3, which suggests that iron was most likely adsorbed on the non-homoionized samples and removed during their homoionization. The K_2O content did not decrease in the homoionic analyses, indicating the presence of interestratified micaceous layers in the smectite particles, which supports their detrital origin.

The structural formula obtained from both the non-homoionized and homoionized samples was used to obtain a ternary plot of the octahedral content (Mg, Fe, Al) of all the analyses, along with the octahedral content of different smectite samples from the literature [38,50,51] (Figure 5). The ternary plot of the octahedral content of non-homoionized samples shows different tendencies for Benavila samples. BEN3A is the most enriched in aluminum, therefore, being the closest to both beidellite and montmorillonite. Samples BEN3B and BEN3C display higher compositional dispersion and show higher contents of iron, which in some cases bring them close to the nontronite octahedral content. However, when the homoionized octahedral content is plotted, it is easily observed that no analyses fall near the notronite octahedral content but rather fall in intermediate fields. There is also an evident trend that can be observed in both ternary plots, which is a progressive transformation towards purest dioctahedral smectites from the BEN3C to the BEN3A sample, due to the progressive loss of magnesium and iron.

The smectites from Benavila had already been classified as Tatatila type montmorillonites [29], meaning they have 15–50% of the charge originating from tetrahedral substitutions and a layer charge > 0.85 [45]. Other studies have simply classified them as montmorillonites, without defining their type [31]. However, recent classifications of dioctahedral smectites [52,53], which consider more variables, such as the layer charge, the octahedral structure, the iron content, and the charge location, allow for giving more descriptive names of the subjects of study. For such purpose, the structural formulae obtained from the homoionized samples were used.

Sample BEN3ACa has a layer charge (LC) of 1.66, a Fe^{3+} content (IC) that represents 18% of the octahedral cations, and a tetrahedral charge (TC) that accounts for 28% of the total charge, and is therefore classified as a high-charged ferrian beidellitic montmorillonite [52]. Sample BEN3BCa has an LC of 1.10, an IC of 33%, a TC of 58% and is classified as a high-charged ferrian montmorillonitic beidellite. Sample BEN3CCa has an LC of 1.30, an IC of 29%, and a TC that is 62% of the total charge, being classified as a high-charged ferrian montmorillonitic beidellite. Therefore, it can be summarized that there are two types of samples: sample BEN3A is a ferrian beidellitic montmorillonite, while BEN3B and BEN3C are ferrian montmorillonitic beidellites. All the samples show high layer charges.

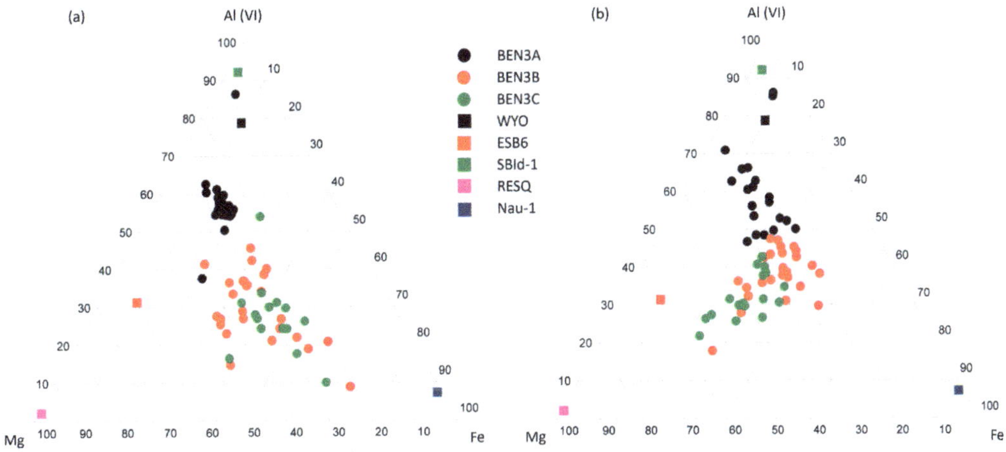

Figure 5. Ternary plot of the octahedral content of the non-homoionized samples (**a**) and the homoionized samples (**b**) and reference samples of montmorillonite (WYO), beidellite (SBId-1), saponite (ESB6), stevensite (RESQ), and nontronite (NAu-1).

4.1.3. TGA

The thermal behavior of Benavila bentonite samples, essential for their possible use in radioactive storage, was determined through thermogravimetric analysis. These analyses also enable an in-depth characterization of the octahedral sheet of the smectites. For that purpose, the homoionized < 2 μm fraction of the three samples from the top level (BEN3A, BEN3B, and BEN3C) was selected. The TGA analyses (Figure 6a–c) show that all the samples present the same tendencies: the most important weight loss (around 15% of their masses) takes place from room temperature up to ~175 °C, due to the dehydration of the absorbed and interlayer water of the smectites. They keep losing small percentages of weight until the range comprised between 550 and 650 °C, where they lose approximately 5% of their mass due to the dehydroxilation of smectites. DTG analyses (Figure 6a–c) give more detailed information. The dehydration of the interlayer water occurs in two peaks due to the presence of different interlayer cations [54], and the dehydroxilation events are marked by defined peaks, although they can be broader or narrower peaks, depending on the samples.

The peaks related to the dehydroxilation of the smectites in the DTG curve (derivative of the thermal analyses) are used to calculate the percentage of cis-vacant (*cv*) and trans-vacant (*tv*) configurations of the octahedral sheet [55]. The two configurations can be differentiated due to their dehydroxylation temperatures. The *cv* varieties dehydroxylate at temperatures between 650 and 700 °C, while *tv* varieties dehydroxylate at temperatures of about 500–550 °C, being the boundary temperature between the two configurations at around 600 °C [55–57]. These percentages were obtained by performing the deconvolution of the DTG curve between 300 and 800 °C using the software package Fityk (v. 1.3.1) [58] (Figure 6d), where it can be seen that a curve is deconvoluted into three curves plus a straight line (the latter is used to subtract the background). Considering the center of the curves obtained from the deconvolution, it is determined whether the dehydroxilation takes place at temperatures below or over 600 °C, and using their respective areas, it is possible to calculate the percentage that *cv* (peaks > 600 °C) and *tv* (peaks < 600 °C) varieties represent

(Table S1) [57,59]. All the samples are classified as *tv/cv*, with a progressive decrease in the *cv* character towards sample BEN3A.

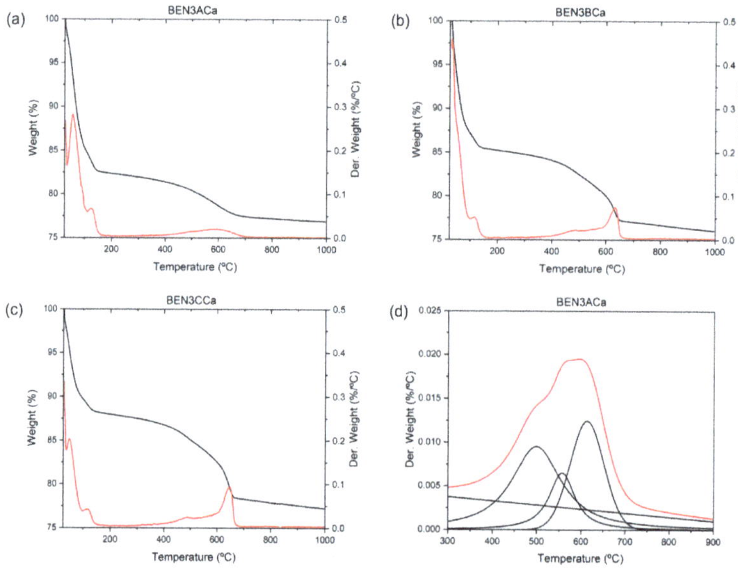

Figure 6. (**a**) TGA (%) (black) and DTA (%/°C) (red) of sample BEN3ACa; (**b**) TGA (%) (black) and DTA (%/°C) (red) of sample BEN3BCa; (**c**) TGA (%) (black) and DTA (%/°C) (red) of sample BEN3CCa; (**d**) Deconvolution of the DTG curve between 300 and 900 °C of sample BEN3ACa, showing the different functions considered (black) and the resulting model (red).

Considering the previous classification from the TEM-AEM analysis of the samples according to recent references [52,53] and focusing on the octahedral configuration, we can conclude that sample BEN3A is a high-charged *tv/cv* ferrian beidellitic montmorillonite, while BEN3B and BEN3C are high-charged *tv/cv* ferrian montmorillonitic beidellites.

4.2. Geochemistry

The major (Table S2) and trace elements (Table S3) contents were obtained by means of ICP-MS for bulk samples and different size fractions. Tables 3 and 4 show the maximum, minimum, and mean values of the chemical analyses of all the granulometric fractions, along with their standard deviation. The trace elements considered were the high field strength elements (HFSE), transition trace elements (TTE), large-ion lithophile elements (LILE) and rare earth elements (REE), as well as other trace elements such as Be, Ga, Ge, As, Y, Mo, Ag, In, Sn, Sb, W and Bi.

The major elements concentrations are in accordance with the mineralogical characterization. Samples with a higher content in phyllosilicates present higher percentages of SiO_2 and Al_2O_3, consistent with their identification as being mainly constituted by dioctahedral smectites, and samples with a higher content in calcite have more CaO (Table S2). This is observed in Figure 7, where the contents of SiO_2, Al_2O_3, and CaO are plotted, and it is possible to observe an enrichment in SiO_2 and Al_2O_3 and a depletion in CaO towards the finer grained fractions (higher content in phyllosilicates, Table 3). The trends observed are corroborated by the high correlation coefficients among SiO_2, Al_2O_3, and CaO shown in Table S4 and by Figure 8, which shows the contents of these elements in the finer grain size fractions normalized with respect to the contents in the bulk rock samples.

Table 3. Maximum (Max), minimum (Min), and mean values of major elements in weight percentage (wt%) of all the granulometric fractions, along with their standard deviation (SD).

	Bulk Sample				<63 μm				<38 μm				<2 μm			
	Max	Min	Mean	SD	Max	Min	Mean	SD	Max	Min	Mean	SD	Max	Min	Mean	SD
SiO_2	51.86	29.76	39.27	6.96	48.76	32.96	42.36	4.57	49.33	34.20	43.31	4.22	50.23	45.27	48.14	1.50
Al_2O_3	13.58	8.26	11.37	2.00	14.39	8.72	12.02	2.24	14.48	8.55	12.14	2.32	16.08	10.38	13.47	1.89
Fe_2O_3	7.81	6.10	6.96	0.62	10.60	5.27	7.56	1.50	11.52	5.30	8.10	1.81	10.53	4.85	7.63	2.02
MnO	0.13	0.06	0.10	0.02	0.09	0.04	0.06	0.02	0.09	0.04	0.07	0.02	0.07	0.02	0.05	0.02
MgO	6.18	4.30	4.93	0.54	5.42	3.99	4.61	0.43	5.44	3.96	4.67	0.47	5.57	4.66	5.14	0.28
CaO	21.65	1.67	13.80	6.32	17.32	1.32	8.15	4.42	15.84	1.38	7.31	3.94	4.47	1.27	2.64	0.93
Na_2O	1.20	0.04	0.34	0.38	0.27	0.05	0.12	0.08	0.24	0.02	0.08	0.07	1.08	0.22	0.55	0.28
K_2O	1.49	0.50	0.87	0.31	0.56	0.29	0.43	0.10	0.52	0.23	0.39	0.09	0.59	0.25	0.35	0.10
TiO_2	1.02	0.40	0.65	0.15	0.75	0.39	0.61	0.12	0.77	0.35	0.61	0.13	0.34	0.12	0.19	0.07
P_2O_5	0.21	0.09	0.15	0.04	0.23	0.03	0.13	0.06	0.26	0.03	0.13	0.07	0.02	0.01	0.01	0.00
LOI	26.20	16.19	20.88	3.32	28.34	22.11	24.20	1.79	26.72	20.87	22.94	1.62	23.78	19.00	21.35	1.26
Total	100.00	98.43	99.31	0.60	100.80	99.18	100.24	0.52	100.80	98.59	99.75	0.72	100.90	98.70	99.53	0.81

Table 4. Maximum (Max), minimum (Min), and mean values of trace elements (μm) of all the granulometric fractions, along with their standard deviation (SD).

	Bulk Sample				<63 μm				<38 μm				<2 μm			
	Max	Min	Mean	SD	Max	Min	Mean	SD	Max	Min	Mean	SD	Max	Min	Mean	SD
Zr	119.00	35.00	72.44	23.22	122.00	49.00	69.67	22.23	127.00	44.00	65.00	24.09	56.00	31.00	40.33	8.16
Nb	6.60	3.40	4.76	0.96	6.10	2.00	4.21	1.26	6.10	2.10	4.24	1.21	1.30	0.20	0.76	0.28
Hf	2.70	1.30	1.89	0.40	3.20	1.70	2.19	0.50	3.40	1.60	2.12	0.55	2.20	0.90	1.49	0.36
Ta	0.74	0.23	0.35	0.15	0.71	0.09	0.30	0.17	1.25	0.52	0.78	0.20	0.15	0.01	0.08	0.04
Tl	0.38	0.16	0.26	0.08	0.21	0.08	0.11	0.04	0.14	0.05	0.09	0.03	0.07	0.05	0.05	0.01
Th	5.22	0.79	2.86	1.83	4.78	0.77	2.54	1.59	5.21	0.87	2.70	1.70	3.27	0.40	1.59	1.17
U	1.22	0.43	0.80	0.30	1.21	0.55	0.71	0.21	1.19	0.50	0.71	0.22	0.24	0.09	0.15	0.04
Sc	33.00	18.00	27.33	4.52	40.00	20.00	30.89	5.26	40.00	21.00	31.22	5.65	42.00	20.00	31.00	5.93
V	265.00	53.00	111.67	61.02	194.00	45.00	89.22	47.79	208.00	46.00	93.89	52.20	171.00	27.00	81.44	48.44
Cr	810.00	20.00	400.00	256.95	850.00	30.00	443.33	282.13	1080.00	40.00	528.89	372.48	890.00	60.00	442.22	302.65
Co	34.00	10.00	21.89	6.35	17.00	11.00	14.33	1.89	22.00	10.00	17.44	3.72	17.00	6.00	11.44	3.59
Ni	120.00	30.00	68.89	34.78	100.00	20.00	51.25	30.18	130.00	20.00	61.25	40.45	120.00	30.00	55.71	32.89
Cu	40.00	20.00	25.00	7.07	30.00	10.00	17.50	6.61	30.00	10.00	20.00	5.00	20.00	10.00	15.00	5.00
Zn	80.00	50.00	64.44	8.31	80.00	40.00	48.89	12.86	90.00	40.00	55.56	14.99	60.00	30.00	43.33	10.54
Rb	68.00	19.00	43.44	17.68	47.00	15.00	24.78	9.59	44.00	13.00	23.67	8.73	16.00	7.00	10.89	2.51
Sr	175.00	81.00	122.56	26.58	98.00	80.00	90.44	5.89	96.00	80.00	88.56	6.45	133.00	93.00	108.78	12.73
Cs	3.70	0.70	2.42	0.95	2.30	0.80	1.51	0.48	2.20	0.60	1.43	0.48	1.50	0.60	1.08	0.27
Ba	604.00	176.00	348.56	145.07	363.00	106.00	225.33	71.50	327.00	98.00	208.67	67.81	556.00	111.00	277.89	153.05
Pb	11.00	5.00	8.25	2.17	27.00	6.00	13.67	9.46	8.00	5.00	6.00	1.22	12.00	5.00	8.50	3.50
La	24.30	6.73	13.97	6.35	21.20	3.11	9.98	5.45	19.20	2.77	9.30	4.89	8.57	0.71	4.27	2.99
Ce	36.80	15.70	24.77	8.06	35.90	10.40	20.39	7.91	33.60	10.40	20.69	7.70	26.20	2.93	12.40	7.09
Pr	6.34	2.47	4.05	1.28	5.27	1.64	2.85	1.11	5.06	1.44	2.77	1.05	1.84	0.18	0.97	0.65
Nd	26.50	10.30	17.42	5.46	23.70	7.65	12.59	4.97	22.60	6.46	11.81	4.72	6.39	0.88	3.61	2.25
Sm	6.34	2.69	4.27	1.29	5.77	2.02	3.09	1.14	5.30	1.73	2.88	1.10	1.19	0.19	0.65	0.37
Eu	1.55	0.67	1.06	0.30	1.35	0.47	0.76	0.26	1.16	0.47	0.71	0.22	0.23	0.05	0.13	0.07
Gd	6.75	2.43	4.06	1.33	5.96	2.00	3.03	1.26	4.84	1.67	2.64	1.04	0.88	0.19	0.47	0.23
Tb	1.09	0.37	0.65	0.21	0.91	0.33	0.49	0.18	0.77	0.30	0.44	0.16	0.12	0.03	0.07	0.03
Dy	6.52	2.20	3.92	1.26	5.34	2.05	2.87	1.09	4.56	1.80	2.62	0.99	0.65	0.15	0.38	0.15
Ho	1.30	0.43	0.76	0.26	1.06	0.39	0.57	0.22	0.90	0.36	0.51	0.19	0.12	0.03	0.07	0.03
Er	3.72	1.30	2.18	0.70	3.12	1.14	1.67	0.63	2.54	1.03	1.47	0.54	0.34	0.10	0.21	0.08
Tm	0.54	0.20	0.33	0.10	0.45	0.17	0.24	0.09	0.36	0.16	0.21	0.07	0.05	0.01	0.03	0.01
Yb	3.29	1.36	2.11	0.55	2.90	1.14	1.59	0.54	2.35	1.07	1.44	0.45	0.33	0.08	0.19	0.07
Lu	0.46	0.21	0.32	0.08	0.44	0.17	0.25	0.08	0.38	0.17	0.23	0.07	0.05	0.01	0.03	0.01
Be	3.00	1.00	1.89	0.57	3.00	1.00	1.67	0.67	2.00	1.00	1.33	0.47	2.00	1.00	1.50	0.50
Ga	15.00	10.00	12.44	1.77	15.00	11.00	12.78	1.47	15.00	12.00	13.11	1.10	14.00	12.00	13.22	0.63
Ge	1.40	0.70	0.94	0.23	0.90	0.50	0.70	0.14	0.90	0.50	0.69	0.14	0.70	0.60	0.65	0.05
Y	39.50	12.30	21.91	7.86	32.90	11.80	16.09	6.94	26.60	10.00	14.46	5.71	3.00	0.80	1.82	0.69
Sn	5.00	1.00	2.33	1.05	6.00	2.00	2.89	1.37	6.00	2.00	3.22	1.40	5.00	2.00	2.78	1.13
Sb	0.30	0.20	0.23	0.04	3.50	0.20	1.54	1.02	0.40	0.20	0.27	0.09	135.00	0.50	38.03	46.11
W	4.00	0.60	1.43	1.03	2.70	0.50	1.26	0.68	8.00	0.90	2.38	2.22	1.90	0.60	1.05	0.42

Focusing on the HFSE, there are no clear differences according to the mineralogy of the samples, as shown by the correlation matrices performed (Table S5). Nb, Tl, Th, and U are significantly correlated with quartz, and it also appears that samples with higher contents in phyllosilicates also have lower total contents in HFSE, showing a progressive depletion towards finer granulometries (Table 4, Figure 8). This is explained by their relative immobility and their association with heavier and more resistant minerals present in the coarser fractions [60–64].

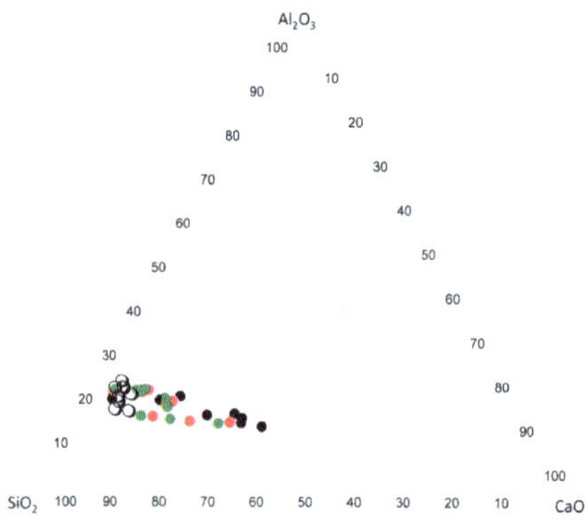

Figure 7. Ternary plot representing the values of SiO_2, Al_2O_3, and CaO. Black: bulk rock samples; Red: <63 μm fraction; Green: <38 μm fraction; Blank: <2 μm fraction.

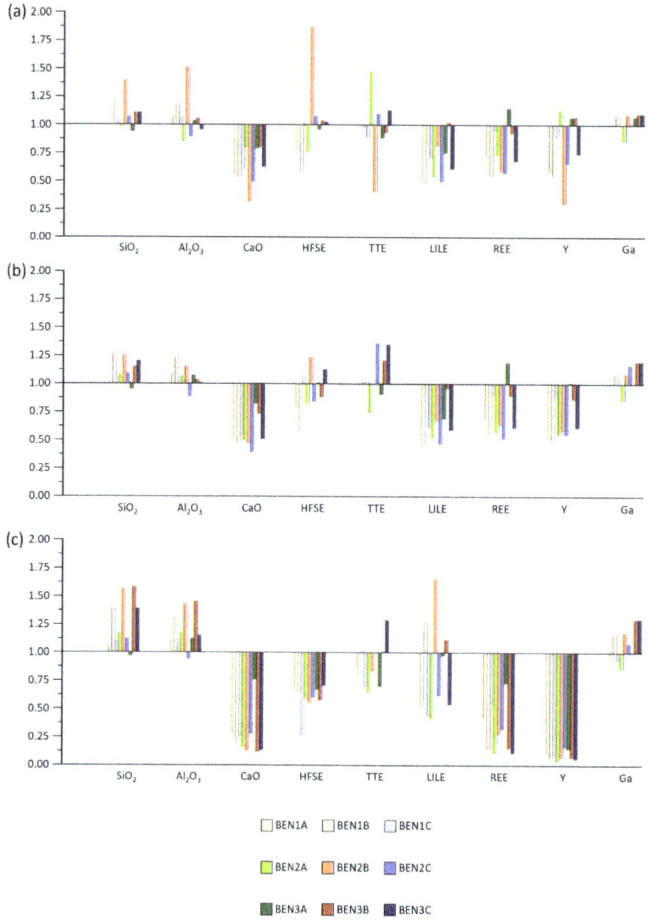

Figure 8. (**a**) Enrichment in the <63 μm fraction respect the bulk rock sample of the represented oxides and elements; (**b**) Enrichment in the <38 μm fraction respect the bulk rock sample of the represented oxides and elements; (**c**) Enrichment in the <2 μm fraction respect the bulk rock sample of the represented oxides and elements.

The TTE shows different correlation coefficients between phyllosilicates and carbonates, being negative in the first case and positive in the latter, considering the total TTE content as well as the content in certain elements within the group (Co, Ni, Cu) (Table S6). However, when observing Figure 8, it is possible to observe that the trend is not as clear as expected, even having an enrichment in TTE in sample (BEN3C), which is almost entirely formed by smectite (98%) and small quantities of calcite (2%). This points to an unclear linkage between TTE and mineralogy.

LILE appear to be less concentrated in samples with higher phyllosilicate contents, shown by their depletion in finer granulometries in comparison to the bulk rock fraction (Table 4, Figure 8), except in the case of the finest fraction (<2 μm). However, the correlation matrix (Table S7) only shows a negative significant correlation of phyllosilicates and Sr, because this element shows higher concentrations in calcite-rich samples due to its similarity to Ca [65–67]. As with the TTE, it is not possible to identify a clear link between LILE and mineralogy, with the exception of Sr and carbonates.

The REE present lower concentrations in the samples with higher contents in phyllosilicates (Figure 8, Table 4 and Table S8), something that can be attributed to their low solubility and their preferential accumulation in the coarser fractions in trace minerals as zircons [68]. However, there is another mechanism that can potentially affect the REE content: their complexation due to the action of carbonate-rich fluids and possible precipitation within calcite veins [69–71]. After chondrite normalization [72] (Figure 9), all samples are enriched both in light REE (LREE) and heavy REE (HREE). The fact that, in the majority of the clay size (<2 μm) fractions, the HREE are much lower than in the other fractions may partially explain the negative correlation between REE and smectite. All the samples studied show negative Eu anomalies. Two samples from the B profile have significant negative Ce anomalies in bulk and silt size fractions. In the clay size fractions, a tendency for positive anomalies was found.

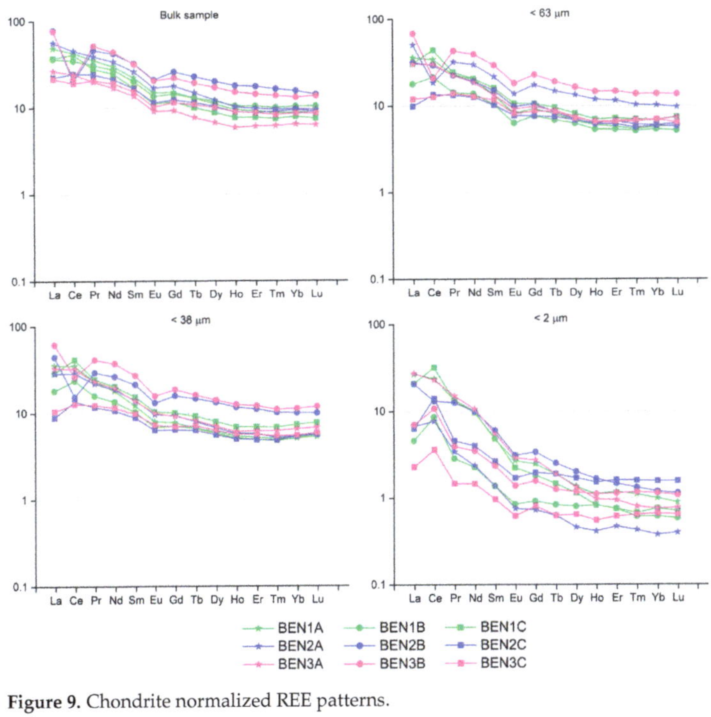

Figure 9. Chondrite normalized REE patterns.

The other trace elements analyzed show that most of them are below detection limits in most samples. Therefore, it is important to focus on the elements above the aforementioned detection limits. Yttrium shows high affinity with the REE and therefore has the same tendency as them (Table S9), with its content being lower within more clay-rich samples (Table S3). On the other hand, Ga is apparently more concentrated in samples with higher contents of phyllosilicates (Figure 8, Table 4 and Table S9), because it can substitute Al^{3+} in the octahedral layer of these minerals [73,74]. In the case of the rest of the other elements analyzed, it is not possible to identify clear links between their content and the mineralogy.

5. Discussion

The mineralogical characterization allows observing a variation in the horizontal direction, suggesting that the alteration process that formed the bentonites followed specific pathways. The presence of amphiboles within the igneous rocks whose alteration gave place to the formation of the clay minerals is cited in the literature [23,24]. Therefore, the presence of those minerals is interpreted as an indicator of a lower degree of weathering of the C profile in comparison with the other vertical profiles. The high content of carbonates in different parts of the sampling grid is related to the circulation of carbonate-rich fluids and the resulting carbonated crusts during the Paleogene–Quaternary [24], events that happened after the weathering that gave place to the bentonite deposit. It can be concluded that the mineral composition of the samples reflects an almost complete weathering of the primary minerals (amphiboles, quartz, feldspars, and micas), as these are present in relatively small proportions. Even if their content was recalculated, obviating the presence of carbonates (deposited after the weathering event), they still represent minor quantities, corroborating their almost complete transformation.

The TEM photographs show an apparent increase in the wavy and pointy-ended faces towards the A profile, therefore evidencing an increase in the development of smectites [46–49]. The TEM-AEM analyses show an increment of the Al_2O_3 and SiO_2 (more abundant within dioctahedral smectites) content from the C profile towards the A profile, as well as a clear decrease in Fe_2O_3 and TiO_2 (less abundant within dioctahedral smectites and more abundant within illite and amphiboles). Additionally, the presence of Fe_2O_3, TiO_2, and K_2O together with the particle morphology is a clear indicator that the precursor phases of the smectites were micas. The results of the smectite crystal-chemical characterization reveal interesting interrelated facts. The first one is that there is a progressive increase in the percentage of *cv* configurations of the octahedral sheet towards sample BEN3C (Table S1). The TEM-AEM study of the samples also enables observing a shift from beidellitic montmorillonite in the A profile to montmorillonitic beidellites in the B and C profiles. In agreement with the interpretations from the XRD characterization of the bulk samples (lower degree of alteration of the C vertical profile deduced from the identification of amphiboles in XRD), it is possible to suggest that lower percentages of *tv* configurations and beidellitic montmorillonites are characteristic of less weathered samples, while higher percentages of *tv* configurations and montmorillonitic beidellites are characteristic of samples which underwent a higher degree of weathering.

Major elements match the expected distribution, in accordance with the mineralogical characterization of the samples. These data were used to calculate two chemical indices, which are adequate for silicate-rich materials [39,40,42,44] and indicate the degree of alteration of the samples (Table 5, Figure 10). These indices are the plagioclase index of alteration (PIA), which considers the content in Al_2O_3, CaO, Na_2O and K_2O [41]; and the chemical index of alteration (CIA), which considers the molecular percentages of Al_2O_3, Na_2O, K_2O and the CaO (CaO*) that is solely contained within the silicates [41,43]. The CaO* was also used in the calculation of the PIA index to avoid the influence of carbonates.

Both chemical indices show a positive correlation between higher values and a higher alteration degree, with the highest values indicating higher clay mineral contents. A correlation matrix analysis (Table S10) reveals strong positive and significant correlations (0.896 and 0.917) with phyllosilicates and negative and significant coefficients with calcite and amphiboles, validating their calculation.

Table 5. Plagioclase index of alteration (PIA) and chemical index of alteration (CIA) of all the samples and granulometric fractions, along with the indices obtained through TEM-AEM.

	Whole Rock		<63 μm		<38 μm		<2 μm		TEM-AEM	
	PIA	CIA	PIA	CIA	PIA	CIA	PIA	CIA	PIA	CIA
BEN1A	50.68	55.80	69.69	74.79	72.16	76.25	81.81	84.14		
BEN1B	33.53	40.18	53.07	60.36	56.95	62.68	72.57	73.37		
BEN1C	51.81	55.59	67.61	72.63	69.83	73.87	82.40	84.69		
BEN2A	49.78	52.89	54.13	61.38	69.85	73.42	84.99	86.00		
BEN2B	30.80	37.44	66.81	71.61	52.23	58.13	75.09	74.16		
BEN2C	35.03	34.18	51.74	55.23	57.36	59.62	60.17	62.38		
BEN3A	76.09	81.27	84.27	88.20	85.09	88.46	85.02	84.47	91.90	90.41
BEN3B	25.46	33.27	31.93	42.86	33.77	42.47	76.89	78.73	86.57	84.61
BEN3C	27.89	30.79	38.92	45.51	44.75	49.05	72.56	74.11	88.73	76.77

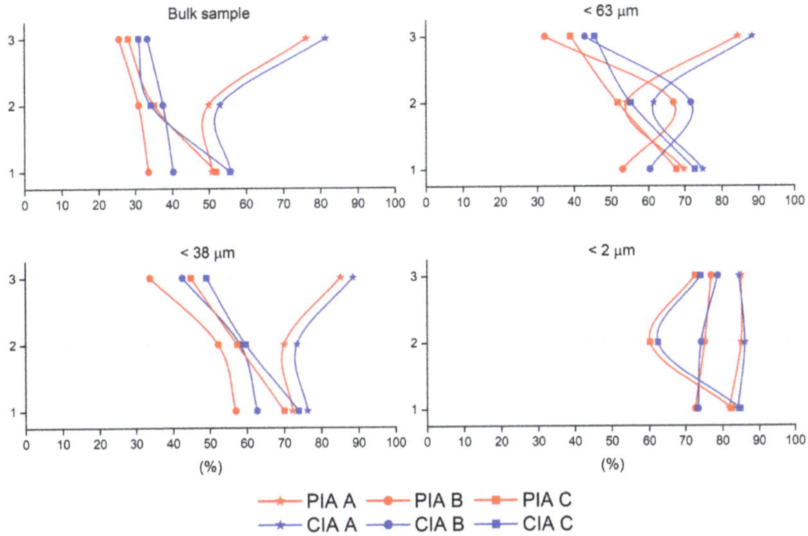

Figure 10. Graphical plot of the PIA (red) and CIA (blue) indices of samples from Benavila.

Despite the expected decline in the PIA and the CIA values from the top to the bottom of the vertical profile due to the field interpretation, it is possible to observe an increase in the values of the indices from top to bottom, indicating more weathered materials at the base of the deposit. However, there is an exception, the A profile, which shows higher values at the top. This can be related to the fracture system running in the NNE–SSW direction that seems to have favored the weathering [26], since it is possible that those fluids came from the NNE and affected the A profile in a slightly different manner. A clear link between the alteration indices and the amount of primary minerals (tectosilicates and amphiboles) is observed. The variation along the horizontal levels shows values that are typically higher towards the A profile, although the B and C profiles have higher variability in their values. The circulation of the carbonated fluids, especially in the B profile, affects the resulting CIA and PIA indices (even after the neglection of the Ca contained within carbonates by the use of the CaO* for their calculation), obscuring their true values and making it challenging to determine the weathering pathways of the deposit (Table 5, Figure 10). When calculating these indices for the finer granulometries, it is possible to observe how the influence of the carbonated fluids is progressively less important (as indicated by their higher values), especially at the <2 μm fraction, which is essentially formed of phyllosilicates. However, these fluids still have some influence on the calculated indices.

To completely avoid the influence of these carbonated fluids on the calculation of the indices and only focus on pure silicates, for which these indices were thought, the results from the TEM-AEM point analyses of the non-homoionized samples were used (Table 5), to avoid the additional input of Ca during the homoionization procedure. The indices calculated this way have slightly higher values than the ones calculated using the ICP-MS analyses of all the granulometric fractions. This is due to the presence of non-silicate minerals (carbonates) within the samples analyzed by ICP-MS, indicating that the corrections performed to completely neglect carbonates to calculate the indices are not 100% efficient. The resulting CIA indices are the most consistent with the previous data (from the mineralogical and crystal-chemical characterizations), showing a higher degree of alteration of the A profile, followed by the B profile, and ending with the C profile. It is especially interesting considering that there is a fracture system which seems to have favored the weathering of the samples running in NNE–SSW direction [26] and that the A profile is located towards the NE and the C profile towards the SW, which can point to a northern origin of the fluids that weathered the samples.

Although clay minerals can adsorb REE or even incorporate them in octahedral positions [75], providing excellent recipients for REE-rich deposits, the richest fractions in phyllosilicates are found to be depleted in such elements. This can be explained by a preferential accumulation of REE in the area within coarser minerals, such as zircons, as well as by their complexation and mobilization by the carbonate-rich fluids and their precipitation within calcite veins [68–71]. This would explain the significant correlation coefficients that REE have both with quartz and with calcite (Table S8). The Ce and Eu anomalies shown in Figure 9 and detailed in Table 6 were calculated as detailed by previous authors [76]. All the samples present negative Eu anomalies. The origin of this anomaly is due to the dissolution of Eu-rich minerals (plagioclase) present on the parent rock by weathering processes [77,78]. On the other hand, there are samples with positive Ce anomalies and others with negative Ce anomalies, indicating the first ones have more oxidizing environments than the second ones [79,80]. In Table S11, a correlation matrix between the values of these anomalies is displayed. It is possible to observe that the Eu anomaly is not correlated with any mineral, probably because there are no major differences within its values. However, the Ce anomaly shows significant positive correlations with the phyllosilicates and a highly significant negative correlation with calcite. This indicates that the weathering that originated phyllosilicates took place in more oxidizing environments, while the negative anomalies are probably caused by more reducing conditions during the precipitation of calcite due to the circulation of carbonate-rich fluids. This corroborates the important role these fluids play in the chemistry of the preexisting materials and how they can impact the indices calculated as well as their interpretations.

Table 6. Ce and Eu anomalies of all the samples and granulometric fractions.

	Whole Rock		<63 μm		<38 μm		<2 μm	
	Ce/Ce*	Eu/Eu*	Ce/Ce*	Eu/Eu*	Ce/Ce*	Eu/Eu*	Ce/Ce*	Eu/Eu*
BEN1A	1.01	0.81	1.11	0.74	1.19	0.82	1.13	0.79
BEN1B	1.04	0.79	1.30	0.71	1.43	0.81	2.27	0.76
BEN1C	1.24	0.78	1.60	0.81	1.59	0.83	1.87	0.74
BEN2A	0.93	0.78	0.42	0.71	1.14	0.84	1.46	0.76
BEN2B	0.31	0.72	1.07	0.78	0.40	0.71	0.77	0.68
BEN2C	1.12	0.80	1.24	0.87	1.44	0.84	2.55	0.75
BEN3A	1.05	0.81	1.14	0.80	1.14	0.88	1.05	0.75
BEN3B	0.33	0.77	0.36	0.70	0.49	0.70	1.85	0.72
BEN3C	0.93	0.77	1.04	0.81	1.18	0.88	1.80	0.70

6. Concluding Remarks

The Benavila deposit is mainly formed by dioctahedral smectites, followed by carbonates (Paleogene–Quaternary carbonated crusts due to the circulation of carbonate-rich fluids) and minor amounts of tectosilicates and other phyllosilicates. The almost complete transformation of primary minerals (as tectosilicates, amphiboles, and micas) suggests an intense weathering of the parent rock. In addition, the presence of amphibole within the C vertical profile is interpreted as indicative of a lower degree of alteration due to the primary character of that mineral.

The crystal-chemical characterization reinforces the importance of performing a homoionization of the samples prior to their analysis. The smectites studied are classified as ferrian montmorillonitic beidellites and beidellitic montmorillonites. Smectites with octahedral configurations with lower percentages of *tv* sites and classified as montmorillonitic beidellites could also indicate a lower degree of weathering within the deposit.

The geochemical analysis of the samples indicates a preferential depletion of the more clayey materials in HFSE, REE, and Y, among other heavy elements. In addition, it also points towards the pathway of weathering along which the bentonite deposit originated. However, the presence of carbonated crusts of Paleogene–Quaternary age due to the circulation of carbonated fluids overprints the deposit in certain parts and hardens the task of obtaining useful weathering indices.

The values of the CIA index are consistent with the aforementioned mineralogical and crystal-chemical criteria regarding the weathering of the parent rocks that originated the Benavila deposit. This study reveals the alteration pathways of the Benavila deposit (NE-SW direction, with the most weathered materials towards the NE) as well as the smectite variability within it. The variations of smectite mainly originate from lower degrees of weathering and the circulation of carbonated fluids that changed the already weathered bentonitic deposit, possibly affecting the properties of the bentonite.

Supplementary Materials: The following supporting information can be downloaded at: https://www.mdpi.com/article/10.3390/min15080836/s1, Table S1: Peak centers and percentages of the areas of the deconvoluted peaks at the DTG between 300 and 800 °C; Table S2: Weight percentage (wt %) of major elements; Table S3: Trace element content (μg/g); Table S4: Correlation matrix of mineralogy and major elements; Table S5: Correlation matrix of mineralogy and HFSE; Table S6: Correlation matrix of mineralogy and TTE; Table S7: Correlation matrix of mineralogy and LILE; Table S8: Correlation matrix of mineralogy and REE; Table S9: Correlation matrix of mineralogy and other trace elements analyzed; Table S10: Correlation matrix of mineralogy and the CIA and PIA indices; Table S11: Correlation matrix of mineralogy and the Ce and Eu anomalies.

Author Contributions: Conceptualization, M.I.D. and I.P.; methodology, J.G.-R., R.M., P.G.F. and E.G.-R.; formal analysis, J.G.-R.; investigation, J.G.-R., E.G.-R., M.S., I.P., P.G.F., R.M. and M.I.D.; data curation, J.G.-R.; writing—original draft preparation, J.G.-R.; writing—review and editing, J.G.-R., E.G.-R., M.S., I.P., P.G.F., R.M. and M.I.D.; supervision, M.I.D. and I.P.; project administration, M.I.D. and M.S.; funding acquisition, M.I.D., I.P. and M.S. All authors have read and agreed to the published version of the manuscript.

Funding: The authors acknowledge financial support of the SA0107P20 (Junta de Castilla y León, Spain, and Fondo Europeo de Desarrollo Regional, FEDER), UID/Multi/04349/2019 (FCT, Portugal) and Work Package ROUTES of EURAD (European Joint Programme on Radioactive Waste Management of the European Union, EC grant agreement n° 847593) projects. Javier García-Rivas thanks the FCT for a postdoctoral fellowship BL233/2019_IST-ID at the C2TN.

Data Availability Statement: The data presented in this study are available on request from the corresponding author.

Conflicts of Interest: The authors declare no conflicts of interest.

References

1. Newman, A.C.D. The Specific Surface of Soils Determined by Water Sorption. *J. Soil Sci.* **1983**, *34*, 23–32. [CrossRef]
2. Środoń, J.; MaCarty, D.K. Surface Area and Layer Charge of Smectite from CEC and EGME/H_2O-Retention Measurements. *Clays Clay Miner.* **2008**, *56*, 155–174. [CrossRef]
3. Woodruff, W.F.; Revil, A. CEC-Normalized Clay-Water Sorption Isotherm. *Water Resour. Res.* **2011**, *47*, W11502. [CrossRef]
4. Suárez, M.; Lorenzo, A.; García-Vicente, A.; Morales, J.; García-Rivas, J.; García-Romero, E. New Data on the Microporosity of Bentonites. *Eng. Geol.* **2022**, *296*, 106439. [CrossRef]
5. Murray, H.H. Bentonite Applications. In *Applied Clay Mineralogy Occurrences, Processing and Application of Kaolins, Bentonites, Palygorskite-Sepiolite, and Common Clays. Developments in Clay Science*; Murray, H.H., Ed.; Elsevier: Amsterdam, The Netherlands, 2006; Volume 2, pp. 111–130, ISBN 978-0-444-51701-2.
6. Harvey, C.C.; Lagaly, G. Industrial Applications. In *Developments in Clay Science*; Elsevier: Amsterdam, The Netherlands, 2013; Volume 5, pp. 451–490, ISBN 978-0-08-099364-5.
7. Eisenhour, D.D.; Brown, R.K. Bentonite and Its Impact on Modern Life. *Elements* **2009**, *5*, 83–88. [CrossRef]
8. Babu Valapa, R.; Loganathan, S.; Pugazhenthi, G.; Thomas, S.; Varghese, T.O. An Overview of Polymer–Clay Nanocomposites. In *Clay-Polymer Nanocomposites*; Elsevier: Amsterdam, The Netherlands, 2017; pp. 29–81. ISBN 978-0-323-46153-5.
9. Ishimaru, S.; Yamauchi, M.; Ikeda, R. A New Semiconducting Organic–Inorganic Nanocomposite, 1,5-Diaminonphthalene–Saponite Intercalation Compound. *Solid State Commun.* **2003**, *127*, 57–59. [CrossRef]
10. Yang, Q.; Saito, T.; Isogai, A. Transparent, Flexible, and High-Strength Regenerated Cellulose/Saponite Nanocomposite Films with High Gas Barrier Properties. *J. Appl. Polym. Sci.* **2013**, *130*, 3168–3174. [CrossRef]
11. Carretero, M.I.; Pozo, M. Clay and Non-Clay Minerals in the Pharmaceutical Industry. *Appl. Clay Sci.* **2009**, *46*, 73–80. [CrossRef]
12. Khurana, I.S.; Kaur, S.; Kaur, H.; Khurana, R.K. Multifaceted Role of Clay Minerals in Pharmaceuticals. *Future Sci. OA* **2015**, *1*. [CrossRef] [PubMed]
13. Gopinath, S.; Sugunan, S. Enzymes Immobilized on Montmorillonite K 10: Effect of Adsorption and Grafting on the Surface Properties and the Enzyme Activity. *Appl. Clay Sci.* **2007**, *35*, 67–75. [CrossRef]
14. Long, L.-H.; Zhang, Y.-T.; Wang, X.-F.; Cao, Y.-X. Montmorillonite Adsorbs Urea and Accelerates Urea Excretion from the Intestine. *Appl. Clay Sci.* **2009**, *46*, 57–62. [CrossRef]
15. Müller, H.-J.; Dobler, D.; Schmidts, T.; Rusch, V. Smectite for Medical Use and Their Toxin Binding Capacity. *J. Food Nutr. Popul. Health* **2019**, *03*. [CrossRef]
16. Ali Khan, S.; Riaz-ur-Rehman; Ali Khan, M. Sorption of Cesium on Bentonite. *Waste Manag.* **1994**, *14*, 629–642. [CrossRef]
17. Bradbury, M.H.; Baeyens, B. A Generalised Sorption Model for the Concentration Dependent Uptake of Caesium by Argillaceous Rocks. *J. Contam. Hydrol.* **2000**, *42*, 141–163. [CrossRef]
18. Durrant, C.B.; Begg, J.D.; Kersting, A.B.; Zavarin, M. Cesium Sorption Reversibility and Kinetics on Illite, Montmorillonite, and Kaolinite. *Sci. Total Environ.* **2018**, *610–611*, 511–520. [CrossRef] [PubMed]
19. Galamboš, M.; Magula, M.; Daňo, M.; Osacký, M.; Rosskopfová, O.; Rajec, P. Comparative Study of Cesium Adsorption on Dioctahedral and Trioctahedral Smectites. *J. Radioanal. Nucl. Chem.* **2012**, *293*, 829–837. [CrossRef]
20. Missana, T.; Benedicto, A.; García-Gutiérrez, M.; Alonso, U. Modeling Cesium Retention onto Na-, K- and Ca-Smectite: Effects of Ionic Strength, Exchange and Competing Cations on the Determination of Selectivity Coefficients. *Geochim. Cosmochim. Acta* **2014**, *128*, 266–277. [CrossRef]

21. Cuevas, J.; Ruiz, A.I.; Fernández, R.; Torres, E.; Escribano, A.; Regadío, M.; Turrero, M.J. Lime Mortar-Compacted Bentonite–Magnetite Interfaces: An Experimental Study Focused on the Understanding of the EBS Long-Term Performance for High-Level Nuclear Waste Isolation DGR Concept. *Appl. Clay Sci.* **2016**, *124–125*, 79–93. [CrossRef]
22. Park, S.-M.; Alessi, D.S.; Baek, K. Selective Adsorption and Irreversible Fixation Behavior of Cesium onto 2:1 Layered Clay Mineral: A Mini Review. *J. Hazard Mater.* **2019**, *369*, 569–576. [CrossRef]
23. Galopim de Carvalho, A.M.; Carvalhosa, A.B. *Carta Geológica de Portugal Na Escala de 1 / 50.000. Notícia Explicativa Da Folha 32-A. Ponte de Sor*; Serviços Geológicos de Portugal: Lisboa, Portugal, 1982.
24. Piçarra, J.M.; Dias, R.P.; Ribeiro, M.L.; Solá, R.; Barbosa, B.; Pais, J.; Vivar, V.; Baltuille Martín, J.M. *Carta Geológica de Portugal Na Escala de 1 / 50.000. Notícia Explicativa Da Folha 32-C. Avis*; LNEG: Lisboa, Portugal, 2009.
25. Pinto Henriques, S. Ensaios de Beneficiação de "Bentonite" de Avis (Portugal) Para Aplicação Em Geologia Médica. Master's Thesis, Universidade de Aveiro, Aveiro, Portugal, 2014.
26. Galopim de Carvalho, A.M.; Alegria, M.F.; Mira Azevedo, M.T. Aspectos de Alteração Em Rochas Dioríticas de Benavila (Avis). *Comun. Geológicas* **1980**, *66*, 67–70.
27. Canilho, M.H. Contribuição Para o Conhecimento Petrográfico e Geoquímico Do Maciço Ígneo de Benavila (Avis). *Ciênc. Terra* **1992**, *11*, 255–273.
28. de Gomes, C.S.F. Industrial minerals: Present situation in Portugal of the commercial clays. In *A Geologia de Engenharia e os Recursos Geológicos. Volume 2: Recursos Geológicos e Formação*; Imprensa da Universidade de Coimbra: Coimbra, Portugal, 2003; pp. 349–366, ISBN 978-989-26-0322-3.
29. Dias, M.I.; Suárez Barrios, M.; Prates, S. Las Bentonitas de Benavila (Portugal). Caracterización Mineralógica y Propiedades. [Bentonites from Benavila (Portugal). Mineralogical Characterisation and Properties]. *Geogactea* **2004**, *35*, 99–102.
30. Dias, M.I. Caracterização Mineralógica e Tecnológica de Argilas Especiais de Bacias Terciárias Portuguesas. Ph.D. Thesis, University of Lisbon, Lisbon, Portugal, 1998.
31. Rebelo, M.; Viseras, C.; López-Galindo, A.; Rocha, F.; da Silva, E.F. Characterization of Portuguese Geological Materials to Be Used in Medical Hydrology. *Appl. Clay Sci.* **2011**, *51*, 258–266. [CrossRef]
32. Guimarães, V.; Bobos, I.; Rocha, F. Acid-Base Properties of Ca-Montmorillonite from Benavila Region (Portugal). *Comun. Geológicas* **2014**, *101*, 791–794.
33. Dziadkowiec, J.; Mansa, R.; Quintela, A.; Rocha, F.; Detellier, C. Preparation, Characterization and Application in Controlled Release of Ibuprofen-Loaded Guar Gum/Montmorillonite Bionanocomposites. *Appl. Clay Sci.* **2017**, *135*, 52–63. [CrossRef]
34. Bastos, C.M.; Rocha, F.; Patinha, C.; Marinho-Reis, P. Bioaccessibility by Perspiration Uptake of Minerals from Two Different Sulfurous Peloids. *Environ. Geochem. Health* **2023**, *45*, 6621–6641. [CrossRef]
35. Gonçalves, F. Estado Actual Do Conhecimento Geológico Do Nordeste Alentejano. In *Proceedings of the IV Curso de Extensão Universitária de Ciências Geológicas*; Faculdade Ciências de Lisboa: Lisbon, Portugal, 1978; pp. 193–213.
36. Pereira, V.B. Prospecção de Bentonites. Referência Aos Trabalhos Realizados e Resultados Obtidos. *Estud. Notas E Trab.* **1993**, *35*, 35–54.
37. Martín Pozas, J.M. Analisis Cuantitativo de Fases Cristalinas Por DRX. In *Método de Debye-Scherrer*; Saja, J., Ed.; ICE Universidad de Valladolid: Valladolid, Spain, 1975.
38. García-Romero, E.; Lorenzo, A.; García-Vicente, A.; García-Rivas, J.; Morales, J.; Suárez, M. On the Structural Formula of Smectites: A Review and New Data on the Influence of the Exchangeable Cations. *J. Appl. Crystallogr.* **2021**, *54*, 251–262. [CrossRef]
39. Bahlburg, H.; Dobrzinski, N. A Review of the Chemical Index of Alteration (CIA) and Its Application to the Study of Neoproterozoic Glacial Deposits and Climate Transitions. *Geol. Soc. Lond. Mem.* **2011**, *36*, 81–92. [CrossRef]
40. Buggle, B.; Glaser, B.; Hambach, U.; Gerasimenko, N.; Marković, S. An Evaluation of Geochemical Weathering Indices in Loess–Paleosol Studies. *Quat. Int.* **2011**, *240*, 12–21. [CrossRef]
41. Fedo, C.M.; Nesbitt, H.W.; Young, G.M. Unraveling the Effects of Potassium Metasomatism in Sedimentary Rocks and Paleosols, with Implications for Paleoweathering Conditions and Provenance. *Geology* **1995**, *23*, 921–924. [CrossRef]
42. Meunier, A.; Caner, L.; Hubert, F.; El Albani, A.; Pret, D. The Weathering Intensity Scale (WIS): An Alternative Approach of the Chemical Index of Alteration (CIA). *Am. J. Sci.* **2013**, *313*, 113–143. [CrossRef]
43. Nesbitt, H.W.; Young, G.M. Early Proterozoic Climates and Plate Motions Inferred from Major Element Chemistry of Lutites. *Nature* **1982**, *299*, 715–717. [CrossRef]
44. Price, J.R.; Velbel, M.A. Chemical Weathering Indices Applied to Weathering Profiles Developed on Heterogeneous Felsic Metamorphic Parent Rocks. *Chem. Geol.* **2003**, *202*, 397–416. [CrossRef]
45. Newman, A.C.D.; Brown, G. The Chemical Constitution of Clays. In *Chemistry of Clays and Clay Minerals*; Newman, A.C.D., Ed.; Monograph No. 6; Mineralogical Society and Longman: London, UK, 1987; pp. 1–128.
46. Fesharaki, O.; García-Romero, E.; Cuevas-González, J.; López-Martínez, N. Clay Mineral Genesis and Chemical Evolution in the Miocene Sediments of Somosaguas, Madrid Basin, Spain. *Clay Miner.* **2007**, *42*, 187–201. [CrossRef]

47. de Santiago Buey, C.; Suárez Barrios, M.; García-Romero, E.; Domínguez Diaz, M.C.; Doval Montoya, M. Electron Microscopic Study of the Illite-Smectite Transformation in the Bentonites from Cerro Del Aguila (Toledo, Spain). *Clay Miner.* **1998**, *33*, 501–510. [CrossRef]
48. Pelayo, M.; Schmid, T.; Díaz-Puente, F.J.; López-Martínez, J. Characterization and Distribution of Clay Minerals in the Soils of Fildes Peninsula and Ardley Island (King George Island, Maritime Antarctica). *Clay Miner.* **2022**, *57*, 264–284. [CrossRef]
49. García-Romero, E.; Suárez, M. A Structure-Based Argument for Non-Classical Crystal Growth in Natural Clay Minerals. *Mineral. Mag.* **2018**, *82*, 171–180. [CrossRef]
50. Post, J.L.; Cupp, B.L.; Madsen, F.T. Beidellite and Associated Clays from the Delamar Mine and Florida Mountain Area, Idaho. *Clays Clay Miner.* **1997**, *45*, 240–250. [CrossRef]
51. Keeling, J.L.; Raven, M.D.; Gates, W.P. Geology and Characterization of Two Hydrothermal Nontronites from Weathered Metamorphic Rocks at the Uley Graphite Mine, South Australia. *Clays Clay Miner.* **2000**, *48*, 537–548. [CrossRef]
52. Emmerich, K. Full Characterization of Smectites. In *Developments in Clay Science*; Bergaya, F., Lagaly, G., Eds.; Elsevier: Amsterdam, The Netherlands, 2013; Volume 5, pp. 381–404. ISBN 978-0-08-099364-5.
53. Emmerich, K.; Wolters, F.; Kahr, G.; Lagaly, G. Clay Profiling: The Classification of Montmorillonites. *Clays Clay Miner.* **2009**, *57*, 104–114. [CrossRef]
54. El-Barawy, K.A.; Girgis, B.S.; Felix, N.S. Thermal Treatment of Some Pure Smectites. *Thermochim. Acta* **1986**, *98*, 181–189. [CrossRef]
55. Drits, V.A.; Lindgreen, H.; Salyn, A.L.; Ylagan, R.; McCarty, D.K. Semiquantitative Determination of Trans-Vacant and Cis-Vacant 2:1 Layers in Illites and Illite-Smectites by Thermal Analysis and X-Ray Diffraction. *Am. Miner.* **1998**, *83*, 1188–1198. [CrossRef]
56. Drits, V.A. An Improved Model for Structural Transformations of Heat-Treated Aluminous Dioctahedral 2:1 Layer Silicates. *Clays Clay Miner.* **1995**, *43*, 718–731. [CrossRef]
57. Wolters, F.; Emmerich, K. Thermal Reactions of Smectites—Relation of Dehydroxylation Temperature to Octahedral Structure. *Thermochim. Acta* **2007**, *462*, 80–88. [CrossRef]
58. Wojdyr, M. *Fityk*: A General-Purpose Peak Fitting Program. *J. Appl. Crystallogr.* **2010**, *43*, 1126–1128. [CrossRef]
59. Wang, X.; Li, Y.; Wang, H. Structural Characterization of Octahedral Sheet in Dioctahedral Smectites by Thermal Analysis. *Minerals* **2020**, *10*, 347. [CrossRef]
60. Deer, W.A.; Howie, R.A.; Zussman, J. *An Introduction to the Rock-Forming Minerals*; Longman: London, UK, 1966.
61. González López, J.M.; Bauluz, B.; Fernández-Nieto, C.; Oliete, A.Y. Factors Controlling the Trace-Element Distribution in Fine-Grained Rocks: The Albian Kaolinite-Rich Deposits of the Oliete Basin (NE Spain). *Chem. Geol.* **2005**, *214*, 1–19. [CrossRef]
62. Middelburg, J.; Vanderweijden, C.; Woittiez, J. Chemical Processes Affecting the Mobility of Major, Minor and Trace Elements during Weathering of Granitic Rocks. *Chem. Geol.* **1988**, *68*, 253–273. [CrossRef]
63. Mosser, C.; Brillanceau, A.; Besnus, Y. Relationship between Sediments and Their Igneous Source Rocks Using Clay Mineral Multi-Element Chemistry: The Cenozoic Lacustrine Anloua Basin (Adamaoua, Cameroon). *Chem. Geol.* **1991**, *90*, 319–342. [CrossRef]
64. Smith, J.V. *Feldspar Minerals 2: Chemical and Textural Properties*; Springer: Berlin/Heidelberg, Germany, 1974.
65. Baker, P.A.; Gieskes, J.M.; Elderfield, H. Diagenesis of Carbonates in Deep-Sea Sediments—Evidence from Sr/Ca Ratios and Interstitial Dissolved Sr2+ Data. *J. Sediment. Petrol.* **1982**, *52*, 71–82.
66. Gabitov, R.I.; Watson, E.B. Partitioning of Strontium between Calcite and Fluid. *Geochem. Geophys. Geosystems* **2006**, *7*, 2005GC001216. [CrossRef]
67. García-Rivas, J.; Suárez, M.; Torres, T.; Sánchez-Palencia, Y.; García-Romero, E.; Ortiz, J.E. Geochemistry and Biomarker Analysis of the Bentonites from Esquivias (Toledo, Spain). *Minerals* **2018**, *8*, 291. [CrossRef]
68. McLennan, S.M.; Nance, W.B.; Taylor, S.R. Rare Earth Element-Thorium Correlations in Sedimentary Rocks, and the Composition of the Continental Crust. *Geochim. Cosmochim. Acta* **1980**, *44*, 1833–1839. [CrossRef]
69. Louvel, M.; Etschmann, B.; Guan, Q.; Testemale, D.; Brugger, J. Carbonate Complexation Enhances Hydrothermal Transport of Rare Earth Elements in Alkaline Fluids. *Nat. Commun.* **2022**, *13*, 1456. [CrossRef] [PubMed]
70. Reece, M.E.; Migdisov, A.A.; Williams-Jones, A.E.; Strzelecki, A.C.; Waters, L.; Boukhalfa, H.; Guo, X. Stability of Aqueous Neodymium Complexes in Carbonate-Bearing Solutions from 100–600 °C. *Commun. Earth Environ.* **2025**, *6*, 353. [CrossRef]
71. Di, J.; Ding, X. Complexation of REE in Hydrothermal Fluids and Its Significance on REE Mineralization. *Minerals* **2024**, *14*, 531. [CrossRef]
72. Boynton, W.V. Cosmochemistry of the Rare Earth Elements: Meteorite Studies. In *Developments in Geochemistry*; Henderson, P., Ed.; Elsevier: Amsterdam, The Netherlands, 1984; Volume 2, pp. 63–114. ISBN 978-0-444-42148-7.
73. Foley, N.K.; Jaskula, B.W.; Kimball, B.E.; Schulte, R.F. *Critical Mineral Resources of the United States—Economic and Environmental Geology and Prospects for Future Supply. Chapter H: Gallium.*; Professional Paper; U.S. Geological Survey: Reston, VA, USA, 2017; p. 862.

74. Eliopoulos, I.-P.D.; Eliopoulos, G.D. Factors Controlling the Gallium Preference in High-Al Chromitites. *Minerals* **2019**, *9*, 623. [CrossRef]
75. Lorenzo, A.; Sánchez-Santos, J.M.; Rivas, M.J.; García-Romero, E.; Suárez, M. Geochemistry of Bentonites: A Statistical Analysis of Trace Element Distribution in Smectites. *Appl. Clay Sci.* **2024**, *257*, 107449. [CrossRef]
76. Shields, G.; Stille, P. Diagenetic Constraints on the Use of Cerium Anomalies as Palaeoseawater Redox Proxies: An Isotopic and REE Study of Cambrian Phosphorites. *Chem. Geol.* **2001**, *175*, 29–48. [CrossRef]
77. Lewis, A.J.; Palmer, M.R.; Sturchio, N.C.; Kemp, A.J. The Rare Earth Element Geochemistry of Acid-Sulphate and Acid-Sulphate-Chloride Geothermal Systems from Yellowstone National Park, Wyoming, USA. *Geochim. Cosmochim. Acta* **1997**, *61*, 695–706. [CrossRef]
78. Karakaya, M.Ç.; Karakaya, N.; Kupeli, S. Mineralogical and Geochemical Properties of the Na- And Ca-Bentonites of Ordu (Ne Turkey). *Clays Clay Miner.* **2011**, *59*, 75–94. [CrossRef]
79. Class, C.; le Roex, A.P. Ce Anomalies in Gough Island Lavas—Trace Element Characteristics of a Recycled Sediment Component. *Earth Planet. Sci. Lett.* **2008**, *265*, 475–486. [CrossRef]
80. Braun, J.-J.; Pagel, M.; Muller, J.-P.; Bilong, P.; Michard, A.; Guillet, B. Cerium Anomalies in Lateritic Profiles. *Geochim. Cosmochim. Acta* **1990**, *54*, 781–795. [CrossRef]

minerals

MDPI

Article

Influence of Geological Origin on the Physicochemical Characteristics of Sepiolites

Leticia Lescano [1], Silvina A. Marfil [1], Luciana A. Castillo [2,3,*] and Silvia E. Barbosa [2,3]

1 Departamento de Geología, Universidad Nacional del Sur. CGAMA (CIC-UNS), Bahía Blanca 8000, Argentina; leticia.lescano@uns.edu.ar (L.L.); smarfil@uns.edu.ar (S.A.M.)
2 Planta Piloto de Ingeniería Química, PLAPIQUI (UNS-CONICET), Bahía Blanca 8000, Argentina; sbarbosa@plapiqui.edu.ar
3 Departamento de Ingeniería Química, Universidad Nacional del Sur, Bahía Blanca 8000, Argentina
* Correspondence: lcastillo@plapiqui.edu.ar

Abstract: In this study the influence of the geological formation environment on the physicochemical properties of two natural sepiolites, as collected, was investigated. The samples analyzed were a lacustrine-derived sample from Tolsa, Spain (ST), and a hydrothermal-derived sample from La Adela, Argentine (SA). Comprehensive characterization was carried out using chemical analysis (XRF), X-ray diffraction (XRD), thermogravimetric analysis (TGA), Fourier-transform infrared spectroscopy (FTIR), scanning electron microscopy (SEM), and evaluations of hydrophobicity/hydrophilicity behavior. The results indicate that the ST sample exhibits a higher SiO_2/MgO ratio and contains amorphous silica impurities, while the SA sample shows a composition more closely aligned with the theoretical stoichiometry of sepiolite. Furthermore, the SA sample demonstrates greater crystallinity compared to the ST sample. Morphological analysis revealed that ST consists of compact, aggregated fibrous structures, while SA is composed of disaggregated, needle-like fibers with high aspect ratios and nanometric diameters. Both samples display predominantly hydrophilic behavior; however, only the SA sample exhibits suspended particles at the interface, suggesting a slightly higher hydrophobic character than ST sample. These findings highlight the significant impact of the geological formation environment on the structural and surface characteristics of sepiolite, which, in turn, influence its performance in applications involving dispersion, adsorption, and interfacial interactions.

Keywords: sepiolite; origin; crystallinity; hydrophobicity/hydrophilicity behavior; chemistry; morphology

1. Introduction

Sepiolite is a natural occurring mineral, belonging to the palygorskite/sepiolite polysomatic series, which can originate in a wide variety of geological environments. It is a magnesium phyllosilicate with the chemical formula $Mg_4Si_6O_{15}(OH)_2(H_2O)_2{\cdot}4H_2O$. In the sepiolite crystal lattice, the octahedral (O) sheet is divided into ribbons that extend along the z-axis, positioned between two continuous, undulating tetrahedral (T) layers. This arrangement arises from a periodic reversal in the orientation of the apical oxygen atoms. These TOT ribbons are alternately connected in a chessboard-like pattern, creating parallel nanochannels with an effective maximum width of approximately 10.6 Å, which run along the c-axis and are typically occupied by weakly held zeolitic water molecules [1]. At the ribbon edges, tightly bound structural water (H_2O) completes the coordination sphere of the octahedral magnesium atoms. These channels and the external surface of the fibers can accommodate small molecules or exchangeable cations, making sepiolite not only industrially valuable but also geochemically informative. Sepiolite typically exhibits a needle-like morphology, often forming porous aggregates and fibrous sheets. Due to its unique structural and physicochemical properties, sepiolite plays a significant role in various geological, environmental, and industrial contexts [1–3].

Besides its physical and chemical characteristics, sepiolite has increasingly been recognized as a valuable indicator of paleoenvironmental conditions. Its formation is highly sensitive to specific geological and climatic settings. Typically, sepiolite forms in lacustrine to palustrine environments, under semi-arid to arid climatic regimes, frequently associated with high evaporation rates [4–10]. Moreover, sepiolite may also form through hydrothermal alteration of rocks and minerals, particularly in volcanic or metamorphic settings. Its formation requirements include a fluid with moderately to highly alkaline pH, sufficient concentrations of magnesium and silica, and minimal aluminum [11]. Giustetto et al. [1] analyzed a sepiolite from Sassello (Ligurian Apennines, Italy) which precipitated from hydrothermal fluids saturated in Mg and silica by the interaction with the host serpentinites. The authors studied its crystal-chemistry and structure, revealing, among other findings, that the fibers are extremely long and flexible, with a length/width aspect ratio significantly greater than 3. Despite this, there are relatively few studies exploring how geological origin impacts sepiolite's physicochemical behavior. The environmental context, whether lacustrine or hydrothermal, can influence parameters such as crystallinity, surface area, trace element distribution, and adsorption capacity. A comprehensive understanding of these contrasts is essential not only for reconstructing the mineral's geological history but also for optimizing its industrial utility.

The objective of this study is to examine how the formation environment of sepiolite influences its physicochemical characteristics. For this purpose, two distinct samples were selected for the analysis: one from a lacustrine setting in the Tajo Basin (Madrid, Spain), and another of hydrothermal origin from the Colorado Basin (province of Río Negro, Argentina). Comprehensive characterization of their physicochemical properties was performed, using X-ray diffraction (XRD) for crystal structure, X-ray fluorescence (XRF) for chemical analysis, Fourier-transform infrared spectroscopy (FTIR) to identify functional groups, and thermogravimetric analysis (TGA) for evaluating their thermal behavior. Morphological features were assessed through scanning electron microscopy (SEM). Additionally, the hydrophobic/hydrophilic behavior of sepiolite samples was qualitatively assessed.

2. Geological Setting

2.1. Tolsa Quarry

The Madrid Basin (Spain) is mainly composed of granitic, metamorphic and sedimentary rocks. During the early and middle Miocene, it was an endorheic sedimentary basin, characterized by central lacustrine and palustrine systems, fringed by alluvial fans and fluvial distributary facies [12]. Figure 1a shows the location of the basin in the center of the Iberian Peninsula, covering an area of more than 10,000 km^2.

The basin's stratigraphy is typically divided into three units: lower, intermediate, and upper [13]. The intermediate unit, dated to the Middle Aragonian–Upper Vallesian, is approximately 60 m thick and is composed of two distinct subunits with different lithologies and depositional environments: the lower, or detrital subunit, primarily consisting of detrital gypsum, and an upper, dolomitic subunit. The facies assemblage within this unit reflects a moderately saline, very shallow lacustrine system with fluctuating water level. Sepiolite deposits are located in marginal lacustrine zones within this unit and at the top of the underlying detrital unit [14]. The mineral occurs in two main stratigraphic levels within these detrital units, interbedded with fine arkosic sands and silts derived from alluvial fans. These deposits may also contain palygorskite and traces of quartz or amorphous silica. Although this sepiolite is acicular and generally poorly crystallized, its fibrous habit is distinctive. Fiber length is typically less than 2 mm, rarely exceeding 5 mm. Fibers are arranged parallel to bedding within fine-grained sediments. Under optical microscopy in thin sections, sepiolite appears colorless to slightly whitish, fibrous, and acicular, with low birefringence, straight extinction, and positive elongation, comparable to descriptions in other lacustrine-evaporitic deposits [14]. From a genetic perspective, the mineralization is interpreted as resulting from chemical precipitation under evaporitic and alkaline conditions within shallow lacustrine settings. The progressive concentration of Mg- and Si-rich waters, coupled with fluctuations in pH and salinity, favored the formation of fibrous Mg-silicates such as sepiolite and palygorskite [11]. These processes were enhanced by the endorheic character of the basin, where limited outflow promoted silica supersaturation and episodic precipitation. The studied deposit occurrence reflects a sedimentary-evaporitic origin, controlled by lacustrine geochemistry and periodic desiccation events. Since 1963, this sepiolite has been extracted in the southeastern area of Madrid municipality, specifically in Vicálvaro. Initially, mining was carried out underground, but since 1970 it has been conducted by open-pit methods [15]. Figure 1b shows the location of the mine.

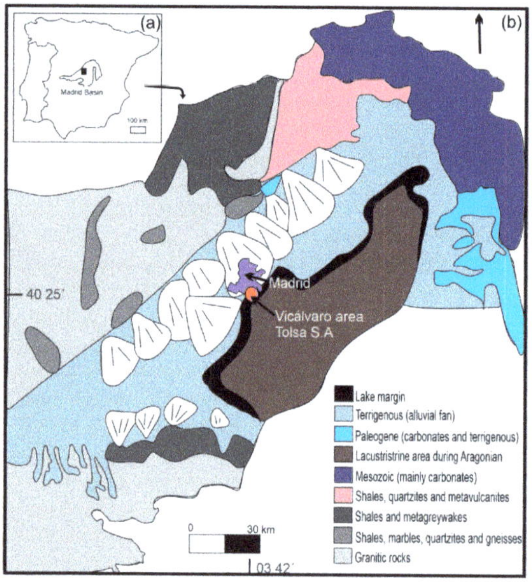

Figure 1. (**a**) Location, and (**b**) geological map of the Madrid Basin (modified from [16]).

2.2. *La Adela Quarry*

The La Adela quarry is located within the Santa Euriciana Ranch, approximately 75 km southwest of the town of San Antonio Oeste, in the province of Río Negro, Argentina (Figure 2a) [17]. The area features around eight small elevations composed mainly of dolomite beds with strike directions ranging from E–W to N 80° W, and dips varying between 45° and 65° to the north. The quarry (currently inactive) was developed on one of these outcrops. It measures approximately 600 m in length and between 70 and 90 m in width, with an estimated thickness of 50 m; the base does not outcrop. It is hosted in pre-Silurian micaceous schists belonging to the Valcheta Group [18]. The country rock consists of a massive, compact dolomite, bluish-white to gray in color, slightly reddish due to the presence of iron oxides. It exhibits a granoblastic texture with medium to coarse grain size, with dolomite crystals reaching up to 0.5 cm [19]. These authors also identified tremolite-actinolite, phlogopite, talc, montmorillonite, and iron oxides. The exploited material is dolomite, containing low amounts of silica and clay minerals as impurities. It was mainly used in the metallurgical processing of iron ore from a nearby Sierra Grande mine. The quarry shows intense jointing, with three predominant directions, variably mineralized. The eastern sector of the quarry is highly dislocated and contains the highest concentration of alteration minerals, the most significant being sepiolite (Figure 2b).

Sepiolite occurs within joints, where it fills the fractures without replacing the host rock, exhibiting exceptional crystal development. Fibers exceed 10 cm in length and occur in bundles up to 1 cm thick. The mineral is white, soft, flexible, and has a waxy luster. It is arranged parallel to the fractures it fills, forming high-purity monomineralic veins, with minor dolomite impurities. Under optical microscopy in thin section, sepiolite appears colorless, fibrous, and acicular, with refractive indices n = 1.512 and n = 1.5222, birefringence of 0.01, straight extinction, and positive elongation. The mineralization was developed along pre-existing fractures and bedding planes in the dolomite, which acted as both structural and lithological controls. The process occurred under hydrothermal conditions, beginning with the precipitation of talc and phlogopite under acidic conditions at ~300 °C. As the pH approached neutrality, illite formed, followed by the crystallization of sepiolite under alkaline conditions, resulting from the alteration of dolomite in the presence of alkali elements. This alteration process is related to Miocene–Pliocene volcanism, to which similar mineralizations in the northern Somuncurá Plateau have been attributed [20,21].

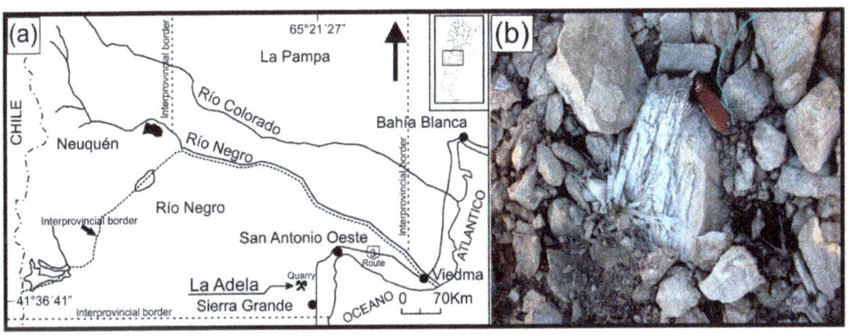

Figure 2. (**a**) Location map of the La Adela quarry within the province of Río Negro (Argentina) (modified from [17]). (**b**) Photograph showing bundles of highly crystalline sepiolite fibers growing within dolomitic host rock.

3. Materials and Methods

Two natural sepiolite samples were analyzed in this study, both previously milled to a uniform fineness of 400 mesh (37 μm). A representative portion of each sepiolite was then selected for analysis following a quartering procedure performed on the ground bulk material. This method ensured the homogeneity and representativeness of the sample used for subsequent physicochemical and mineralogical characterizations. The samples are referred to as ST (from the Tolsa quarry, Madrid, Spain, supplied by Tolsa S.A.) and SA (from La Adela quarry, province of Río Negro, Argentina). These samples were analyzed according to chemical, mineralogical, and physical characterization.

Chemical composition was determined by XRF using a PANalytical PW 4400/40 Axios with CrKα radiation (Malvern Panalytical, Almelo, The Netherlands). Mineralogical analysis was conducted by XRD, using a Rigaku D Max III-C X-ray diffractometer (Rigaku, Tokyo, Japan) operating at 35 kV and 15 mA, with CuKα radiation (wavelength of 0.154060 nm), and a graphite monochromator in the secondary diffracted beam was used to eliminate CuKβ radiation. Diffractograms were recorder over a 2θ ranges of 3–60° 2θ, in 0.04° 2θ increments with a counting time of 1 s per increment. Samples were mounted on dimpled glasses and pressed with a flat glass plate to obtain a uniform surface. Phase identification was performed using JADE software version 8 (Materials Data Inc., Livermore, CA, USA), with intensities expressed in counts per second (CPS). Raw, unsmoothed diffractograms were used for all measurements. FTIR analysis was carried out using a Nicolet 520 FT-IR spectrophotometer (Thermo Fisher Scientific, Waltham, MA, USA) equipped with a DTGS detector. A 2 mg sample was mixed with KBr at 1:49 ratio (sepiolite/KBr) and pressed under 120 MPa until a disk was obtained. Spectra were recorder from 4000 cm^{-1} to 400 cm^{-1}, using 32 scans at resolution of 4 cm^{-1} under a nitrogen atmosphere. Thermal behavior was evaluated by thermogravimetric analysis (TGA) and derivative thermogravimetry (DTG) using a thermogravimetric analyzer (TGA 5500, Discovery Series, TA Instruments, New Castle, DE, USA). In total, 25 mg of sample was placed in a ceramic crucible and heated at 10 °C/min up to 800 °C under a nitrogen atmosphere (25 mL/min). Morphological characterization was performed using a SEM (LEO EVO 40 XVP, Carl Zeiss AG, Cambridge, UK) equipped with an energy-dispersive X-ray spectroscopy (EDS) detector (Oxford X-Max 50, Oxford, Oxford, UK), operating at an accelerating voltage of 15 kV and under a vacuum of 2.1×10^{-5} Torr. The optimum working distance for sepiolite morphology observation was 7 mm. Samples were prepared by gently dispersing the particles with the aid of a dry air stream onto a double-sided aluminum adhesive tape mounted on bronze stubs. Then, they were coated with a 30 Å gold layer argon plasma PELCO 91000 coater (Ted Pella, Redding, CA, USA), in order to ensure conductivity. Particle length and width were determined by analyzing hundreds of particles by using AnalySIS2.1 software (Soft Imaging System GmbH, Münster, Germany) on high-resolution SEM micrographs. The hydrophobic/hydrophilic behavior of the samples was qualitatively evaluated by dispersing 1 g of each sample in a biphasic system composed of octane and distilled water. The settling behavior was visually assessed after 1 h, and photographic documentation was recorded. To complement this test, UV-Vis spectroscopy was performed to verify the presence of particles at the interface. The experiment was replicated in a quartz cuvette with identical proportions of components. A specific window was created in the cuvette to allow UV–Vis light to pass through the interfacial region. Spectra were acquired using a Shimadzu UV-160 spectrophotometer (Shimadzu Corporation, Kyoto, Japan), in the range of 200 nm–600 nm, with a spectral resolution of 1 nm.

4. Results and Discussion

The influence of the formation environment on the physicochemical characteristics of the sepiolite samples investigated in this study was assessed through integral comprehensive chemical, mineralogical, and physical characterization.

Table 1 shows the chemical composition of the two samples (ST and SA), along with the theoretical values derived from the ideal formula for sepiolite (S) $Mg_4Si_6O_{15}(H_2O)_2$•$4H_2O$ [22]. The SiO_2/MgO ratio is 2.65 for ST and 2.21 for SA, the latter being very close to the theoretical value of 2.23. The high ratio observed in the ST sample is attributed to the higher silica content, which is consistent with a lower loss on ignition (LOI), likely due to the presence amorphous silica as an impurity. The calcium oxide content in both samples is associated with the presence of calcite ($CaCO_3$) or dolomite ($CaMg(CO_3)_2$) as accessory minerals. According to [23], the Mg/(Mg + Al) atomic ratio must exceed 0.6 for a mineral to be classified as sepiolite. The calculated values for this ratio are 0.96 for ST and 0.93 for SA, confirming their classification as sepiolite.

Table 1. Chemical composition of analyzed samples from Tolsa (ST) and La Adela (SA) compared with ideal sepiolite (S).

Samples	SiO_2 (%)	Al_2O_3 (%)	Fe_2O_3 (%)	MgO (%)	CaO (%)	Na_2O (%)	K_2O (%)	LOI (%)
ST	63.10	1.08	0.27	23.80	0.49	0.09	0.21	10.96
SA	53.15	1.83	1.31	24.08	0.90	0.10	0.08	18.54
S	56.26	-	-	25.27	-	-	-	18.57

Figure 3 shows the XRD patterns of the analyzed samples. The principal reflections corresponding to sepiolite were identified in both diffractograms according to the ICDD 26-1226 reference card [24]. Although the overall diffractograms are comparable, several differences exist between the ST and SA samples, attributable not only to the presence of impurities but also to variations in their degree of crystallinity. In the ST sample, the elevation of the background, between 20° and 30° 2θ, suggests the presence of amorphous material, likely corresponding to amorphous silica, as previously reported in the literature [14]. This interpretation is consistent with the elevated SiO_2 content and the lower LOI observed in the chemical analysis. In contrast, the SA sample exhibits reflections associated with dolomite impurities, the most intense being located at 0.289 nm. The crystallinity of each sepiolite sample can be inferred from the presence, shape, and intensity of specific reflections as well as the full width at half maximum (FWHM) parameter, measured in degrees 2θ, which reflects the mineral crystallinity [25]. The diffraction peaks in the ST sample are broader and less defined, whereas those in the SA sample are sharper and more intense, indicating a higher degree of crystallinity and a more ordered structural arrangement. These differences are particularly noticeable in the reflections corresponding to the d-spacings of 0.374 nm, 0.334 nm, 0.318 nm, 0.261 nm, and 0.256 nm, although additional variations are also observed. Quantitatively, FWHM values for the SA sample range from 0.095° 2θ to 0.331° 2θ, while those for the ST sample range from 0.422° 2θ to 1.065° 2θ. The broader peaks and higher FWHM values in the ST sample confirm its relatively lower crystallinity.

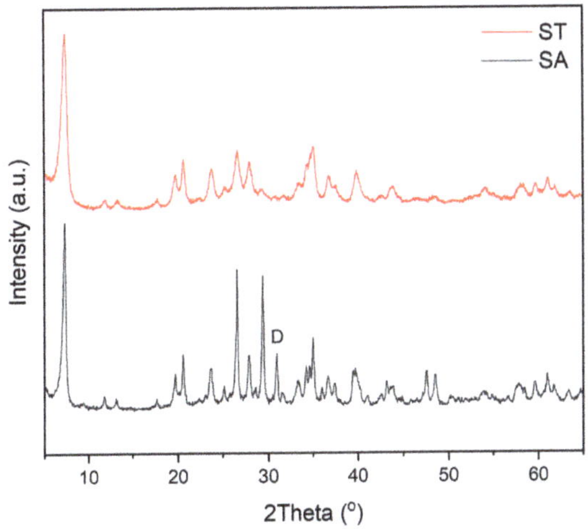

Figure 3. X-ray diffractograms of ST and SA samples. D: dolomite.

Figure 4 presents the thermogravimetric curves of the two sepiolite samples. Both exhibit broadly similar slopes, suggesting a comparable thermal behavior. At ~100 °C, the hydration water is released, with mass loss values of 7.5% and 5.0% for the ST and SA samples, respectively. The total mass loss values are consistent with those obtained from the chemical composition analysis. The lower total mass loss observed in the ST sample is likely due to the presence of impurities identified (amorphous silica). The mass loss observed at approximately 700 °C corresponds to two overlapping thermal decomposition processes. On one hand, water molecules coordinated to Mg^{2+} cations in octahedral sites within the sepiolite structure are released at this temperature [26,27]. On the other hand, this event coincides with the decomposition of dolomite, which occurs within the same temperature range [28,29]. The DTG curves (Figure 4b) provide further insight into the thermal events underlying the overall mass loss. Both samples exhibit an initial peak around 40–45 °C, associated with the release of physically adsorbed water. The slightly higher intensity in the ST sample could indicate a higher affinity for surface water. A second thermal event is observed at 245–248 °C, corresponding to the loss of loosely bound water within the sepiolite channels. The similarity in temperature between both samples indicates comparable structural features, although the peak intensity in ST is higher, suggesting a greater amount of intracanalicular water. At higher temperatures, a distinct peak at 528 °C is evident in the ST sample, indicating the dehydroxylation of structural hydroxyl groups. In SA, this event is greatly attenuated or absent, which could indicate a lower proportion of structural hydroxyl groups and/or a higher crystallinity. Only the SA sample presents a clear signal at 683 °C, which may be attributed to the decomposition of dolomite impurities, as indicated by the thermogravimetrical analysis [28,29].

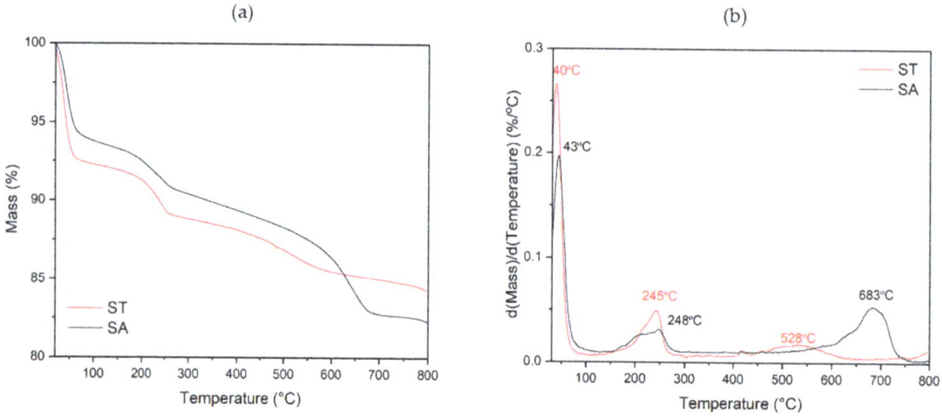

Figure 4. (**a**) Mass loss curve, and (**b**) first derivative of the mass loss curve of ST and SA samples from TGA.

Figure 5 presents the FTIR spectra of both sepiolite samples. Most absorption bands in both SA and ST, identified with stripped lines, correspond to characteristic sepiolite vibration. The band at 3689 cm^{-1} is assigned to the stretching vibrations of MgOH. The band observed at 3566 cm^{-1} corresponds to the OH stretching vibrations of bound water, while the band at 3621 cm^{-1} is associated with zeolitic water. The OH bending vibration corresponding to the bound water can be observed at 1657 cm^{-1}. The SiOSi bands at 1209 cm^{-1} and 1016 cm^{-1} are due to SiO vibrations. A minor band at 785 cm^{-1} is assigned to the OH bending vibration of MgFeOH. The bands observed at 690 cm^{-1} and 646 cm^{-1} correspond to the bending vibration of MgOH. The band at 470 cm^{-1} is attributable to the SiOSi bending vibration and 442 cm^{-1} arising from the SiOMg of the octahedral–tetrahedral linkage [30–37]. In the SA sample, additional bands at 1437 cm^{-1}, 877 cm^{-1} and 728 cm^{-1}

were detected, indicating the presence of carbonate impurities [38–40], consistent with TGA and XRD results.

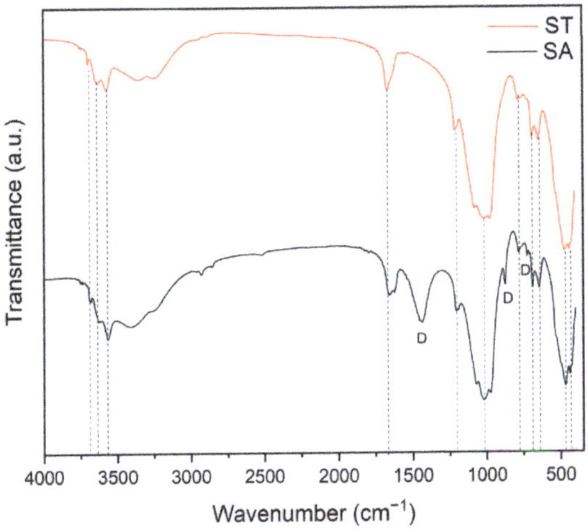

Figure 5. FTIR spectra of ST and SA samples. Ref: D: dolomite.

The fibrous morphology of the ST and SA milled samples is revealed in the SEM micrographs shown in Figure 6. The sample ST exhibits densely aggregated fibrous structures, forming entangled mats and compact bundles with a homogeneous appearance (Figure 6a). This aggregation is likely promoted by hydrogen bonding between surface SiOH groups and a small amount of adsorbed water, which facilitates network formation [41]. The ST fibers appear oriented in planar arrangements, consistent with their sedimentary depositional origin. Despite the fact that the fibers are mostly densely and compactly aggregated, it was possible to determine the length and width of loose fibers among these fiber aggregates (e.g., circles in Figure 6a). The length and width values are 1.51 ± 0.47 µm and 0.12 ± 0.04 µm, respectively, resulting in a length-to-width ratio of 13 ± 5. However, a more representative size of ST sample is the average length of the aggregated fibers (4.5 µm). These aggregates were not included in the individual fiber statistical analysis. From a textural perspective, no discernible porosity is evident in the ST, at the used observation conditions, as the fibers form a compact and planar structure. This reduced porosity may be attributed to the presence of amorphous silica impurities, identified in previous analyses, which could fill voids between fibers or promote a denser packing, thereby modifying the material's overall texture and fiber arrangement. In contrast, the SA sample (Figure 6b) displays elongated needle-like and rod-shaped particles, along with short disaggregated fibers dispersed throughout the field of view. Individual long fibers, exceeding 5 microns in length with notable flexibility, are clearly visible (see arrows in Figure 6b). Statistical analysis of measurable loose fibers revealed a bimodal particle distribution, associated with two distinct size populations. One of them consists of short fibers (~70%) with length and width values of 1.16 ± 0.39 µm, and 0.10 ± 0.02 µm, respectively, resulting in a length-to-width ratio of 11.22. Meanwhile, ~30% is composed of long and flexible fibers, with a length of 7.08 ± 2.83 µm and a width of 0.18 ± 0.07 µm, yielding a length-to-width ratio of 44. Taking into account the relative abundance of each population, the weighted mean fiber length for the SA sample is 2.94 µm. The nanometric thicknesses and high aspect ratio of these needle-like crystals are characteristic of the anisotropic crystal growth of the SA sample. Importantly, the overall mean fiber length for SA, weighted by the relative abundance of the two size populations is ~3 µm, which remains on the same order of magnitude as the average length of the aggregated fibers in ST (4.5 μm).

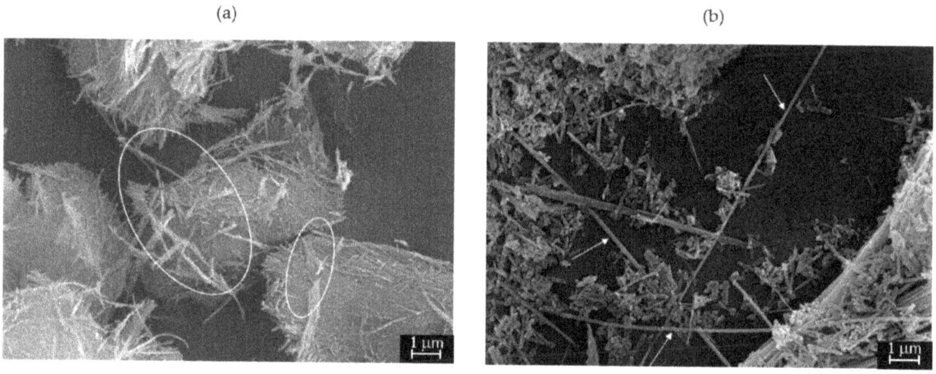

Figure 6. SEM micrographs (20,000×) of (**a**) ST sample, showing densely aggregated fibrous structures with a few loose fibers (enclosed in circles), and (**b**) SA sample displaying short disaggregated fibers and individual long fibers (indicated by arrows).

Although both samples exhibit a similar degree of fineness, the fibers display different dispersion/aggregation behavior, likely associated with their origin. These morphological features have been explored in various studies using different defibrillation methods, without affecting sepiolite length and flexibility [42]. Moreover, the potential to obtain tridimensional (3D) networks and bidimensional (2D) layered structures from hydrogels containing high-aspect-ratio sepiolite nanofibers has been demonstrated [43]. In addition, Giustetto et al. also analyzed hydrothermal-origin sepiolite, similar to SA, finding extremely long (up to several centimeters) and flexible fibers, composed of bundles of thinner fibrils [1]. These previous works revealed the influence of geological formation environment on sepiolite, highlighting that sepiolite crystallized from hydrothermal solution presents large size of the fibers and high degree of crystallinity.

To further investigate potential differences in surface properties, a simple hydrophobic/hydrophilic test was conducted on both sepiolite samples, ground under identical conditions. Figure 7 shows photographs of octane–water systems one hour after the addition of the samples. Upon introduction, particles predominantly settled, indicating a primarily hydrophilic nature. This is consistent with the expected behavior of sepiolite, which is inherently hydrophilic due to its composition as a hydrated magnesium silicate and its high surface area, rich in silanol groups (SiOH). Nevertheless, distinct interfacial behaviors were observed since the system containing ST samples exhibits an interface sharp and clear, whereas for SA, a cloudy interfacial region appeared. Under higher magnification, small air bubbles and fine particles were observed suspended at the interface for the system containing the SA sample.

To confirm the presence and nature of particles at the interface, the hydrophobic/hydrophilic test was replicated directly in a UV-Vis cuvette, limiting UV-Vis exposure to the interfacial region. Figure 8 displays the resulting UV-Vis spectra obtained for each sample. The ST sample shows no significant absorption consistent with visual observations of a clear interface. In contrast, the SA sample exhibits a characteristic absorption band between approximately 242 nm and 262 nm [44], delimited with stripped lines; confirming the presence of sepiolite at the interface. This interfacial behavior is likely related to morphological and textural differences: the SA sample comprises smaller, disaggregated particles with a higher specific surface area compared to ST. As a fibrous mineral, sepiolite has greater lateral surface area than basal surface area. Furthermore, the chemical composition of these surfaces depends on particle size and exposed surface orientation [45], with the lateral surface being more hydrophilic than the basal surface. Consequently, the hydrophilic nature of the particles varies according to their size, with very small SA particles exhibiting lower hydrophilicity than ST particles.

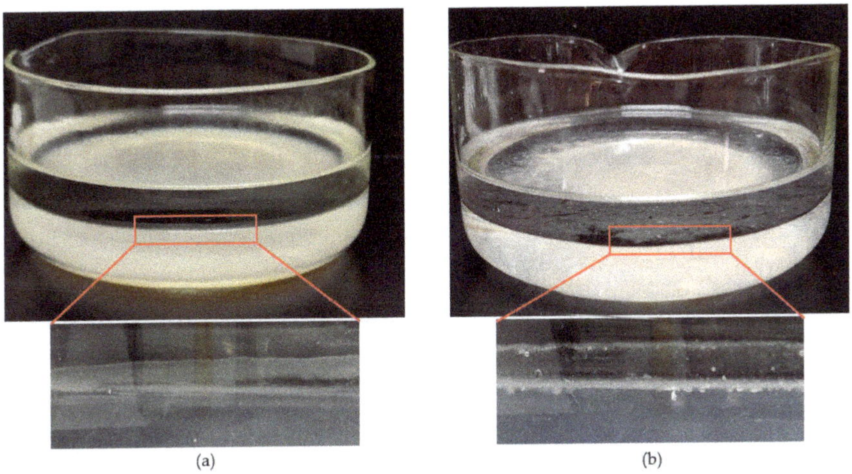

Figure 7. Photographs of hydrophilicity/hydrophobicity test for: (**a**) ST and (**b**) SA.

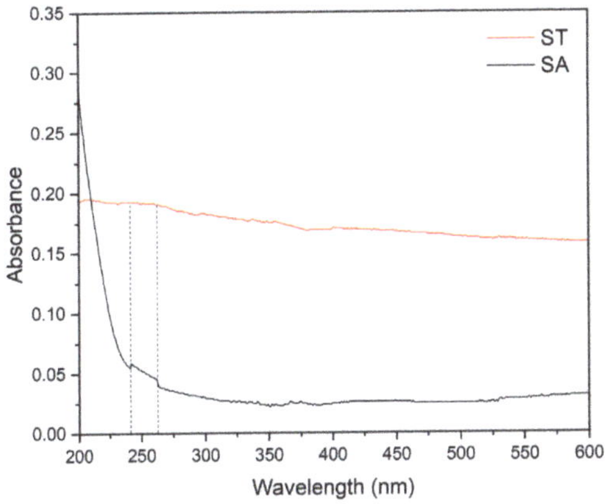

Figure 8. UV-vis spectra at the water–octane interfaces of the systems containing ST and SA particles.

5. Industrial and Environmental Applications

Sepiolite from Tolsa has been commercially distributed worldwide since 1957. Its applications are diverse, ranging from functional additives (biocidal, fungicidal, algicidal, and virucidal agents, synergistic agents for flame retardants, photocatalytic decontamination materials, and self-cleaning additives) to industrial solutions including paints and coatings, bitumen and asphalt formulations, drilling muds, construction material additives, foundry aids, and paper industry inputs. Additionally, it is used in environmental applications such as absorbents, bleaching earths, filtration, purification, and clarification. In the life sciences sector, this sepiolite is also employed in animal feed and agricultural additives, while in the pet care sector, it is widely used in cat litter and other hygiene products [46].

Regarding the SA sample, characterized by its disaggregated fibers, higher aspect ratio, and greater crystallinity than ST, it exhibits properties that could be advantageous for adsorption and interfacial processes. Its enhanced surface accessibility and higher specific surface area make it particularly suitable for wastewater treatment applications, such as the removal of heavy metals or organic contaminants. It is important to mention that the high-aspect sepiolite fibers of the SA sample allow obtaining 3D networks and 2D layered nanostructures [43]. These sepiolite-based nanoarchitectures show promising potential for the development of inorganic membranes with pore size and pore size distribution appropriated for ultrafiltration process. Moreover, the incipient hydrophobicity observed in the interfacial tests suggests possible applications in emulsion stabilization, or as a carrier for active agents in catalytic or controlled-release systems. The SA sample's well-defined structure and surface reactivity also support its potential use as a catalyst support or matrix for functional nanocomposites. These application pathways highlight the importance of geological origin in tailoring sepiolite for specific technological uses, expanding its utility beyond traditional sectors.

6. Conclusions

The geological origin of the sepiolite samples, lacustrine-derived (ST) and hydrothermal-derived (SA), has an impact on their physicochemical properties. Comparative analysis reveals differences in chemical composition, crystallinity, thermal behavior, morphology, and hydrophobic/hydrophilic behavior. The ST sample exhibited elevated SiO_2 content, likely due to amorphous silica impurities, consistent with its lower crystallinity, and higher structural defects. In contrast, the SA sample presented a chemical composition aligned with ideal sepiolite stoichiometry, accompanied by sharper XRD reflections and lower FWHM values, indicating a greater structural order than ST. SEM observations further underscore these contrasts. ST displays dense, aggregated fiber bundles forming planar textures, associated with the sedimentary depositional origin. SA, by contrast, reveals disaggregated needle-like fibers with high aspect ratios and nanometric thickness, characteristics indicative of anisotropic hydrothermal crystal growth. Hydrophilicity/hydrophobicity tests for both samples confirm their predominantly hydrophilic nature; however, SA exhibits additional interfacial behavior since fine particles suspended at the octane–water boundary, suggesting relatively less hydrophilicity or incipient hydrophobicity. This is attributable to the high specific surface area and surface orientation effects in the SA sample, which enhance interfacial activity. Thus, the sepiolite formation environment critically defines the mineral's structural, morphological, and surface characteristics, directly influencing its interaction with fluids, adsorption capacities, and dispersion behavior, essential for both geological interpretation and industrial utility. These characteristics confer a wide range of applications on sepiolite. ST is mainly used as an adsorbent, especially for wastewater treatment, whereas SA enables the formation of 3D networks that could be applied in the fabrication of inorganic membranes.

Author Contributions: Conceptualization, L.A.C. and S.E.B.; methodology, L.A.C., L.L.; software, L.A.C., L.L.; validation, L.A.C., L.L.; formal analysis, L.A.C., L.L.; investigation, L.A.C., L.L., S.A.M. and S.E.B.; resources, S.A.M., S.E.B.; data curation, L.A.C., L.L.; writing—original draft preparation, L.A.C., L.L.; writing—review and editing, L.A.C., L.L., S.A.M. and S.E.B.; visualization, L.A.C., L.L.; supervision, S.E.B.; project administration S.A.M., S.E.B. All authors have read and agreed to the published version of the manuscript.

Funding: This research received no external funding.

Data Availability Statement: Data are contained within the article.

Acknowledgments: The authors thank the Consejo Nacional de Investigaciones Científicas y Técnicas (CONICET), the Centro de Geología Aplicada, Agua y Medio Ambiente (CGAMA, CIC-UNS) and the Universidad Nacional del Sur for their support.

Conflicts of Interest: The authors declare no conflict of interest.

References

1. Giustetto, R.; Macaluso, L.; Berlier, G.; Ganjkhanlou, Y.; Barale, L. Characterisation and possible hazard of an atypical asbestiform sepiolite associated with aliphatic hydrocarbons from Sassello, Ligurian Apennines, Italy. *Mineral. Mag.* **2019**, *83*, 209–222. [CrossRef]
2. Jones, B.F.; Galán, E. Sepiolite and palygorskite. *Rev. Mineral. Geochem.* **1988**, *19*, 631–674.
3. Ruiz, A.I.; Ruiz-García, C.; Ruiz-Hitzky, E. From old to new inorganic materials for advanced applications: The paradigmatic example of the sepiolite clay mineral. *Appl. Clay Sci.* **2023**, *235*, 106874. [CrossRef]
4. Stoessell, R.K.; Hay, R.L. The geochemical origin of sepiolite and kerolite at Amboseli, Kenya. *Contrib. Mineral. Petrol.* **1978**, *65*, 255–267. [CrossRef]
5. Herranz, J.E.; Pozo, M. Sepiolite and other authigenic Mg-clay minerals formation in different Palustrine environments (Madrid Basin, Spain). *Minerals* **2022**, *12*, 987. [CrossRef]
6. Tateo, F.; Sabbadini, R.; Morandi, N. Palygorskite and sepiolite occurrence in Pliocene lake deposits along the River Nile: Evidence of an arid climate. *J. Afr. Earth Sci.* **2000**, *31*, 633–645. [CrossRef]
7. Mokatse, T.; Prud'Homme, C.; Vainer, S.; Adatte, T.; Shemang, E.; Verrecchia, E.P. Sepiolite as a multifactorial indicator of paleoenvironments in the Chobe Enclave (northern Botswana). *Sediment. Geol.* **2023**, *454*, 106459. [CrossRef]
8. Kadir, S.; Erkoyun, H.; Eren, M.; Huggett, J.; Önalgil, N. Mineralogy, geochemistry, and genesis of sepiolite and palygorskite in Neogene lacustrine sediments, Eskişehir Province, west central Anatolia, Turkey. *Clays Clay Miner.* **2016**, *64*, 145–166. [CrossRef]
9. Calvo, J.P.; Pozo, M. Geology of magnesian clays in sedimentary and non-sedimentary environments. In *Magnesian Clays: Characterization, Origin and Applications*; Pozo, M., Galán, E., Eds.; Digilabs: Bari, Italy, 2015; pp. 123–174.
10. Draidia, S.; Ouahabi, M.E.; Daoudi, L.; Havenith, H.B.; Fagel, N. Occurrences and genesis of palygorskite/sepiolite and associated minerals in the Barzaman formation, United Arab Emirates. *Clay Miner.* **2016**, *51*, 763–779. [CrossRef]
11. Baldermann, A.; Mavromatis, V.; Frick, P.M.; Dietzel, M. Effect of aqueous Si/Mg ratio and pH on the nucleation and growth of sepiolite at 25 °C. *Geochim. Cosmochim. Acta* **2018**, *227*, 211–226. [CrossRef]
12. Cañaveras, J.C.; Calvo, J.P.; Ordóñez, S.; Muñoz-Cervera, M.C.; Sánchez-Moral, S. Tectono-Sedimentary Evolution of the Madrid Basin (Spain) during the Late Miocene: Data from Paleokarst Profiles in Diagenetically-Complex Continental Carbonates. *Geosciences* **2020**, *10*, 433. [CrossRef]
13. Reijmer, J.J.; Blok, C.N.; El-Husseiny, A.; Kleipool, L.M.; Hogendorp, Y.C.; Alonso-Zarza, A.M. Petrophysics and sediment variability in a mixed alluvial to lacustrine carbonate system (Miocene, Madrid Basin, Central Spain). *Depos. Rec.* **2022**, *8*, 317–339. [CrossRef]
14. Herranz, J.E.; Pozo, M. Authigenic Mg-clay minerals formation in lake margin deposits (the Cerro de los Batallones, Madrid Basin, Spain). *Minerals* **2018**, *8*, 418. [CrossRef]
15. Pozo, M.; Galán, E.; González, J.L. Sepiolite and palygorskite from the Madrid Basin, Spain: Mineralogy and genetic relations. *Appl. Clay Sci.* **2014**, *95*, 31–45.
16. Ordóñez, S.; Calvo, J.P.; García del Cura, M.A.; Alonso Zarza, A.M.; Hoyos, M. Sedimentology of sodium sulphate deposits and special clays from the Tertiary Madrid Basin (Spain). In *Lacustrine Facies Analysis*; Anadón, P., Cabrera, L., Kelts, K., Eds.; Blackwell Scientific Publications: Oxford, UK, 1991; pp. 39–55.
17. Cortelezzi, C.R.; Marfil, S.A.; Maiza, P.J. A sepiolite of large crystalline growth from "La Adela" mine province of Río Negro, Argentina. *N. Jb. Miner. Mh.* **1994**, *4*, 157–166. [CrossRef]
18. Caminos, R.; Llambías, E.J. El basamento cristalino. In Proceedings of the Geología y Recursos Naturales de la Provincia de Río Negro, IX Congreso Geológico Argentino, Bariloche, Argentina, 5–9 November 1984; Ramos, V., Ed.; pp. 37–63.
19. Maiza, P.; Marfil, S.A. Diaclasas mineralizadas con sepiolita de cantera La Adela, provincia de Río Negro, Argentina. In Proceedings of the XII Congreso Geológico Argentino y II Congreso de Exploración de Hidrocarburos, Mendoza, Argentina, 10–15 October 1993; pp. 82–86.
20. Dominguez, E.A.; Maiza, P.J. Yacimientos no metalíferos y rocas de aplicación. In Proceedings of the Geología y Recursos Naturales de la Provincia de Río Negro, IX Congreso Geológico Argentino, Bariloche, Argentina, 5–9 November 1984; Ramos, V., Ed.; pp. 611–628.
21. Cordenons, P.D.; Remesal, M.B.; Salani, F.M.; Cerredo, M.E. Temporal and spatial evolution of the Somún Curá magmatic province, northern extra-andean Patagonia, Argentina. *J. S. Am. Earth Sci.* **2020**, *104*, 102881. [CrossRef]

22. International Mineralogical Association (IMA). 2025. Available online: https://cnmnc.units.it/files/editor/IMA_Master_List_(2025-07).pdf (accessed on 15 August 2025).
23. Velde, B. *Introduction to Clay Minerals: Chemistry, Origins, Uses and Environmental Significance*; Chapman and Hall Ltd.: London, UK, 1992; Volume 198, 198p.
24. JCPDS, International Centre for Diffraction Data (ICDD). *Mineral Powder Diffraction File Databook*; ICDD: Swarthmore, PA, USA, 1993; pp. 614–615.
25. Pozo, M.; Calvo, J.P.; Pozo, E.; Moreno, A. Genetic constraints on crystallinity, thermal behavior and surface area of sepiolite from the Cerro de los Batallones deposits (Madrid Basin, Spain). *Appl. Clay Sci.* **2014**, *91–92*, 30–45. [CrossRef]
26. Dikmen, S. Zeta potential study of natural- and acid-activated sepiolites in electrolyte solutions. *Can. J. Chem. Eng.* **2012**, *90*, 421–427. [CrossRef]
27. Locatelli, D.; Pavlovic, N.; Barbera, V.; Giannini, L.; Galimberti, M. Sepiolite as reinforcing filler for rubber composites: From the chemical compatibilization to the commercial exploitation. *Kautsch. Gummi Kunststoffe* **2020**, *73*, 34–41.
28. Zhou, F.; Ye, G.; Gao, Y.; Wang, H.; Zhou, S.; Liu, Y.; Yan, C. Cadmium adsorption by thermal-activated sepiolite: Application to in-situ remediation of artificially contaminated soil. *J. Hazard. Mater.* **2022**, *423A*, 127104. [CrossRef]
29. Shahraki, B.; Mehrabi, B.; Dabiri, R. Thermal behavior of Zefreh dolomite mine (Central Iran). *J. Min. Metall.* **2009**, *45B*, 35–44. [CrossRef]
30. Suárez, M.; García-Rivas, J.; García-Romero, E.; Jara, N. Mineralogical characterisation and surface properties of sepiolite from Polatli (Turkey). *Appl. Clay Sci.* **2016**, *131*, 124–130. [CrossRef]
31. Largo, F.; Haounati, R.; Ouachtak, H.; Hafid, N.; Jada, A.; Addi, A.A. Design of organically modified sepiolite and its use as adsorbent for hazardous Malachite Green dye removal from water. *Water Air Soil Pollut.* **2023**, *234*, 183. [CrossRef]
32. Wei, S.; Wang, L.; Wu, Y.; Liu, H. Study on removal of copper ions from aqueous phase by modified sepiolite flocs method. *Environ. Sci. Pollut. Res.* **2022**, *29*, 73492–73503. [CrossRef]
33. Jiang, X.; Wang, S.; Ge, L.; Lin, F.; Lu, Q.; Wang, T.; Lu, B. Development of organic–inorganic hybrid beads from sepiolite and cellulose for effective adsorption of malachite green. *RSC Adv.* **2017**, *7*, 38965–38972. [CrossRef]
34. Zhou, X.; Li, H.; Liu, Y.; Hao, J.; Liu, H.; Lu, X. Improvement of stability of insecticidal proteins from Bacillus thuringiensis against UV-irradiation by adsorption on sepiolite. *Adsorpt. Sci. Technol.* **2018**, *36*, 1233–1245. [CrossRef]
35. Yang, L.; Deng, Y.; Gong, D.; Luo, H.; Zhou, X.; Jiang, F. Effects of low molecular weight organic acids on adsorption of quinclorac by sepiolite. *Environ. Sci. Pollut. Res.* **2021**, *28*, 9582–9597. [CrossRef]
36. Alves, L.; Ferraz, E.; Santarén, J.; Rasteiro, M.G.; Gamelas, J.A. Improving colloidal stability of sepiolite suspensions: Effect of the mechanical disperser and chemical dispersant. *Minerals* **2020**, *10*, 779. [CrossRef]
37. Zhang, G.; Liu, L.; Shiko, E.; Cheng, Y.; Zhang, R.; Zeng, Z.; Zhao, T.; Zhou, Y.; Chen, H.; Liu, Y.; et al. Low-price MnO_2 loaded sepiolite for Cd^{2+} capture. *Adsorption* **2019**, *25*, 1271–1283. [CrossRef]
38. Caicedo-Pineda, G.A.; Prada-Fonseca, M.C.; Casas-Botero, A.E.; Martínez-Tejada, H.V. Effect of the tryptone concentration on the calcium carbonate biomineralization mediated by Bacillus cereus. *Dyna* **2018**, *85*, 69–75. [CrossRef]
39. Nguyen, M.B.; Le, G.H.; Pham, T.T.; Pham, G.T.; Quan, T.T.; Nguyen, T.D.; Vu, T.A. Novel Nano-Fe_2O_3-Co_3O_4 Modified Dolomite and Its use as highly efficient catalyst in the ozonation of ammonium solution. *J. Nanomater.* **2020**, *2020*, 4593054. [CrossRef]
40. Sánchez-Sánchez, A.; Cerdán, M.; Jordá, J.D.; Amat, B.; Cortina, J. Characterization of soil mineralogy by FTIR: Application to the analysis of mineralogical changes in soils affected by vegetation patches. *Plant Soil* **2019**, *439*, 447–458. [CrossRef]
41. Ruiz-Hitzky, E.; Ruiz-García, C.; Fernandes, F.M.; Lo Dico, G.; Lisuzzo, L.; Prevot, V.; Darder, M.; Aranda, P. Sepiolite-hydrogels: Synthesis by ultrasound irradiation and their use for the preparation of functional clay-based nanoarchitectured materials. *Front. Chem.* **2021**, *9*, 733105. [CrossRef] [PubMed]
42. Lescano, L.; Castillo, L.; Marfil, S.; Barbosa, S.; Maiza, P. Separation and purification of sepiolite fibers. Contribution to special processing for industrial use. *Appl. Clay Sci. Intern. J. Appl. Technol. Clays Clay Miner.* **2014**, *93*, 378–382.
43. Castillo, L.; Lescano, L.; Sirvent, L.; Barbosa, S.; Marfil, S.; Maiza, P. Separación y purificación de fibras de sepiolita: Contribución al procesamiento de arcillas especiales para uso industrial. *Geoacta* **2011**, *36*, 113–127.
44. Fajdek-Bieda, A.; Wróblewska, A.; Miądlicki, P.; Szymańska, A.; Dzięcioł, M.; Booth, A.M.; Michalkiewicz, B. Influence of technological parameters on the isomerization of geraniol using sepiolite. *Catal. Lett.* **2020**, *150*, 901–911. [CrossRef]
45. Tian, G.; Han, G.; Wang, F.; Liang, J. Sepiolite nanomaterials: Structure, properties and functional applications. In *Nanomaterials from Clay Minerals*; Wang, A., Wang, W., Eds.; Elsevier: Amsterdam, The Netherlands, 2019; pp. 135–201.
46. Tolsa. Available online: https://www.tolsa.com/es/ (accessed on 15 August 2025).

minerals MDPI

Article

Composition, Genesis, and Adsorption Properties of Smectite–Palygorskite Clays (Lower Carboniferous, Russia)

Sergey Zakusin [1,2], Olga Zakusina [1], Tatiana Koroleva [1], Ivan Morozov [1], Mikhail Chernov [2] and Victoria Krupskaya [1,2,*]

1 Institute of Ore Geology, Petrography, Mineralogy and Geochemistry, Russian Academy of Science (IGEM RAS), 35 Staromonetny Pereulok, 119017 Moscow, Russia; zakusinsergey@gmail.com (S.Z.); o.zakusina@yandex.ru (O.Z.); tanakoro@yandex.ru (T.K.); ivan.morozov@yandex.ru (I.M.)
2 Geological Faculty, M.V. Lomonosov Moscow State University, GSP-1, Leninskie Gory, 119991 Moscow, Russia; miha.chernov@yandex.ru
* Correspondence: krupskaya@ruclay.com

Abstract

Infrared spectroscopic analysis of palygorskite clay from the Dashkovskoye and Borshchevskoye deposits yielded key insights into the sedimentation conditions prevailing in the study area. In this paper, the composition, structure, and adsorption properties of smectite–palygorskite clays from the Steshevian sub-stage of the Lower Carboniferous (Russia) are investigated. The study applied X-ray diffraction, infrared spectroscopy, scanning electron microscopy, assessment of cation exchange capacity by adsorption of $[Cu(trien)^{2+}]$, assessment of Cs sorption, and particle size analysis. It is demonstrated that the Al–palygorskite of the Dashkovskoye deposit was formed by sedimentation from suspended matter in a shallow-water basin in the Steshevian sub-age, despite a different genesis (chemogenic in the case of the palygorskites, clastic/redeposited in the case of the smectites). The palygorskites of the Borschovskoye deposit have a complex terrigenous genesis and were formed from redeposited chemogenic Al–palygorskites transported into the basin from the surrounding region of the Dashkovskoye deposit. With increasing depth of the basin in the Steshevian sub-age, the volume of incoming palygorskite material decreases, and the proportion of smectite material increases. The Fe–palygorskites entered the Borschovskoye deposit due to the redeposition of sediments from soils upstream of water flows. All the studied clays have considerable adsorption properties (32–49 mg-eq/100 g) and can be used in various industries.

Keywords: bentonite; cation exchange capacity; genesis; microstructure; paleoreconstruction; palygorskite; smectite; surface area

1. Introduction

Smectite and palygorskite clays of the Moscow and Kaluga regions have been studied since the 1960s because they have high sorption capacity, colloidal properties, and good heat resistance. As is typical in geological research, particular attention was paid to the stratigraphic correlation and interpretation of the paleogeographic features of the Moscow syneclise [1–6]. The mineral composition was also considered in the works mentioned above. In particular, the study [6] employed an integrated approach with consideration of the mineralogical and technological characteristics of smectite and palygorskite clays. However, despite a detailed examination of the occurrences of the smectite and palygorskite outcrops, some aspects of the structure of these clay minerals were not covered in these

 https://doi.org/10.3390/min16010070

studies. The current state of research equipment enables a fresh examination of the composition and structure of these clays, offering new insights into the genesis of the palygorskite and smectite clay deposits.

Smectite belongs to layered silicates. An elementary 2:1 layer of smectites consists of two tetrahedral sheets and an octahedral sheet enclosed between them (Figure 1a). A negative charge of the layer comes from isomorphic substitutions of cations in the octahedral sheet and, to a lesser extent, in the tetrahedral sheets. This negative charge is then neutralized by exchangeable interlayer cations (Na^{+}, Ca^{2+}, Mg^{2+}, etc.), usually in a hydrated form [7–9]. Smectite is formed in various geological settings. It is most frequently the product of exogenous processes, such as the weathering of acidic eruptive rocks (predominantly volcanic ash) in an alkaline environment, as well as in various marine and continental conditions, in soils on various parent materials, etc. [10–12].

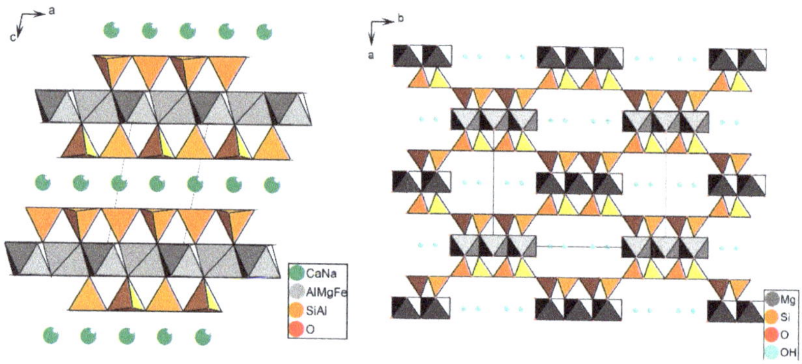

(a) (b)

Figure 1. Schematic presentation of the structure of smectite (**a**) and palygorskite (**b**). The dotted line defines the boundaries of the unit cell.

Palygorskite is a clay mineral with a ribbon-like and fibrous structure [13–15]. The basis of its structure consists of double silicon-oxygen tetrahedron chains, which are turned toward each other by their vertices (Figure 1b). These silicon-oxygen chains are connected into ribbons by means of the cations Al^{3+}, Mg^{2+}, Fe^{3+}, and Fe^{2+}, which have octahedral coordination [16–20]. This structural configuration of palygorskite results in the formation of channels. These channels have a cross-section of 3.7×6.4 Å and are typically filled with water molecules [13]. Palygorskite has a predominantly authigenic origin. Its formation typically occurs in closed basins enriched with Si, Al, and Mg [21,22].

The detailed composition and structure of palygorskite clays of the Moscow and Kaluga regions are less studied than those of smectitic clays and therefore deserve priority attention in this work. Authigenic palygorskite is far more common in geological formations than allothigenic palygorskite, which is rarely preserved. Therefore, the study of the peculiarities of the clay's composition and structure makes it possible to identify the features of the formation of the studied deposits.

The main method for studying the crystal structure of clay minerals is usually X-ray diffraction. However, methods of infrared spectroscopy have emerged over the last few years. This makes it possible to obtain new data on isomorphic substitutions in the structure of palygorskites. The studies [23–26] provide extensive summary material, including data on palygorskites of various chemical compositions and geneses. They also introduce a new methodological approach proposed at the Theoretical and Physical Chemistry Institute (National Hellenic Research Foundation, Greece). This method makes it possible to determine the chemical composition of the octahedral sheets in the presence

https://doi.org/10.3390/min16010070

of possible mineral impurities (smectite and hydroxides of Al, Mg, and Fe). Such impurities can be used to clarify the genesis of palygorskite clays.

Smectite and palygorskite clays are known for their high adsorption properties. This leads to their widespread use as sorbents and isolation materials for highly toxic (including radioactive) wastes [27–36]. Although smectite–palygorskite mixed clays are of limited use, the central location of their deposits in the Moscow and Kaluga regions makes their use promising for a number of regional industries.

The main aim of the present study is to ascertain the genesis of the clays of the Steshevian sub-stage based on a detailed study of their mineral structure, as well as to evaluate their adsorption properties for industrial applications.

2. Geological Position of the Considered Deposits

The studied smectite–palygorskite mixed clays of the Russian Platform (Moscow and Kaluga regions), which belong to the Steshevian sub-stage of the Serpukhovian stage of the Lower Carboniferous, are overlain with nonconformity by argillaceous sandstones and dolomites of the Protvian and Zapaltyubian sub-stages of the Serpukhovian stage.

The Serpukhovian is developed on the East European Platform and belongs to the "Russian" stages of the international stratigraphic scale. The bed boundary of this stage and its paleogeographic interpretation are still not fully elucidated. The stratigraphic relationships of the known sections of the Serpukhovian stage and the problems associated with the correlation of their boundaries are described in detail [5]. The identification of the sedimentation features in the Steshevian sub-stage is important for the purpose of clarifying environmental processes. This importance stems from [37], who identified this period as marking a transition to high-amplitude, glacioeustatic sea-level fluctuations. Through a comparative analysis of the lithological and mineral composition, as well as the fossil record from the three most complete sections of the Serpukhovian stage of the East European Platform, the detailed work [5] makes important conclusions about the existence of a relatively shallow water basin with strong terrigenous migration. This migration is characterized by flowing rather than stagnant hydrodynamic regimes. In contrast, other earlier works reported the predominance of lagoonal conditions [38].

The Steshevian sub-stage is a stratum of clays with smectite predominant in the lower part, which is chocolate to dark gray and black in color. In the upper part, there is a relatively high content of palygorskite. The clay stratum has a light red to grayish yellow color with interlayers of dolomites. Generally, in the middle part of the section, there are variegated clays of mixed composition with interlayers of dolomites and limestones varying from several millimeters up to half a meter thick. The clays occur subhorizontally. The thickness of the layers varies from 2 to 4 m, reaching a total of 15 m. The boundaries with the overlying and underlying rocks are quite clearly traced by the change in the mineral and grain size composition. The features of changes in the mineral composition between the Dashkovskoye deposit in the Moscow region and the Borschovskoye deposit in the Kaluga region can be considered in more detail (Figure 2a). These deposits are in sufficient proximity to each other (approximately 100 km apart) to allow this comparison.

The Dashkovskoye deposit was exploited from the early 1970s for the extraction of smectite and palygorskite raw material. The quarry is now closed, and its operation has ceased. The bulk of the products made from the clay material were bentonite raw material for the production of drilling mud, as well as iron ore pellets.

The study of the Borschovskoye deposit also started in 1970. At that time, its explored reserves were estimated at seven million tons. Geological exploration carried out by the Lafarge Company in 2007 showed that the deposit's reserves exceed ten million tons [6].

https://doi.org/10.3390/min16010070

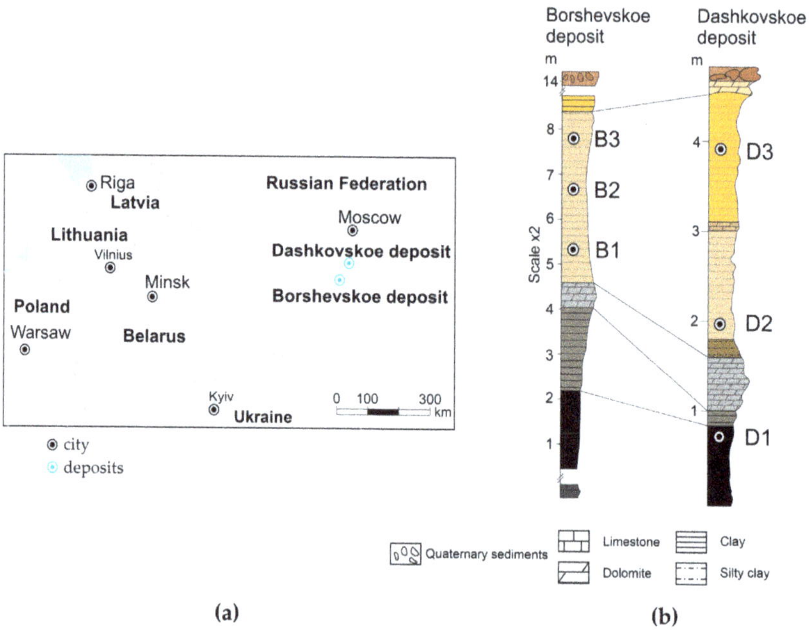

Figure 2. Location of the studied deposits of smectite–palygorskite clays (**a**) and the lithological section of the Steshevian sub-stage of the Lower Carboniferous of the Moscow Oblast in the mining area of the Borschovskoye and Dashkovskoye deposits (**b**). The color matches the color of the rocks.

3. Materials and Methods

Samples from the Dashkovskoye (D) deposit were collected from cleared quarry walls during several field expeditions. For mineralogical studies, samples were taken vertically at intervals of 20–30 cm, maintaining natural moisture. Samples from the Borschovskoye (B) deposit were provided by the company "Lafarge SA" in the form of core material from different boreholes in sealed packaging.

In total, about 40 samples taken from the Dashkovskoye field and about 20 from the Borschovskoye field were studied. Based on the research results, 6 samples were selected that most fully characterize the various types of clays found in the deposits (Figure 2, Table 1).

Table 1. Description of the samples selected for detailed studies.

Sample Name	Clays, Deposit	Sampling Depth, m
	Dashkovskoye deposit	
D-1	predominantly smectite clay	0.8
D-2	smectite–palygorskite clay	1.9
D-3	predominantly palygorskite clay	3.8
	Borschovskoye deposit	
B-1	predominantly smectite clay	7.8
B-2	smectite–palygorskite clay	6.6
B-3	predominantly palygorskite clay	5.2

Determination of grain size distribution was carried out using sedimentation. Sample preparation was carried out according to the method of P.F. Melnikov (rubbing with sodium pyrophosphate) as described in [39].

https://doi.org/10.3390/min16010070

Particle size distribution analysis in the fine fraction of clay samples was carried out using an ANALYSETTE 22 MicroTec Plus (Fritsch GmbH, Idar-Oberstein, Germany) laser analyzer in cooperation with the Moscow representative office of Fritsch. Before imaging with the laser analyzer, the samples were pre-dispersed by ultrasound for 5 min at a frequency of 44 kHz. For each sample, the measurements were repeated 20 times at intervals of several seconds, during which the samples were exposed to additional sonication for 10 s directly in the container of the device.

The microstructure of the clay samples was studied using a scanning electron microscope (SEM) LEO 1450VP (Carl Zeiss, Oberkochen, Germany). To preserve the natural structure of the samples, they were frozen in liquid nitrogen and subsequently vacuum freeze-dried. An electrically conductive thin film of gold (10–20 nm thick) was applied to investigate the surface of the finished samples using IB-3 (Eiko Engineering, Hitachinaka, Japan). This helps to avoid any charging effects on the sample during SEM analysis.

X-ray diffraction patterns were obtained using a Rigaku Ultima-IV X-ray diffractometer (Rigaku, Tokyo, Japan) (Cu-Kα, detector D/Tex-Ultra, scan range 3–65° 2θ) from textured and untextured preparations. To obtain oriented samples, the suspension (fraction < 2 μm) was applied to a glass slide. The analysis of the results was carried out according to the recommendations described in [40]. Quantitative mineral analysis of bulk samples was carried out by the Rietveld method [41] in the PROFEX GUI software, 5.2.8 version package for BGMN [42].

The features of the crystal-chemical structure of the clay samples were studied using near- and mid-infrared (NIR and MIR) spectroscopy. The methods of NIR spectroscopy are the most sensitive for minerals with a fibrous structure, such as palygorskites, sepiolites, etc. The optimal temperature for the dehydration of zeolitic water from the channels in the structure of palygorskites is in the range of 80–100 °C. After the loss of this water, the structure contracts ("folds"), and a shift of the bands in the NIR spectra can be expected. This is the basis of the principle of studying palygorskites using dehydration, which was developed at the Theoretical and Physical Chemistry Institute [25]. NIR spectra were obtained using a Vector 22N FT-IR spectrometer (Bruker Optics, Billerica, MA, USA), equipped with an external integrating sphere. The study sample was placed directly on a quartz window of the outer sphere and was protected from external influences by a box in the presence of P_2O_5. The use of a strong magnifying glass and a conventional incandescent lamp makes it possible to simulate heating conditions in the sample box. This leads to the loss of zeolitic water from palygorskite. Spectra are collected every minute until dehydration stabilizes. Some of the absorption bands in the NIR spectrum of palygorskites and other clay minerals, as well as oxides and hydroxides, are superimposed on each other or are in close spectral regions. As the sample dehydrates, the absorption bands from the changing structure of palygorskites shift, while the bands from the structure of other clay minerals present (in the case of the present study, smectites) only decrease in intensity. Thus, it is possible to confidently study the crystal-chemical features of the palygorskites mixed with other minerals (both clayey and non-clayey, such as iron oxides). The absorption bands in the near range have wide profiles; therefore, the second derivative of the absorption feature is used for a more correct determination of the position of the bands. The spectroscopy research experiments were carried out at the Theoretical and Physical Chemistry Institute, National Hellenic Research Foundation (Athens, Greece).

The cation exchange capacity (CEC) was determined by the method of adsorption of methylene blue and a triethylene tetramine complex of copper (II) ($[Cu(trien)]^{2+}$) [43–49].

The determination of the absorption properties with respect to Cs^+ was carried out under static conditions on powdered samples with a particle size of <0.25 mm. The experiments were carried out in two parallel repetitions. Solutions of cesium nitrate ($CsNO_3$)

https://doi.org/10.3390/min16010070

were prepared for the experiment at cesium concentrations of 50, 200, 350, 500, 700, 800, 900, 1000, and 1100 mg/L. Then, 200 mL of each solution of the aforementioned corresponding concentrations were poured into a series of flasks each containing a weighed quantity of the study sample weighing 2 g. The flasks with the resulting solutions were periodically mixed during the day. It was experimentally found that the optimal duration of contact of the adsorbent with the solutions of varying concentrations of cesium nitrate is one day (24 h). After the reaction time, solutions were filtered using blue ribbon filter paper. Measurement of the Cs concentrations in the original and equilibrium solutions was carried out on an inductively coupled plasma mass spectrometer Thermo Scientific Element 2 (Thermo Fisher Scientific, Bremen, Germany). Adsorption isotherms were plotted after measurements.

4. Results

4.1. Mineral Composition

The content of smectite and palygorskite in the samples of the considered deposits is 20%–57% and 5%–55%, respectively (Figures 2 and 3). Several patterns can be observed based on the detailed section of the Dashkovskoye deposit. The ratio of smectite and palygorskite varies along the section, from the predominance of smectite in the lower part (up to 82%) to the predominance of palygorskite in the upper part of the section (Figures 2 and 3; Table 2). Smectite is present throughout the section. Simultaneously with an increase in smectite content, small amounts of illite impurity appear. Quartz, dolomite, calcite, and feldspars were noted as non-clay minerals. An admixture of quartz is present throughout the section, with its content fluctuating between 10% and 40%. Meanwhile, the content of quartz and palygorskite in the dolomite interbeds decreases sharply, while smectite is present in only trace amounts.

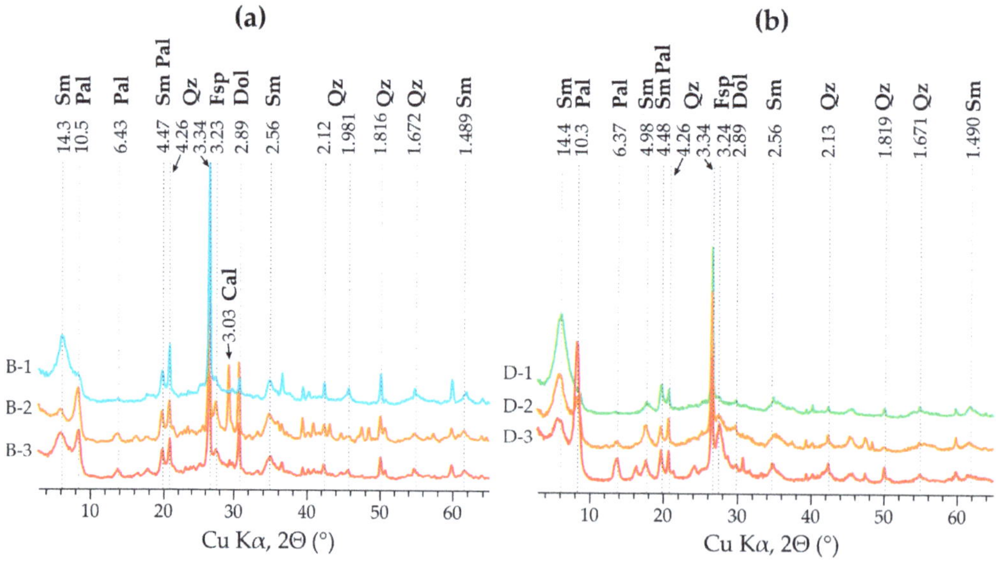

Figure 3. Typical X-ray diffraction patterns of clay samples from the Steshevian sub-stage of the Dashkovskoye (**a**) and Borschovskoye (**b**) deposits.

https://doi.org/10.3390/min16010070

Table 2. Mineral composition of samples from typical clay layers.

Sample	Smectite	Palygorskite	Illite	Kaolinite	Quartz	Field Spars	Dolomite	Calcite
Dashkovskoye deposit								
D-1	57	6	10	6	18	2	-	1
D-2	36	50	1	2	11	-	-	-
D-3	27	55	2	-	13	-	3	-
Borschovskoye deposit								
B-1	55	6	-	-	37	1	1	-
B-2	41	25	-	-	25	2	6	1
B-3	20	31	-	-	30	-	11	8

The mineral composition of the clays from the Borschovskoye deposit is more heterogeneous (Table 2, Figure 3). This is reflected in a high content of quartz (25%–37%) and dolomite (up to 11%), calcite (up to 8%), and feldspars (up to 2%). The content of clay minerals is not constant and varies both laterally in different boreholes of the Borschovskoye deposit and vertically within the layers in the same borehole. The lithological differences identified at the Dashkovskoye deposit are also not regularly traced. This mineral composition refutes the hypothesis of the predominant chemogenic genesis of palygorskites in the area of the Borschovskoye deposit and suggests a redeposited origin of these clays.

The variation in mineral associations indicates a change in the conditions of sedimentation at the Steshevian sub-stage. Along with the prevalence of smectite, trace amounts of quartz, illite, and kaolinite are also present in the lower parts of the section. This may be the result of an intensification of the action of terrigenous migration. Although illite is found in small amounts, it clearly shows a correlation with the content of smectite and quartz: In strata with an increase in smectite, the content of quartz, illite, and kaolinite is also increased. Such an association of minerals indicates an intensification of the action of terrigenous outwash of material. At the same time, with an increase in the content of palygorskite, the presence of other clay minerals is suppressed. This is the result of a change in the conditions of sedimentation.

The MgO content changes in the same direction, which directly depends on the concentration of palygorskite in the clays. Such a significant increase in the MgO content in the paleobasin provides an opportunity for the chemogenic synthesis of palygorskite. With a continued increase in the MgO content and a decrease in the SiO_2 content in the aquatic environment of the paleobasin, the conditions for the synthesis of palygorskite become unfavorable and the formation of chemogenic dolomite begins. In lithological terms, such dolomites are massive, uniform granular interbeds, similar in structure to poorly cemented sands. This was observed in the Dashkovskoye deposit quarry.

4.2. Crystal-Chemical Structure of Clay Minerals

Pure smectite or palygorskite was not detected in any of the analyzed samples from the deposits. Instead, all samples were composed of a mixture of different minerals. Obtaining fine fractions of different sizes (from 1 to 0.1 μm) also did not allow us to achieve mono-mineral samples. IR spectroscopy studies were carried out to clarify the crystal-chemical characteristics, such as the occupancy of octahedral and tetrahedral positions of palygorskites and smectites.

Investigations of the samples in the MIR spectrum allowed the confirmation of the conclusion about their mineral composition made on the basis of diffractometric analysis. The position of the Al–Al–OH bands in the range from 912 to 914 cm^{-1} (Figure 4) indicates the dioctahedral structure of both smectite and palygorskite [24,50]. Against the

https://doi.org/10.3390/min16010070

background of distinct bands of the matrix and in the region of hydroxyl groups, the bands of adsorbed water (~3400 and 1650 cm^{-1}) stand out in the spectrum of the predominantly smectite sample (red line). The characteristic bands of zeolitic water in the structure of palygorskite are found in the range of 3270–3385 cm^{-1}. Palygorskite can be characterized as low-magnesium (band at 3616 cm^{-1}), dioctahedral, with a predominance of aluminum in octahedral positions (band at 3627 cm^{-1}), and with a small amount of iron substitution in octahedral positions.

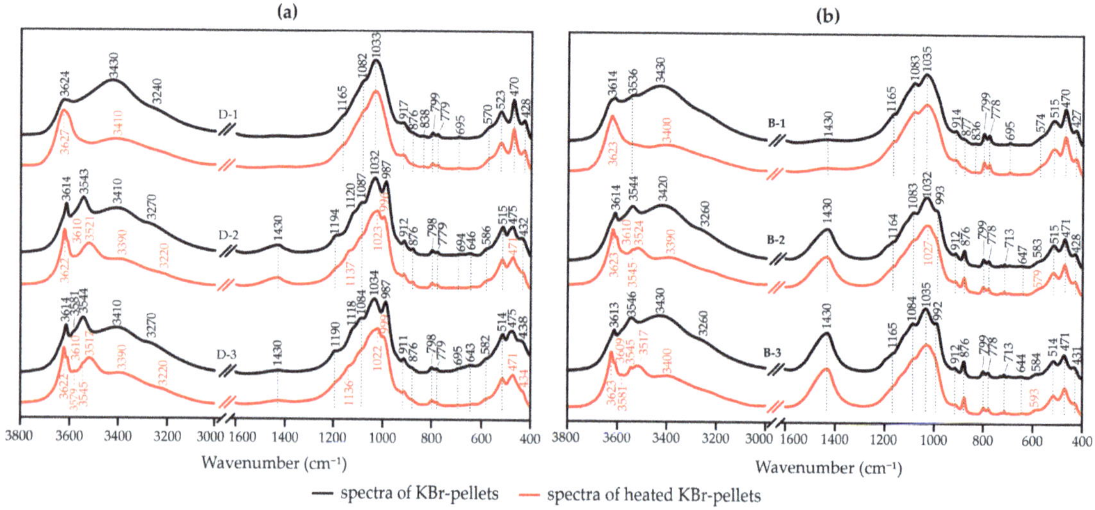

Figure 4. Typical IR spectra of clay samples from the Steshevian sub-stage of the Dashkovskoye (**a**) and Borschovskoye (**b**) deposits of palygorskite and smectite samples.

NIR spectroscopy studies of palygorskites make it possible to analyze a clay's crystal-chemical composition, in particular, the isomorphic substitutions of Fe and Al in the structure of palygorskites. The relationship between zeolitic and coordination water and the nature of their behavior during heating is clearly illustrated in Figure 5. The Al–Al–OH bands are detected in the regions of 7075 and 4514 cm^{-1}; Al–Fe–OH at 7012 and 4444 cm^{-1}; and Fe–Fe–OH at 6946 and 4361 cm^{-1}, respectively [23]. The ratio of the intensities of these bands allows us to assess the features of the distribution of iron in the octahedral positions. It turned out that in the palygorskites D-3, octahedral sites are mainly occupied by Al^{3+}, while the content of isomorphic Fe^{3+} is insignificant. This indicates the homogeneity of all the studied palygorskites from this deposit. The palygorskites B (Figure 5b) belong to the aluminum–ferruginous variety, with a relatively high content of isomorphic Fe in the octahedral sheets. This is demonstrated by the relatively strong absorption bands of Fe–Fe–OH and Al–Fe–OH.

Based on the ratios of the intensities of Fe–Fe–OH, Al–Al–OH, and Al–Fe–OH bands, their relative abundances were calculated (assuming that the sum of the intensities I (Al–Al–OH), I (Al–Fe–OH), and I (Fe–Fe–OH) equals 1). The results are plotted on a triangular diagram (Figure 6). According to [25], the distribution of Al^{3+} and Fe^{3+} in the octahedral positions of all natural palygorskites follows a particular type of curve. It would be reasonable to assume that the samples that do not fall within this curve represent mechanical mixtures.

https://doi.org/10.3390/min16010070

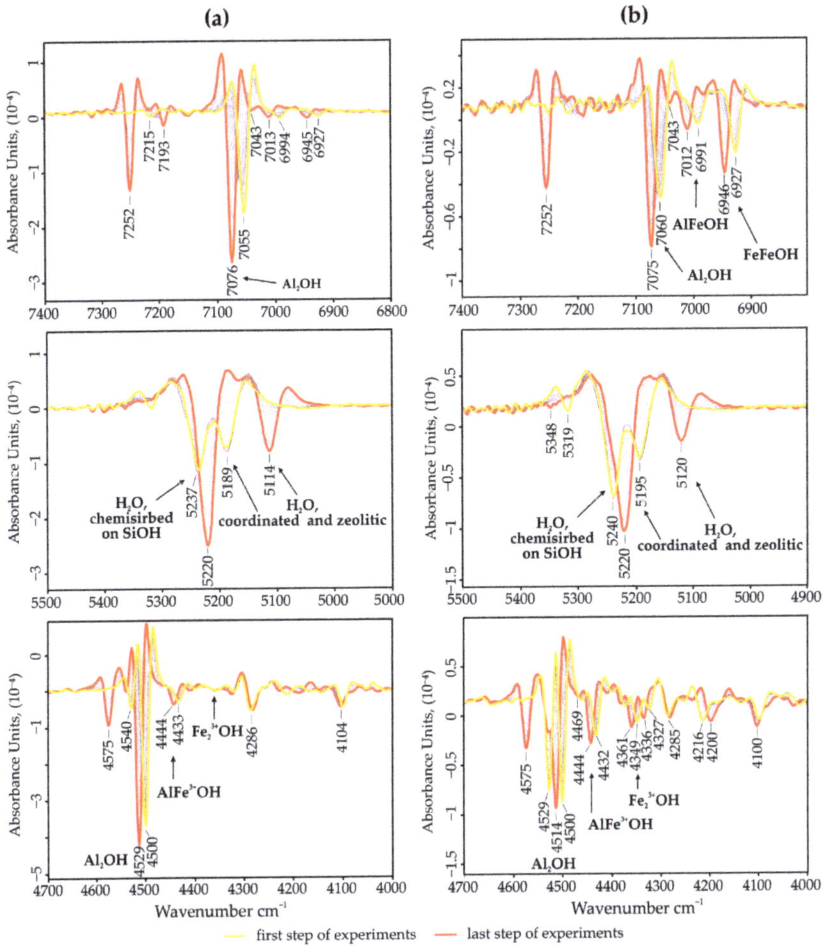

Figure 5. The second derivative of the absorption spectra in the near-IR region of palygorskite-rich samples from the deposits: (**a**)—Dashkovskoye, (**b**)—Borschovskoye.

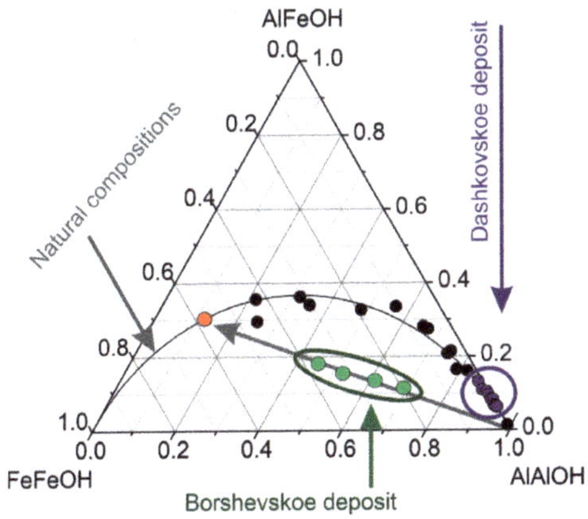

Figure 6. Diagram of the ratio of Fe–Fe–OH, Al–Fe–OH, and Al–Al–OH bonds in palygorskite.

https://doi.org/10.3390/min16010070

The palygorskites from the Dashkovskoye deposit follow this curve, meaning they represent pure natural Al–dioctahedral varieties without the presence of a significant admixture of other palygorskite varieties (of a different genesis). Based on the stability of the crystal-chemical composition of palygorskites along the section during the entire period of formation of the layer up to the Vereiskian time, we can conclude that the conditions for the chemogenic synthesis and deposition of clay material in the Steshevian sub-age in a shallow-water basin were stable.

In contrast to the palygorskites from the Dashkovskoye deposit, which were described above, all the compositions of the palygorskites from the Borschovskoye deposit are below the curve of natural compositions. By analyzing the locations of the Borschovskoye deposit points on the curve, it can be concluded that there are mainly Al–natural palygorskites and Fe–Al–palygorskites, with a significant predominance of isomorphic iron. Al–palygorskites were most likely formed under conditions similar to those of the Dashkovskoye deposit, and they likely have a chemogenic origin. Fe–Al–palygorskites have a distinctly different origin and were possibly formed in pedogenic conditions. Both of them were apparently transported to the sedimentation basin of the Borschovskoye deposit. Taking into account that the grain size distribution of the samples studied, it can be concluded that different sources of material were transported into the sedimentation basin. It is likely that the palygorskites in question came from different sources via different migration paths.

4.3. Grain Size Composition

Despite the different mineral composition, the grain size composition of the D clays is uniform throughout the section. Specific convex cumulative curves are prevalent (Figure 7a), which correspond to slow sedimentation from a unidirectional water flow. These cumulative curves are characteristic of marine hemipelagic sediments on the continental slope, as well as of sediments in the deep-water part of deltas and lagoon basins with medium hydrodynamics [51].

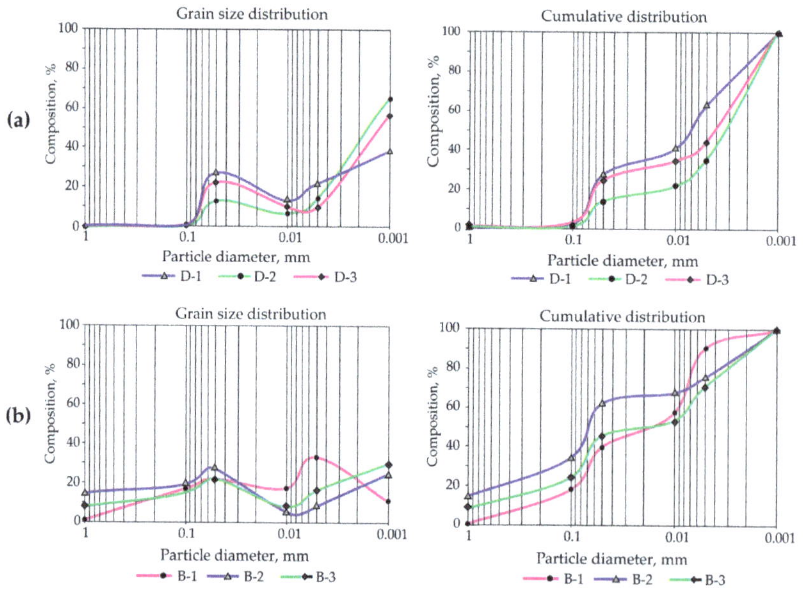

Figure 7. Grain size distribution and cumulative curves of the studied clays: (**a**)—Dashkovskoye deposit, (**b**)—Borschovskoye deposit.

https://doi.org/10.3390/min16010070

The content of particles in the clay fraction (<5 μm) in samples D is more than 50% (Table 3). Particle size distribution across all three samples exhibits a consistent bimodal pattern, with a predominant mode in the fine clay fraction range. Nevertheless, there are clear regularities for different samples: in the predominantly smectite clay (D-1), the fine-silty mode is more pronounced, whereas the clay-fraction mode is less prominent. Conversely, the predominantly palygorskite clay exhibits a weaker fine-silty mode, but a stronger mode in the fine clay fractions. The smectite–palygorskite clay is characterized by the middle position of the fine-silt clayey mode (Figure 7).

Table 3. Grain size composition of the studied clays.

Sample	Grain Size Composition (%), Size Fraction in mm					
	1.0–0.1	0.1–0.05	0.05–0.01	0.01–0.005	0.005–0.001	<0.001
			Dashkovskoye deposit			
D-1	0.0	0.3	26.4	13.2	21.6	38.4
D-2	0.0	0.9	22.3	10.6	9.4	56.9
D-3	0.0	0.4	13.1	6.9	14.0	65.6
			Borschovskoye deposit			
B-1	14.5	19.5	27.5	5.5	8.2	24.7
B-2	7.9	15.1	22.0	8.2	16.5	30.2
B-3	0.8	16.8	22.0	16.5	33.0	11.0

The cumulative curves also show a high degree of similarity and appear as concave lines with a clear «plateau» in the silt-size area and a steep slope towards the fine fraction. Such a distribution of fractions in the D clay indicates the preservation of hydrodynamic conditions in the paleobasin during the formation of the deposits. At the same time, the gentle shapes of the cumulative curves indicate slow sedimentation and the action of a single factor in the sedimentation process. This could take place in a shallow basin, where fine terrigenous material is introduced by slow-flowing water, and the process of chemo-deposition of palygorskite minerals occurs.

The grain size composition of B clays exhibits specific differences, characterized primarily by a polymodal distribution with modes in the sandy, silty, and clay fractions, as well as by raised cumulative curves with numerous inflections. This pattern reflects the redeposited genesis of the sediments in the Borschovskoye deposit. The most prominent among the others is the stratum rich in palygorskite material, which came from a different source than the other clay and non-clay materials.

For a detailed study of the grain size composition of the finely dispersed part using laser diffraction, fractions < 5 μm from samples D-3 (predominantly palygorskite clay) and D-1 (predominantly smectite clay) were selected. The distribution of particles of the clay component in the samples D is characterized by a monomodal log-normal distribution, with an average particle diameter of 7.3 μm for the predominantly palygorskite sample and 7.7 μm for the predominantly smectite sample (Figure 8).

Particles of clay minerals are sensitive to any mechanical treatment, including ultrasonic treatment, which is well evidenced by changes in the behavior of the spectra of the fractional composition (Figure 8). During several repetitions with additional sonication for 10 minutes, the prevailing average particle size (mode size) decreases from 7.3 to 5.6 μm for the predominantly palygorskite clay sample (D-3) and from 7.7 to 5.9 μm for the predominantly smectite clay sample (D-1). The number of particles < 0.3–0.5 μm in size also increases. With an increase in the time of ultrasound exposure to 20 minutes or more, the number of particles in the range of 5–7 μm decreases, and a second mode appears in the spectrum with an average diameter of 440–830 μm. This behavior of the spectra

https://doi.org/10.3390/min16010070

indicates that, under short exposure to ultrasound (10 min), there is a partial destruction of relatively low-strength large microaggregates ranging in size from 8 to 20 μm, resulting in the enrichment of the specimen with primary particles. Meanwhile, with an increase in the duration of ultrasound treatment, aggregation occurs with the formation of large secondary aggregates due to an increase in the static charge on the surface of the particles.

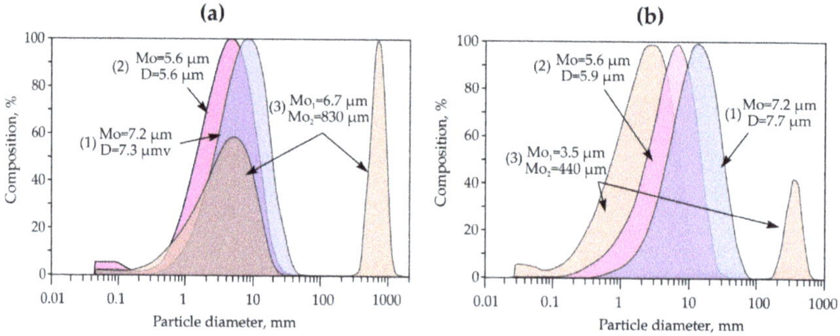

Figure 8. Features of the distribution of the grain size composition of the studied clays: (**a**)—sample of predominantly smectite clay D-1, (**b**)—sample of predominantly palygorskite clay D-3; Mo—mode, D—average size. 1—start of measurements, 2—measurements after 10 min of exposure to ultrasound, 3—measurements after 20 min of ultrasound.

4.4. Micromorphology of Clays

The sample D-1 is characterized by the prevalence of smectite and by a predominantly laminar microstructure with elements of turbulence, which is clearly seen in Figure 9(a1,a2). The microstructure is represented by anisometric sheet microaggregates with a thickness of 2–10 μm and a length of 10 μm or more. The microaggregates are in contact according to the edge-to-edge type, merge into each other along the layering, and are composed of sharply anisometric sheet particles 30–50 nm thick and 1–4 μm long, also contacting according to the edge-to-edge type. The edges of the particles are ragged and curved. The void space of the sample is mainly represented by elongated intra-aggregate pores and, to a lesser extent, inter-microaggregate pores. The microstructure of sample B-1 was found to have many similarities with sample D-1. Nevertheless, Figure 9(b1) shows that B-1 contains substantially smaller microaggregates with less distinct particle features. Morphologically, the particles are less pronounced in B-1 than in D-1. Such differences in smectite micromorphology suggest that the B-1 clay is of redeposited origin.

The microstructure of the predominantly palygorskite clays (D-3 and B-3) (Figure 9(a2,a3)) is characterized by more noticeable laminarity. The size of the palygorskite particles in cross-section is 50–100 nm. Due to the tangled fibers, it is difficult to estimate the length of the particles precisely; the apparent length of individual particles reaches 1–10 μm. Elongated interaggregate pores are prevalent in the palygorskite samples. Furthermore, the proportion of intra-microaggregate pores is much lower than in the predominantly smectite clays. The main difference between the particles of palygorskite B is that the particles (aggregates of particles) are smaller, with less distinct morphology, and are often characterized by a shorter length when compared to the particles of palygorskite D.

Characterizing the microstructure of the samples studied in general, it should be noted that they are similar at relatively low magnifications (up to 1000×). Most likely, this is due to the fact that the sediment had an almost identical dispersed composition and underwent diagenetic transformations of the same nature. This is likely responsible for the similarity of the microstructure.

https://doi.org/10.3390/min16010070

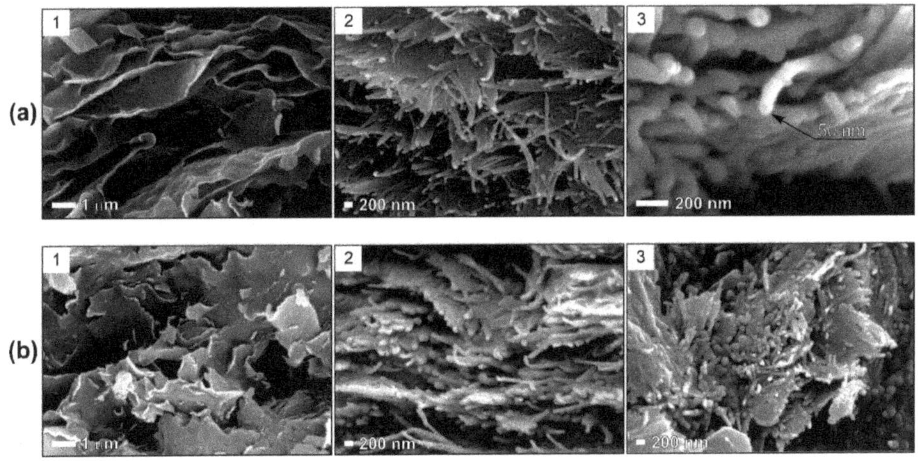

Figure 9. Microstructure of samples of the Steshevian sub-stage, Moscow region: (**a**)—Dashkovskoye, (**b**)—Borschovskoye deposits. The predominant minerals are: 1—smectite, 2, 3—palygorskite.

4.5. Adsorption Properties

The ability of clay materials to absorb various cations (including technogenic ones) is often used in industry to obtain various sorbents. Based on this, it is reasonable to evaluate the sorption properties of materials that can be used in industrial conditions. The clays of the Dashkovskoye deposit were previously mined in an open-pit operation and have the potential to be mined again; they are easily available on the mineral raw materials market and are distinguished by a relatively uniform mineral composition with a predominance of smectite or palygorskite. Therefore, they were chosen for the adsorption experiments presented below.

Determination of the CEC has several conventions that make the application of different methods fairly subjective [49]. For the selected samples, the CEC value was measured by adsorption of methylene blue (MB) and by $[Cu(trien)]^{2+}$. Additionally, measurements were made for the clay fraction of the samples using the MB adsorption method (Table 4).

Table 4. Indices of the sorption properties of the studied clays.

Sample	1	2	3	4	5	6
	S_{BET}, m^2/g	CEC (Bulk), $[Cu(trien)]^{2+}$, mg-eq/100 g	CEC (Bulk), MB, mg-eq/100 g	CEC (<1 μm), MB, mg-eq/100 g	A_{max}, mg-eq/100 g	K, mmol/L
D-1	32	49	65	92	43	1.13
D-2	65	42	51	56	31	0.27
D-3	161	32	35	39	26	0.08

1—S_{BET}—specific surface area; 2–4—CEC values determined by the adsorption of $[Cu(trien)^{2+}]$ in a bulk sample (2), by adsorption of methylene blue in a bulk sample (3), and in the fraction <1 μm (4); 5—maximum adsorption of Cs+; 6—Cs+ adsorption equilibrium constant.

According to Table 4, the CEC values determined by the MB adsorption method are higher than those obtained with $[Cu(trien)]^{2+}$. This is attributed to the polymerization of MB during the short interaction period. Although this process overestimates adsorption values, it simultaneously facilitates a more thorough characterization of sorption capacity in clay minerals, predominantly of the smectite group. Regardless of the analytical method employed, the CEC value of the studied samples is predominantly controlled by smectite

https://doi.org/10.3390/min16010070

content [46,52]. It can be expected that the relatively large MB molecules, being in an aggregated state, are unable to penetrate the channels of palygorskite; therefore, low CEC values tend to be observed in predominantly palygorskite clays.

At the same time, the specific surface area significantly increases in strata enriched in palygorskite, which is explained by the more developed inner surface of palygorskites (Table 4).

Experiments on the adsorption of Cs^+ cations from aqueous solutions made it possible to reveal differences in the sorption properties of the selected samples. Three characteristic areas can be distinguished on the graphs (Figure 10):

1. A linear part at low concentrations of the occluded cation;
2. A non-linear transition section at intermediate concentrations;
3. A curve that flattens off at high concentrations, corresponding to almost complete saturation of the adsorbent.

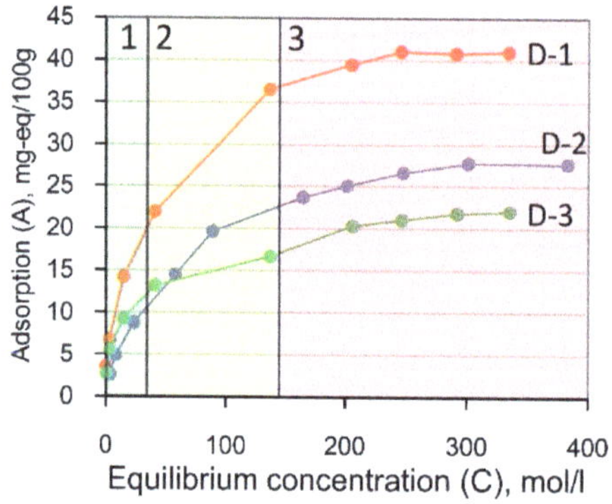

Figure 10. Isotherms of Cs^+ adsorption on clay samples from the Dashkovskoye deposit.

Thus, the isotherm of adsorption can be satisfactorily described within the framework of Langmuir's theory of monomolecular adsorption. As a first approximation, the sample with the highest content of smectite (D-1) has the highest adsorption capacity. At the same time, adsorption capacity decreases with a decrease in the content of smectite. In order to calculate the value of the maximum adsorption (A_{max}) more accurately, it is necessary to graphically solve the Langmuir equation, for which the isotherms were plotted in the coordinates A/Cequiv = f(Cequiv).

Although CEC and A_{max} values show slight divergence, both decrease with declining smectite content. Higher CEC values should be associated with the tendency of MB cations to undergo multilayer adsorption on the surface of clay minerals. This is confirmed by spectrometric studies of the kinetics of its sorption, as well as by other studies [51]. In contrast, judging by the nature of the obtained isotherms, the Cs^+ cations are prone to monolayer adsorption on these samples. The values of the adsorption equilibrium constant K characterize the interaction energy of the adsorbate with the adsorbent; in other words, the chemical affinity between them. Thus, we can conclude that the interaction of the Cs^+ cations is energetically more favorable with the surface of smectite than that of palygorskite.

5. Discussion

The integration of all the results mentioned above made it possible to significantly supplement the history of the geological development of the region in the Steshevian sub-age (Figure 11).

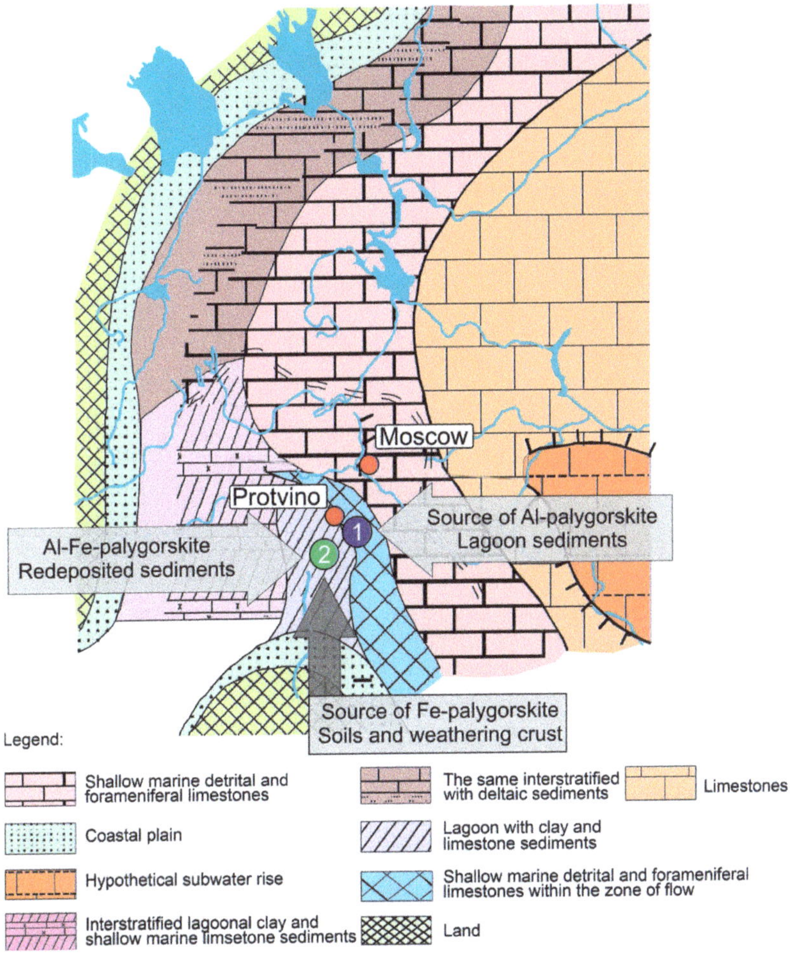

Figure 11. Scheme of the distribution of precipitation and fauna during the Steshevian sub-age: 1—Dashkovskoye deposit, 2—Borschovskoye deposit, ref. [53] with additions.

An open sea covered the eastern part of the Russian platform (Volga–Ural region) throughout the Carboniferous period, while the Moscow syneclise constituted its western marginal zone, separated by islands or underwater uplifts. Increasing crustal uplift intensified the isolation of the Moscow syneclise basin, ultimately transforming it into a large lagoon [53]. The Steshevian sub-age was marked by significant changes in the conditions of sedimentation in the area under consideration. A stagnant lagoon with a relatively higher salinity formed in the southern part of the Moscow syneclise basin. It accumulated clay and dolomite silts. The lagoon was isolated from the open sea by shoals. Pronounced climatic aridization also impacted sedimentation: It affected the sedimentary dolomites in the lagoon, the palygorskite clay composition, and, according to [53], there was a decrease in TiO_2 content in clays and the cessation of coal formation.

In the Steshevian sub-age, the basin was very shallow and turbid due to a large amount of incoming clayey matter. This statement is supported by specific faunal species found in sections of the same age in the Moscow region [4,5]. The introduction of smectite material prevailed at the initial stage of the formation of the basin. The terrigenous genesis

https://doi.org/10.3390/min16010070

of the layers with increasing smectite is confirmed by the simultaneous increase in quartz and illite.

Due to the intensity of the climactic aridity and the proportion of MgO, the active formation of palygorskite began. The clay minerals were deposited simultaneously and formed dense aggregates, which are clearly visible by SEM. With an increase in aridity and, as a consequence, shallowing of the basin, the supply of smectite ceased, and palygorskite predominated in the upper part of the section. Over time, as the aridity continued to increase, the MgO content significantly prevailed over Al_2O_3 and SiO_2. As a result of this, layers of chemogenic dolomite were formed. By the end of the Steshevian sub-age, the basin had become extremely shallow throughout; the conditions of sedimentation changed, and terrigenous sedimentation began to prevail.

Following the end of the Steshevian sub-age, during the subsequent Protvian time, calcareous sediments accumulated in the southern part of this area, while sedimentary dolomites formed in the northwestern part [53]. Then calcareous marine sediments began to be deposited almost everywhere. At the turn of the Early and Middle Carboniferous epochs, the region dried up again for a long time. Lower Carboniferous carbonate strata underwent prolonged weathering and erosion. At an early stage of the Bashkirian age, the basal surface of the Lower Namyurian (Protvian sub-stage) deposits was probably covered by the sea. However, it is unknown how widely this basin was spread.

In the last few years, based on the results of a detailed study of paleontological remains, the features of sedimentation in the Lower Carboniferous have been clarified. The "Steshevian basin" was probably formed by a moderate input of thin terrigenous material ahead of the delta front and was not essentially a "lagoonal" basin [5]. The main difference in the opinion of the above authors is that the basin was not stagnant, as previously assumed by many authors, but had quite good mixing, with the inflow of oxygen and nutrients for the favorable development of organisms. At the end of the Steshevian sub-age, the Steshevian basin was practically filled in. At the same time, the continuing aridization of the climate increased the concentration of Mg and Si by evaporation in the shallow clay basin, which led to the deposition of palygorskite material.

A detailed study of the crystal-chemical structure of palygorskites (in particular the amount of isomorphic iron) made it possible to detail the scheme of the formation of palygorskites in the Moscow basin. It was revealed that Al varieties of the clay were formed as a result of chemical synthesis in a shallow saline basin, while Fe varieties were probably formed in soils and transferred to the sedimentation basin by rivers and water streams.

Thus, based on the studied clays, some refinements can be made to the previously described sedimentation schemes [5,6,53]. In addition, there were basins with a mixed type of sediments. As a result of the development of soil formation processes (as, for example, described for other basins of a similar age in [1,2,5]), palygorskites with a high Fe content in octahedral positions could form. Such palygorskites can be defined as end members in a series of mixtures of palygorskites of different composition in the Borschovskoye deposit. While most of the clay material in the sediments of the Dashkovskoye deposit has a local in situ origin, the clays of the Borschovskoye deposit were formed by sediment mixing with Al–palygorskites, which were carried in from the lagoon areas, and Fe–Al–palygorskites eroded from the surrounding soils. At the same time, the directions of the flows that brought the material could also differ quite significantly (Figure 11). Material enriched in Fe–Al–palygorskites has been transported by river flows from the nearby southern land, while material enriched in Al–palygorskite has come from the west as a result of seasonal fluctuations in sea level.

https://doi.org/10.3390/min16010070

6. Conclusions

Comprehensive studies of the composition, structure, and properties of smectite–palygorskite clays of the Russian Platform (using sections of the Steshevian sub-stage of the Dashkovskoye and Borschovskoye deposits as examples) made it possible to detail their genesis and evaluate the prospects for their use as sorbents.

The content of smectite in the lower parts of the Steshevian sub-stage reaches 55%–57%. Homogeneous particle size distribution, micromorphological features, and an increase in the proportion of quartz and illite indicate that the smectite has a predominantly terrigenous genesis.

At the same time, palygorskites in the studied area are distinguished by various types of genesis:

1—the predominantly Al–palygorskites are characterized by chemogenic synthesis and would have formed in situ in lagoon conditions;

2—the Fe–(Fe–Al)–palygorskites also have a chemogenic genesis, but they were formed in pedogenic conditions in the presence of microbial communities.

In the sections of the Dashkovskoye deposit, only Al–palygorskites are found, and thus, it can be argued that this basin served as a source of chemogenic palygorskite. Meanwhile, palygorskites of mixed composition are encountered in the sections of the Borschovskoye deposit.

IR spectroscopy studies in the near-IR region allowed us to conclude that these palygorskites are represented by a mixture of Al–palygorskites and Fe–(Fe–Al)–palygorskites. That is, the clays of this basin have a terrigenous genesis, and the material came from different sources: from the erosion of material from the coastal areas of the lagoons with Al–palygorskites and from the erosion of soil areas with Fe–(Fe–Al)–palygorskites carried in by water flows.

It should be noted that the studied clays are quite old for bentonite and, despite this, retain a considerable sorption capacity. These clays could be utilized as sorbents in the future, but their properties are not yet comparable to those of established bentonite deposits (e.g., those in the Russian Federation and Kazakhstan [32,44]). The advantage of using smectite–palygorskite clays is convenient logistics and proximity to nearby enterprises located in the central part of the country.

Author Contributions: Conceptualization: V.K., S.Z. and O.Z.; Methodology: S.Z., O.Z. and V.K.; Formal analysis and investigation: S.Z., O.Z., M.C., I.M. and T.K.; Writing—original draft preparation: S.Z., O.Z. and V.K.; Writing—review and editing: M.C., O.Z. and T.K.; Funding acquisition: V.K.; Resources: V.K.; Supervision: V.K. All authors have read and agreed to the published version of the manuscript.

Funding: The experimental studies were partially performed using equipment acquired with the funding of Moscow State University Development Program (X-ray Diffractometer Ultima-IV, Rigaku and Scanning Electron Microscope LEO 1450VP, Carl Zeiss). Mineral composition was funded by the Russian Science Foundation (project 16-17-10270). Investigation of adsorption properties was performed under a state assignment conducted by the Institute of Ore Geology, Petrography, Mineralogy and Geochemistry, Russian Academy of Science.

Data Availability Statement: The original contributions presented in this study are included in the article. Further inquiries can be directed to the corresponding author.

Acknowledgments: The authors are grateful to JSC "Keramzit" https://www.zao-keramzit.com (accessed on 11 January 2026) and Lafarge SA for their assistance in conducting the work and in collecting samples for research. The authors thank Vassilis Gionis and Georgios Chryssikos (Theoretical and Physical Chemistry Institute, National Hellenic Research Foundation) for studying the samples using near-IR spectroscopy and for providing valuable comments on the text of the article;

https://doi.org/10.3390/min16010070

Vyacheslav N. Sokolov (Geological Faculty of M.V. Lomonosov Moscow State University) for his help in investigations of microstructure; Boris V. Pokidko (IGEM RAS) for his help in conducting research on the dispersion of clays; Ekaterina A. Tyupina (Mendeleev University of Chemical Technology of Russia) for her help in investigating surface properties, and Margarita L. Kuleshova and Zoya P. Malashenko for their help in the Cs adsorption experiments.

Conflicts of Interest: The authors declare that they have no competing interests.

Abbreviations

The following abbreviations are used in this manuscript:

SEM	Scanning electron microscope
NIR	Near-infrared
MIR	Mid-infrared
FT-IR	Fourier transform infrared spectroscopy
IR	Infrared spectroscopy
CEC	Cation exchange capacity
MB	Methylene blue

References

1. Alekseeva, T.V.; Alekseev, A.O.; Kabanov, P.B.; Zolotareva, B.N.; Alekseeva, V.A.; Gubin, S.V. Carboniferous paleosols of the Moscow syneclise: Humic substances, mineralogical and geochemical properties. In *Paleosols and Indicators of Continental Weathering and Biosphere History*; Series "Geo-biological systems in the past"; PIN RAS: Moscow, Russia, 2010; pp. 76–94. (In Russian)
2. Alekseeva, T.V.; Alekseev, A.O.; Gubin, S.V. Paleosol Complex in the Uppermost Mikhailovian Horizon (Viséan, Lower Carboniferous) in the Southern Flank of the Moscow Syneclise. *Paleontol. J.* **2016**, *50*, 319–335. [CrossRef]
3. Alekseeva, T.V.; Alekseev, A.O.; Gubin, S.V.; Kabanov, P.B.; Alekseeva, V.A. Palaeoenvironments of the Middle—Late Mississippian Moscow Basin (Russia) from multiproxy study of palaeosols and palaeokarsts. *Palaeogeogr. Palaeoclimatol. Palaeoecol.* **2016**, *450*, 1–16. [CrossRef]
4. Kabanov, P.B.; Alekseeva, T.V.; Alekseeva, V.A.; Alekseev, A.O.; Gubin, S.V. Paleosols in Late Moscovian (Carboniferous) marine carbonates of the East European Craton revealing "Great calcimagnesian plain" Paleolandscapes. *J. Sediment. Res.* **2010**, *80*, 195–215. [CrossRef]
5. Kabanov, P.B.; Alekseeva, T.V.; Alekseev, A.O. Serpukhovian stage (Carboniferous) in type area: Sedimentology, mineralogy, geochemistry, and section correlation. *Stratigr. Geol. Correl.* **2012**, *20*, 15–41. [CrossRef]
6. Nasedkin, V.V.; Boeva, N.M.; Belousov, P.E.; Vasiliev, A.L. Geology, mineralogy, and genesis of palygorskite clay from Borshchevka deposit in the Kaluga region and outlook for its technological use. *Geol. Ore Depos.* **2014**, *56*, 208–227. [CrossRef]
7. Guggenheim, S.; Adams, J.M.; Bain, D.C.; Bergaya, F.; Brigatti, M.F.; Drits, V.A.; Formoso, M.L.L.; Galan, E.; Kogure, T.; Stanjek, H. Summary of recommendations of Nomenclature Committees relevant to clay mineralogy: Report of the Association Internationale Pour L'etude des Argiles (AIPEA) nomenclature committee for 2006. *Clays Clay Miner.* **2006**, *54*, 761–772. [CrossRef]
8. Wilson, M.J. Rock-forming minerals. In *Sheet Silicates: Clays Minerals*; The Geological Society: London, UK, 2013; p. 724.
9. Brindley, G.W.; Brown, G. *Crystal Structures of Clay Minerals and Their Identification*; The Mineralogical Society of Great Britain and Ireland: Twickenham, UK, 1980; p. 496.
10. Belousov, P.; Chupalenkov, N.; Christidis, G.E.; Zakusina, O.; Zakusin, S.; Morozov, I.; Chernov, M.; Zaitseva, T.; Tyupina, E.; Krupskaya, V. Carboniferous bentonites from 10Th Khutor deposit (Russia): Composition, properties and features of genesis. *Appl. Clay Sci.* **2021**, *215*, 106308. [CrossRef]
11. Belousov, P.E.; Krupskaya, V.V. Bentonite clays of Russia and neighboring countries. *Georesursy* **2019**, *21*, 79–90. [CrossRef]
12. Christidis, G.E. Genesis and compositional heterogeneity of smectites. Part III: Alteration of basic pyroclastic rocks—A case study from the Troodos Ophiolite Complex, Cyprus. *Am. Mineral.* **2006**, *91*, 685–701. [CrossRef]
13. Post, J.E.; Heaney, P.J. Synchrotron powder X-ray diffraction study of the structure and dehydration behavior of palygorskite. *Am. Mineral.* **2008**, *93*, 667–675.
14. Jones, B.F.; Galan, E. Sepiolite and palygorskite. In *Hydrous Phyllosilicates*; Baily, S.W., Ed.; Walter de Gruyter: Berlin, Germany, 1988; Volume 19, pp. 631–674.
15. Pozo, M.; Calvo, J.P. An overview of authigenic magnesian clays. *Minerals* **2018**, *8*, 520. [CrossRef]
16. Christ, C.L.; Hathaway, J.C.; Hostetler, P.B.; Shepard, A.O. Palygorskite: New X-ray data. *Am. Mineral.* **1969**, *54*, 198–205.
17. Drits, V.A.; Sokolova, G.V. Structure of palygorskite. *Crystallography* **1971**, *16*, 183–185.

https://doi.org/10.3390/min16010070

18. Chisholm, J.E. Powder diffraction patterns and structural models for palygorskite. *Can. Mineral.* **1992**, *30*, 61–73.
19. Artioli, G.; Galli, E. The crystal structure of orthorhombic and monoclinic palygorskite. *Mater. Sci. Forum* **1994**, *166–169*, 647–652. [CrossRef]
20. Galan, E.; Carretero, I. A new approach to composition limits for sepiolite and palygorskite. *Clays Clay Miner.* **1999**, *47*, 399–409. [CrossRef]
21. Allouche, F.; Ammous, A.; Tlili, A.; Kallel, N. Geological setting, geochemical, textural, and genesis of palygorskite in Eocene carbonate deposits from Central Tunisia. *Carbonates Evaporites* **2024**, *39*, 31. [CrossRef]
22. Arefiev, M.P.; Shchepetova, E.V.; Pokrovskaya, E.V.; Shkurskii, B.B.; Nurgalieva, N.G.; Batalin, G.A.; Gareev, B.I. Palygorskite Mineralization in the Induan Sediments of the Moscow Syneclise as an Effect of Regional and Global Paleogeographic Changes at the Permian–Triassic Boundary. *Dokl. Earth Sci.* **2024**, *519*, 2196–2207. [CrossRef]
23. Chryssikos, G.D.; Gionis, V.V.; Kacandes, G.H.; Stathopoulou, E.T.; Suárez, M. Octahedral cation distribution in palygorskite. *Am. Mineral.* **2009**, *94*, 200–203. [CrossRef]
24. Suarez, M.; Garcia-Romero, E. FTIR spectroscopic study of palygorskite: Influence of the composition of the octahedral sheet. *Appl. Clay Sci.* **2006**, *31*, 154–163. [CrossRef]
25. Chryssikos, G.D.; Gionis, V.V. On the structure of palygorskite by mid- and near-infrared spectroscopy. *Am. Mineral.* **2006**, *91*, 1125–1133.
26. Gionis, V.; Kacandes, G.H.; Kastritis, I.D.; Chryssikos, G.D. Combined near-infrared and X-ray diffraction investigation of the octahedral sheet composition of palygorskite. *Clays Clay Miner.* **2007**, *55*, 543–553. [CrossRef]
27. Selin, P.; Leupin, O.X. The use of clay as an engineered barrier in radioactive-waste management—A review. *Clays Clay Miner.* **2013**, *61*, 477–498. [CrossRef]
28. Zhang, Q.; Zhao, Y.; Qin, L.; Liang, W.; Chen, K.; Li, K.; Yan, R. Adsorption properties of cesium by natural Na-bentonite and Ca-bentonite. *J. Radioanal. Nucl. Chem.* **2024**, *333*, 5347–5361. [CrossRef]
29. Oskoueian, F.; Hosseini, S.F.; Seddighian, H.H.; Abdi, S.; Jalalian, Y.; Mashhadi, Y.B.; Oskoueian, E.; Karimi, E.; Jahromi, M.F.; Shokryazdan, P.; et al. Biological properties of activated bentonite vs. non-activated bentonite in mice fed an aflatoxin-contaminated diet: A comparative investigation. *Mycotoxin Res.* **2025**, *41*, 339–348.
30. Liu, W.; Liang, D.; Yang, Z.; Gao, C.; Xie, J. Study on the influence of thermal aging on the swelling properties of GMZ bentonite. *Environ. Earth Sci.* **2025**, *84*, 315. [CrossRef]
31. Krupskaya, V.; Novikova, L.; Tyupina, E.; Belousov, P.; Dorzhieva, O.; Zakusin, S.; Kim, K.; Roessner, F.; Badetti, E.; Brunelli, A.; et al. The influence of acid modification on the structure of montmorillonites and surface properties of bentonites. *Appl. Clay Sci.* **2019**, *172*, 1–10. [CrossRef]
32. Krupskaya, V.; Zakusin, S.; Zakusina, O.; Belousov, P.; Pokidko, B.; Morozov, I.; Zaitseva, T.; Tyupina, E.; Koroleva, T. On the Question of Finding Relationship Between Structural Features of Smectites and Adsorption and Surface Properties of Bentonites. *Minerals* **2025**, *15*, 30. [CrossRef]
33. Sun, Y.; Zhang, P.; Liu, Y.; Wu, S.; Yang, Q. An efficient and recyclable palygorskite-supported palladium catalyst for Suzuki–Miyaura coupling reactions in water at room-temperature. *React. Kinet. Mech. Catal.* **2024**, *137*, 163–176. [CrossRef]
34. Perez-Matu, R.R.; Avilés, F.; Gonzalez-Chi, P.I. Water absorption kinetics of palygorskite nanoclay/polypropylene composite foams. *Polym. Bull.* **2024**, *81*, 4149–4174. [CrossRef]
35. Tyupina, E.; Kozlov, P.; Pryadko, A.; Krupskaya, V. Radioiodine Sorption on AgCl-modified Bentonite and its Stability in different Environments. *Open Chem. Eng. J.* **2025**, *19*, e18741231403108. [CrossRef]
36. Hussein, A.F.; Al-Bidry, M.A.; Wasiti, A.A. Investigation and evaluation of palygorskite microstructure following acid pretreatment and its potential use as an adsorbent for copper. *Sci. Rep.* **2025**, *15*, 3813. [CrossRef]
37. Bishop, W.J.; Montanez, I.P.; Gulbranson, E.L.; Brenckle, P.L. The onset of mid-Carboniferous glacio-eustasy: Sedimentologic and diagenetic constraints, Arrow Canyon, Nevada. *Palaeogeogr. Palaeoclimatol. Palaeoecol.* **2009**, *276*, 217–243. [CrossRef]
38. Zchus, I.D. Palygorskite from the Steshevsky horizon of the Moscow region basin. *Dokl. Akad. Nauk SSSR* **1956**, *107*, 5. (In Russian)
39. *GOST 12536-79*; Soils. Methods of Laboratory Granulometric (Grain-Size) and Microaggregate Distribution. Gosstandart of the USSR: Moscow, Russia, 1980.
40. Moore, D.M.; Reynolds, R.C., Jr. *X-ray Diffraction and the Identification and Analysis of Clay Minerals*, 2nd ed.; Oxford University Press: Oxford, UK; New York, NY, USA, 1999; p. 378.
41. Post, J.E.; Bish, D.L. Rietveld refinement of crystal structures using powder X-ray diffraction data. *Rev. Mineral. Geochem.* **1989**, *20*, 277–308.
42. Doebelin, N.; Kleeberg, R. Profex: A graphical user interface for the Rietveld refinement program BGMN. *J. Appl. Crystallogr.* **2015**, *48*, 1573–1580. [CrossRef] [PubMed]
43. Kaufhold, S.; Dohrmann, R.; Ufer, K.; Meyer, F.M. Comparison of methods for the quantification of montmorillonite in bentonites. *Appl. Clay Sci.* **2002**, *22*, 145–151. [CrossRef]

https://doi.org/10.3390/min16010070

44. Belousov, P.E.; Pokidko, B.V.; Zakusin, S.V.; Krupskaya, V.V. Quantitative methods for quantification of montmorillonite content in bentonite clays. *Georesursy* **2020**, *22*, 38–47. [CrossRef]
45. Bujdák, J.; Komadel, P. Interaction of Methylene-blue with reduced charge montmorillonite. *J. Phys. Chem. B* **1997**, *101*, 9065–9068. [CrossRef]
46. Czímerová, A.; Bujdák, J.; Dohrmann, R. Traditional and novel methods for estimating the layer charge of smectites. *Clay Sci.* **2006**, *34*, 2–13. [CrossRef]
47. Dohrmann, R.; Kaufhold, S. Three new, quick CEC methods for determining the amounts of exchangeable cations in calcareous clays. *Clays Clay Miner.* **2009**, *57*, 338–352. [CrossRef]
48. Lorenz, P.; Meier, L.; Kahr, G. Determination of the cation exchange capacity (CEC) of clays minerals using the complexes of copper (II) ion with triethylenetetramine and tetraethylenepentamine. *Clays Clay Miner.* **1999**, *47*, 386–388. [CrossRef]
49. Dohrmann, R.; Genske, D.; Karnland, O.; Kaufhold, S.; Kiviranta, L.; Olsson, S.; Plötze, M.; Sandén, T.; Sellin, P.; Svensson, D.; et al. Interlaboratory CEC and exchangeable cation study of bentonite buffer materials: I. Cu(II)-triethylenetetramine method. *Clays Clay Miner.* **2012**, *60*, 162–175. [CrossRef]
50. Russel, J.D.; Fraser, A.R. Chapter 2. Infrared methods. In *Clay Mineralogy: Spectroscopic and Chemical Determinative Methods*; Wilson, M.J., Ed.; Chapman & Hall: London, UK, 1994; pp. 11–67.
51. Krupskaya, V.V.; Andreeva, I.A.; Sergeeva, E.I. Clay minerals in bottom sediments of the Medvezhii Island region, Norwegian Sea. *Lithol. Miner. Resour.* **2004**, *39*, 31–40. [CrossRef]
52. Ching-Hsing, Y.; Newton, S.Q.; Norman, M.A.; Miller, D.M.; Schafer, L.; Teppen, B.J. Molecular dynamics simulation of the adsorption of methylene blue at clay mineral surfaces. *Clays Clay Miner.* **2000**, *48*, 665–681. [CrossRef]
53. Osipova, A.I.; Belskaya, T.N. On facies and paleogeography of the Serpukhov time in the Moscow basin. *Lithol. Miner. Resour.* **1965**, *5*, 3–17.

https://doi.org/10.3390/min16010070

MDPI AG
Grosspeteranlage 5
4052 Basel
Switzerland
Tel.: +41 61 683 77 34

Minerals Editorial Office
E-mail: minerals@mdpi.com
www.mdpi.com/journal/minerals

www.ingramcontent.com/pod-product-compliance
Lightning Source LLC
Chambersburg PA
CBHW061437210926
51850CB00008B/1534
9783725877171